Genetics of Infectious Disease Susceptibility

Tjeerd G. Kimman
Research Laboratory for Infectious Diseases,
RIVM-National Institute for Public Health and the Environment,
Bilthoven, The Netherlands

Kluwer Academic Publishers
BOSTON   DORDRECHT   LONDON

RIJKSINSTITUUT VOOR VOLKSGEZONDHEID EN MILIEU
NATIONAL INSTITUTE OF PUBLIC HEALTH AND THE ENVIRONMENT

Colophon

Graphic design	Karin Janmaat and Marjan Kramer / RIVM
Photography cover	Shehzad Noorani / LINEAR BV (Soeda cattle)
	Ron Giling / LINEAR BV (Liber portrait girl)
	KINA (sheep)
	Ernst Rozendal / RIVM (mouse)
Print	Wilco b.v., Amersfoort

Library of Congress Cataloging-in-Publication Data is available.

ISBN 978-90-481-5763-1

Published by Kluwer Academic Publishers,
P.O. Box 17, 3300 AA Dordrecht, The Netherlands.

Sold and distributed in North, Central and South America
by Kluwer Academic Publishers,
101 Philip Drive, Norwell, MA 02061, U.S.A.

In all other countries, sold and distributed
by Kluwer Academic Publishers,
P.O. Box 322, 3300 AH Dordrecht, The Netherlands.

Printed on acid-free paper

Table of contents

Nothing in biology makes sense except in the light of evolution
Theodosius Dobzhansky

Preface

Infectious diseases are still the world's most important cause of death. They are faced with emerging interest and their study flourishes. Traditionally infectious diseases have been studied from many perspectives, each with their own advantages and disadvantages. While many texts are devoted to the properties of micro-organisms, their epidemiological spread, the pathogenesis of disease, immunity of the host, vaccination, or public health policy, this text is mainly directed to the question why some individuals (or races or breeds or species) are inherently more susceptible to infection than others. We probably all remember infections spreading in households, classrooms, or countries, wondering why some get sick while others remained healthy. The purpose of this text is to promote and exploit that curiosity.

Susceptibility or resistance to infectious disease thus shows marked variation within and between genetically diverse populations and we can exploit that variation to gain knowledge of infectious diseases and improve their management, treatment, and prevention. Infectious disease is clearly the result of a complex interplay between multiple human and microbial genes and a strong environmental component, each contributing an additive risk. Somewhat oversimplified, the severity of infectious disease depends on genetic properties of the pathogen, the amount of invading micro-organisms (influenced by environmental and social factors, such as population density, hygiene and herd immunity), and resistance of the host (influenced by its genetic make-up, age, immune and nutritional status, and stress). Host genetic factors are thus some of the many determinants of the outcome of infectious diseases. Complete understanding of infectious diseases includes knowledge of host genes that influence susceptibility to pathogens and that regulate the host's response to them. In this text, I will restrict myself to host genetic factors that influence the severity of disease and the spread of the micro-organisms. Notwithstanding that restriction, the result in no way claims a comprehensive overview of all knowledge in the field. The choice is mainly personal. For several years I had the privilege to work on many, both theoretical and practical, aspects of infectious diseases, while (like many others) more or less neglecting the genetic make-up of the host. However, over the past years there is increasing evidence that host genetic factors play a major role in determining susceptibility to and outcome of infectious disease. A complete study of infectious diseases should therefore include the genetic approach. I have no doubt that understanding them will require a complex integration of knowledge from host, pathogen, environment, and their mutual interactions (see Figure 1).

The study of host genetic factors influencing the course of infectious diseases may have many scientific and practical implications.

- It will provide insight into the genes and the physiological mechanisms involved in resistance or tolerance against infectious disease. While some genes may be disease specific, others may give resistance to several infections. Based on knowledge of pathways of infectious diseases new ways to treat or prevent infections may become available. While vaccines and antibiotics are very powerful tools to combat infections, there are still no vaccines against major diseases, such as AIDS or malaria. Moreover increasing bacterial resistance against antibiotics is threatening their use.

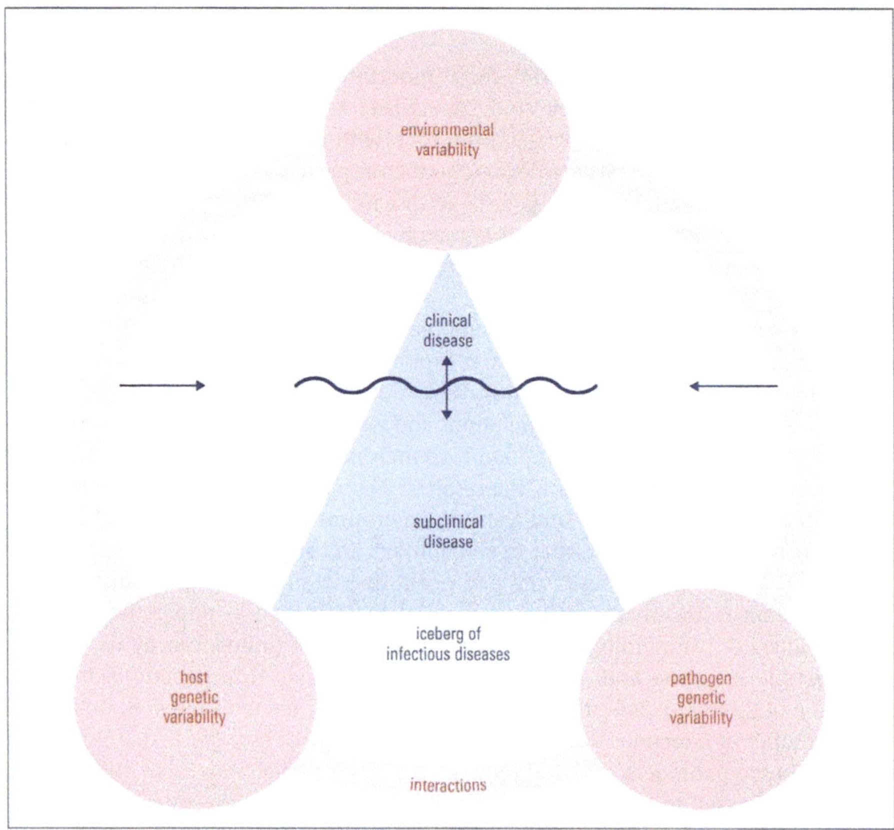

Figure 1. According to the iceberg concept of infectious diseases, most infections run a subclinical course, and only a minority of the infections result in overt clinical disease. Still a smaller proportion results in hospital admission or a fatal outcome. The environment, host and pathogen genetic factors, and their mutual interactions influence the ratio between clinical and subclinical disease.

- Studying the many interactions of genes from host and pathogen may provide further insight into important pathogenic pathways exploited by the pathogen, as well as into mechanisms used by the host to overcome the pathogen, and their relative importance.
- We may understand the contribution of genetic variation to the variation in infectious disease patterns in populations. Especially when infections appear to contribute to chronic diseases, such as gastric ulcer disease, cancer or diabetes mellitus, the impact of host genes appears to be large. Estimating the contribution of genetic variation to the variation in disease expression can help in making public health policy. By estimating the host genetic impact on infection (even or notably when it is small or absent), we will also be able to more fully appreciate the influence of genetic differences of pathogen strains and environmental factors.
- By including host genetics in studying infectious disease we will more fully understand the evolutionary pressures that influenced and still influence populations of humans, animals, and micro-organisms. Infectious diseases reflect the battle of genes from host and pathogen. Micro-organisms have been major agents of selec-

tion in the evolution of genetic diversity. Although evolutionary forces from the past are somewhat difficult to fully appreciate, such forces are still in work today in some populations and their study may thus facilitate their understanding.

- Studying host genetics will add in understanding their complex interactions with environmental and social factors. Humans and animals have been fine-tuned by evolution in a long battle with other organisms in a strong selective environment. Changing the environment and social conditions may drastically alter their vulnerability to disease. Notably some polymorphisms that evolved from natural selection against infection in the past may nowadays be involved with susceptibility to allergic, autoimmune, or other disease.

- Insight into the risks of individuals or breeds to particular infections may lead to disease management tailored to the need of the individual. Understanding the basis of variable disease susceptibility of individuals or groups may lead to specific or additional interventions or prophylaxis. In animal husbandry genetic improvement may lead to animals with increased disease resistance or enhanced response to vaccination.

- Pharmacogenetics in the field of infectious disease may give insight into an individual's drug metabolism and may thus optimize the use of new and existing drugs. This may help in preventing ineffective or even toxic treatments and the development of antibiotic resistance.

While I find this approach useful, interesting, and intellectually challenging, it is obvious that the result is in no way a comprehensive overview of infectious diseases, molecular microbiology, epidemiology, or genetics. I will not deal with many of the basic concepts of disease and defense mechanisms. I therefore refer to excellent textbooks, which I read with joy and consulted with great admiration (Mims et al 1995, Mims et al 1993, Roitt 1991, Strachan and Read 1996).

This time is fascinating. The completion of the Human Genome Project, the Human Genome Diversity Project, the sequencing of several other complex genomes, the DNA chip technology and increasing competence in bioinformatics will strongly stimulate progress in the field of infectious disease genetics and genomics. Essential is that knowledge from various disciplines is being combined: molecular microbiology, microbial pathogenesis, molecular genetics, and population genetics theory. In the next years we will therefore see a great wealth of new information on host genes (initial estimates of the number of human genes exceeded 150,000, but recent estimates predict only 26,000 to 38,000 genes [Pennisi 2000, Venter et al 2001]), their functions, their regulation, and their complex interactions influencing susceptibility to infectious disease. This will soon outdate this text, but I hope that the basic principles and concepts will hold. "The work is in progress".

References

Mims, C.A., Dimmock, N.J., Nash, A., Stephen, J. (1995). Mims' pathogenesis of infectious disease, 4th ed. Academic Press.

Mims, C.A, Playfair, J.H.L., Roitt, I.M., Wakelin, D., Williams, R. (1993). Medical microbiology, Mosby-Year Book Europe Ltd.

Pennisi, E. (2000). And the gene number is ......? Science 2888, 1146-1147.

Roitt, I.M. (1997). Essential immunology, 9th ed. Blackwell scientific publications.

Strachan, T., Read, A.P. (1996). Human molecular genetics. BIOS scientific publishers Ltd.

Venter, J.G. et al (2001). The sequence of the human genome. Science 291, 1304-1351.

Part I   General principles and methods

Chapter 1.   What is the contribution of host genetics to infectious disease?

Usually infectious diseases only affect a proportion of those invaded by the pathogen. An example is respiratory syncytial virus (RSV) infection, causing disease in young children and infants. RSV-associated disease may vary from completely unnoticed to severe respiratory distress followed by death. An illustrating exception at the other side of the spectrum is infection with rabies virus, which appears to cause nearly 100 % lethality in naturally infected individuals. In this chapter I will try to describe whether and to what degree the genetic make up of the host influences susceptibility to infectious disease. I will do so by briefly treating different approaches used to answer this question. Beforehand it is clear that the answer to this question may differ with place and time in history, and the pathogens and host populations that are examined. Species differences in susceptibility to infectious disease for example illustrate that the genetic constitution of the host may completely determine susceptibility, but this insight is of little or no use for the practicing physician or public health officer. Often mortality to infectious disease is high just after the introduction of a new pathogen in a virgin population or species. In time the pathogen may lose virulence and the host population may develop a degree of genetic resistance to this invader. During this evolutionary period towards a state of balanced pathogenicity, the genetically determined variability in disease expression varies. I will here review some approaches that have been used to determine the extent of the contribution of the host genetic make up to the variation in severity of infectious diseases. Difficulties, of course, are that the exposure of the infectious agent to the members of a population may vary considerably in space and time, and that genetically linked persons often share the same environment.

Historical studies
McNeill (1976) made It plausible, in his famous study "Plagues and Peoples", that throughout history new, previously unrecognized, infections appeared locally and strongly influenced the course of history. Changes that promoted infectious disease include population growth, migration, animal husbandry, and more recently the expansion of transportation and trade networks (Strassmann and Dunbar 1999). Especially when two previously separated groups met, infections new to one of them, notably smallpox, measles and cholera, could easily spread in the so-called virgin population. After time the newly invaded populations developed both immunological and genetic resistance, and developed the medical and hygienic methods to deal with the new threat. Over the 100 years following the Spanish conquest in Mexico, the indigenous population in Central and South America has been reduced by a factor of at least 20 mostly due to infectious disease (Dobyns 1966). Infectious diseases have therefore likely been the most significant agents of natural selection on human and animal populations, and have produced and maintained genetic differences in different sub-populations (Haldane 1949). Nonetheless, demographic data of the influence of infections on population growth and their influence as selective force are scarce.

Exact demographic data only begin to emerge from the middle of the 19th century. Despite this short period they illustrate the demographic influence of infections. For example, in The Netherlands the structural declining development in mortality is disturbed by incidental increases in mortality associated with cholera epidemics in 1853, 1854, 1855, 1859, 1866, and 1867, smallpox epidemics in 1858 and 1871, and the influenza epidemic in 1918. During these epidemic years the mortality in children and young adults increased at least two-fold (Poppel and Imhoff 2000).

There is further ample evidence from several sources that genetic determinants affected susceptibility to specific infectious diseases in the course of history, for example tuberculosis. The present day European population shows considerable resistance to this disease, probably because the great epidemics of pulmonary tuberculosis in Europe in the 17th, 18th, and 19th centuries weeded out genetically susceptible individuals. However, tuberculosis did not exist or was rare amongst the indigenous people of South Africa until the end of the 19th century. Upon introduction, the clinical picture of tuberculosis in Black and Khoi San peoples was initially very acute in nature and highly lethal, reminiscent of infection in a virgin population. Subsequently mortality felled, maybe due to the elimination of susceptibles in the early stages of the epidemic and the creation of a genetically resistant group. A similar observation was made in Qu'Apelle Indians, in which tuberculosis was introduced in 1890. Initially tuberculosis caused an annual mortality rate of 10 % of the population, which declined to 0.2 % within 40 years (Cummins 1929, Donald 1998, Motulsky 1960).

Even recently such selective forces were in action. In 1973 an epidemic of influenza killed 27 % of a community of Yanomamö Indians in Southern Venezuela (Chagnon and Melancon 1980). In the 1980s, the initial Pacific "island-hopping" of Ross River virus from its original niche in Australia caused virgin-soil epidemics of arthritis-myalgia syndrome in Fiji and Samoa (Murphy 1998). Without doubt such epidemics apply strong selective forces and have a major impact on the genetic make-up of populations.

Although the point appears obvious, much of the information on the role of infections as selective force early in history is anecdotal. There is therefore little data to allow exact historical reconstructions, perhaps with the exception of the well-known examples of sickle cell anemia and the myxomatosis case. Actual proof of the role of genes in protecting a population in the past against an infectious disease cannot be obtained in a population study at this moment. We cannot therefore exactly define the historical contribution of genetic variability in infectious disease resistance and vice versa. However the tool of phylogenetic analysis of host and microbial genomes allows some retrospective inferences on the history of infectious disease to be made. Genomes preserve an archaeological record of passed evolutionary events, such as natural selection, genetic drift, and mode of population growth. In addition, phylogenetic data can be correlated with other biological variations, such as host-pathogen interactions, geographical locations, and time in history. The statistical parameters of evolutionary models are typically estimated by maximum-likelihood and maximum parsimony methods, although the latter does not take into account the lengths of the branches and therefore tends to underestimate the amount of change in long branches. Through phylogenetic analysis it is possible to estimate when in time a particular pathogen emerged or started to spread epidemically, it is possible to estimate the timing of evolutionary events, and it is possible to estimate probable characteristics of ancestral species (Holmes 1999, Pagel 1999). For example, molecular data have suggested that rapid increases in the number of dengue virus and human T-cell lymphot-ropic virus–1 (HTLV-1) lineages were associated with increased spread of these viruses made possible by the rapid population growth in the Americas, which began in the fifteenth century (Dekaban et al 1995, Zanotto et al 1996). Phylogeny is thus based on the concept that present variation in genomes is due to descent with modification.

Inherited disorders

Familial forms of infectious disease and high rates of parental consanguinity in affec-ted persons may be strong indications for a genetic influence on disease susceptibility and a recessive mode of inheritance. Moreover they allow further studies into the genetic basis of enhanced susceptibility. Homozygosity mapping, without a need for family studies, is very suitable to detect a disease locus in such instances thanks to the fact that the adjacent region will be preferentially homozygous by descent in inbred children (Lander and Botstein 1987). Because inbred children are overrepresented among those with a recessive disorder, it is often easier to gather enough inbreds for homozygosity mapping than to gather enough families with multiple affected mem-bers for family linkage approaches. Severe and fatal infections with the attenuated strain of *Mycobacterium bovis* used for the Bacillus Calmette-Guérin vaccine, and other atypical *Mycobacteriae*, in some consanguineous family members pointed to rare mutations in the gene encoding the interferon-γ receptor-1 (IFN-γR1). Using a linkage study based on homozygosity mapping, the genetic defect was mapped to chromoso-me 6q22-q23, where the gene encoding the IFN-γR1 is located. Four affected Maltese children were subsequently found to be homozygous for a nonsense mutation (i.e. resulting in a stop codon), and another child was homozygous for a frameshift muta-tion in this gene. Functional studies in these patients indicated the absence of IFN-γR1 at the surface of macrophages and other immune cells, the absence of macrophage upregulation in response to IFN-γ, decreased macrophage mycobactericidal activity, and impaired granuloma formation (Newport et al 1996, Jouanguy et al 1996, Casanova et al 1997). This genetic defect therewith clearly demonstrates the essential role of IFN-γ and macrophages for resisting mycobacterial infections. The findings also indicate a strong selective power against this defect. However, further studies should indicate whether other, perhaps more subtle polymorphisms in the IFN-γ receptor gene also affect disease susceptibility to mycobacterial and other infections. Furthermore, while this IFN-γ receptor deficiency clearly indicates a role for a host's genetic influence on susceptibility to mycobacterial infections, it does not explain how much of the total variability of mycobacterial disease expression is genetically determined.

Population studies

Association studies in populations directly allow determining the contribution of gene variants to infection. These are typically case-control studies examining the dis-tribution of a polymorphic allele between unrelated affected and unaffected individu-als. For example, after the identification of CCR-5 as co-receptor for human immuno-deficiency virus (HIV) and a 32 base pair (bp) deletion mutation, population studies have shown the complete absence of homozygotes for the mutation in seropositive individuals compared with seronegative individuals. Possession of one copy of the mutant allele appeared to afford partial protection (Samson et al 1996, Dean et al 1996). These studies do not provide information on genetic variation in HIV-susceptibility in populations where the ΔCCR-5 mutation is absent, such as the African and Japanese populations. It is interestingly therefore that another association study found increased risk of HIV infection and shorter survival time after AIDS diagnosis in homozygous carriers of mannose-binding protein gene variants (Garred et al 1997). Other examples are discussed in chapter 19. Thus, while this approach is powerful for one or more single genes, they do not provide an estimation of the total genetic contribution to disease variability.

Familial studies

Simple approaches using for example odds ratios (ORs) applied to pairs of related individuals may be used to assess familial clustering as a first step to indicate genetic predisposition. Familial clustering has been reported for many infectious diseases. For example, it has been repeatedly observed that intense infections with *Ascaris lumbricoides* and *Trichuris trichiura* are aggregated in some individuals, which subsequently tend to be aggregated within certain families. The families themselves are predisposed to above average intensity of infections. The individuals are also predisposed to reacquire intense infections even after successful treatment (Chan et al 1994). However, not only genetic predisposition but also persistent environmental exposure may likely determine such long-term effects. Williams-Blangero et al (1999) presented a genetic epidemiological analysis of *A. lumbricoides* infection in a single pedigree of the Jiel population in eastern Nepal. Variance component analysis provided unequivocal evidence for a strong genetic component accounting for 30 to 50 % of the variation in worm burden, both before and after treatment (i.e. reinfection). Shared environmental effects (such as a common household) accounted for 3 – 13 % of the total phenotypic variation.

Familial clustering has been expressed as sibling relative risk or λ_s value. λ_s is the risk ratio of the risk of disease found in a sibling of a case compared to the general population risk. This λ_s may be expressed as an overall risk ratio that summarizes the collective effect of all disease loci or it may be expressed as the relative risk attributable to a specific locus. Thus the higher λ_s is, the stronger is the genetic effect. Because in many infectious diseases family members likely have a greater risk of exposure than the general population, the λ_s values likely overestimate the importance of host genetics. Statistical techniques as multiple linear regression, multivariate logistic regression, and regressive modeling may adjust for such risk factors that also aggregate in families, and other covariates. In addition, various genetic models may be considered using regressive models. λ_s values have been estimated from epidemiological surveys and from twin studies. For most infectious diseases λ_s values lie between one and ten. For comparison, in insulin-dependent diabetes mellitus λ_s is about 15, and in multiple sclerosis, Crohn's disease and systemic lupus erythematodes disseminatus (SLE) λ_s is about 20 (Hill 1998, Irfan et al 1999, Todd and Farral 1996, Weeks and Lathrop 1995).

Several twin studies have indicated the genetic basis to susceptibility to infectious disease, including poliomyelitis, tuberculosis, leprosy, *Toxoplasma gondii*, and *Helicobacter pylori* infection (Chakravati and Vogel 1973, Comstock 1978, Couvreur et al 1976, Malaty et al 1994, Vogel and Motulsky 1986). These studies compare the concordance rates in monozygotic and in dizygotic twins. Monozygous twins share all genes and dizygous twins share half of their genes. If disease susceptibility is genetically determined, monozygous twins resemble each other more than dizygous twins. In that case monozygotes have a higher concordance rate (i.e. both have the disease) than do dizygotes, assuming that both types of twins share their environments to equivalent degrees. To dissect the relative importance of genetic and environmental influences, it is necessary to calculate intraclass correlations for monozygous twins reared apart and together and dizygous twins reared apart and together. Although twin studies may suffer from many biases, they are able to detect both single and multiple gene-controlled disease susceptibility and also allow a direct measure of penetrance. Penetrance is the proportion of carriers of a genotype who also express it phenotypically. In twin studies penetrance can be estimated from the proband concordance rate, or the proportion affected among co-twins of affected monozygotes (Bulmer 1970, Wakelin and Blackwell 1988). A complicating factor, however, may occur if exposure to the infective organism is low.

Twin studies are one way to estimate the heritability of a disease. Heritability (h^2) is defined as the proportion of the total phenotypic variance in a trait attributable to segregation of a gene or genes. It may refer to a locus or to all loci combined. Heritability can range from 0, indicating a negligible genetic contribution, to 1, indicating a complete genetic influence. Hence the heritability of a trait is not universal, but specific to a particular population at a particular moment in a particular environment. Because heritability indicates the relative importance of genetic effects, the remaining variance is accounted for by shared rearing environmental factors, non-shared environmental factors, and strain differences. Evidently estimates of heritability may differ among populations due to genetic and environmental heterogeneity. Populations that are genetically homogeneous will yield lower heritability estimates than populations that are genetically heterogeneous. Likewise a large environmental variability will yield a low heritability estimate. Important factors are also measurement variation and repeatability. Evidently, heritability estimates may yield low values when measurement variation is high and repeatability of the measurement is low. In twin studies heritability can be estimated quantitatively from the formula:

$$\frac{\text{variance within pairs of dizygotic twins } - \text{ variance within pairs of monozygous twins}}{\text{variance within pairs of dizygotic twins}}$$

Thus heritability may be estimated from comparisons of variance within monozygotic and dizygotic twins, and further from the correlation coefficient for monozygotic twins that are reared apart, from model fitting analyses, and from threshold animal models (Detilleux et al 1995, Graham et al 1994), Malaty et al 1994). To compensate for shared environmental factors, the correlation coefficient for monozygotic twins reared apart provides the best single estimate of the relative importance of genetic effects (h^2).

Higher concordance rates have repeatedly been found in monozygotic twins for tuberculosis, leprosy, *Helicobacter pylori*, and several other infections. For example, monozygotic twins have a 36 % concordance for paralytic poliomyelitis compared with 6 –7 % shown by dizygotic twins and other sibs (Herndon and Jennings 1951). Malaty et al (1994) found a significantly higher concordance for infection with *H. pylori* in monozygotic than in dizygotic twin pairs, yielding a heritability estimate of 0.6. In studies examining presence or absence of tuberculosis in monozygotic twins and dizygotic twins, markedly higher concordance rates have repeatedly been observed in monozygotic twins. The studies suggest 30 – 60 % penetrance for the predisposing phenotype (Fine 1981). In a sib-pair linkage analysis, the heritability of a single locus on chromosome 5q31-q33 controlling *Plasmodium falciparum* blood infection levels was approximately 0.45 (Rihet et al 1998). Interestingly, concordance rate for hepatitis B persistence, but not for hepatitis B infection, was higher in monozygotic twins (Lin et al 1989), suggesting (not contra imagination) that genetic influence on disease variability differs in various steps of the pathogenesis. Data from the Dutch twin register estimated the heritability of physician-diagnosed bronchitis, bronchiolitis, and pneumonia between 0.3 and 0.4 (Boomsma 2000). Using a maximum likelihood variance decomposition approach, the maximum likelihood of the heritability of *Trypanosoma cruzi* infection in pedigrees was 0.56 ± 0.27 (mean $\pm$ standard error [SE]), indicating that genetic factors account for more than half of the observed variation. An additional 23 % of the variation was attributable to the effects of shared environment (Williams-Blangero et al 1997).

Some reported heritabilities of infectious diseases are given in Table 1.
An interesting possibility is the use of twins in experimental studies to demonstrate a host genetic influence on infection. For example, human immunodeficiency virus-1 (HIV-1) shows a remarkable similar pattern of replication kinetics in monocytes and macrophages obtained from identical twins when compared with the same cells obtained from non-identical twins and unrelated donors (Chang et al 1996). Subsequent studies using cells from twins may elucidate the mechanisms by which genetic effects influence HIV-1 replication *in vitro*, as well as the significance of the findings for the *in vivo* situation.

Veterinary species allow estimating heritability after experimental inoculation. However, while this approach may clearly indicate a genetic influence under the specific experimental circumstances, extrapolating the results to the natural disease under field conditions may be difficult if not impossible. For example, Matheron et al (1987) estimated a heritability of 0.49 – 0.85 for resistance of Guadeloupe goats to *Cowdria ruminantium* infection. Experimental or well-observed natural inoculations further allow examining more in detail the heritabilities of different disease-associated phenotypes. In Romney sheep, Bisset and Morris (1996) demonstrated that the genetics of resistance to nematode infection is not identical with the genetics of resilience (the ability to maintain production despite infection). In fact, lambs with low worm burdens did not appear to have increased production advantages, such as growth rate, when they were grazed together with lambs with high worm burdens. The heritabilities ($\pm$ SE) of various measures of resilience appeared relatively low, ranging between 0.1 ± 0.03 and 0.19 ± 0.04. Interestingly, heritability of resilience appeared to increase with increasing severity of the challenge.
A relatively high level of heritability ($h^2 = 0.49$) was observed in Hungarian Merino breed lambs experimentally inoculated with *Haemonchus contortus*. These authors suggested that the heritability estimate was applicable towards a wide range of nematode species (Sreter et al 1994). Studies in New Zealand and Australia have estimated that nematode resistance, measured by counting nematode egg counts in feces, has a heritability of about 0.3 in Romney and Merino sheep (Douch et al 1996). Detilleux et al (1995) estimated heritability for susceptibility to two bovine retroviral infections, bovine leukosis virus and bovine immunodeficiency virus, close to zero.

Table 1. *Estimated heritabilities of some infectious disease phenotypes*

Disease phenotype	Heritability (h^2)	Population	Study design	Ref.
Bovine retroviral infections	~0	Holstein cattle	pedigree study	Detilleux et al 1995
Susceptibility to *Mycobacterium avium subsp. paratuberculosis*	0.06 - 0.09	Dutch dairy cattle	pedigree study	Koets et al 2000
Nematode resilience (e.g. growth rate)	0.1 - 0.19	Romney sheep	pedigree studies	Bisset and Morris 1996
Salmonella typhimurium disease manifestations	0.06 (survival time) 0.09 (cecal carriage) 0.12 (mortality)	Chickens	experimental inoculation	Janss and Bolder 2000
Marek's disease manifestations	0.1 - 0.2	Chickens	pedigree studies	Van der Zijpp 1987
Nematode egg counts	0.3	Romney and Merino sheep	pedigree studies	Douch et al 1996
Plasmodium falciparum blood levels	0.45 (one locus on 5q31-q33)	humans	sib-pair linkage analysis	Rihet et al 1998
Plasmodium falciparum immune responses (several)	0.04 - 0.8	humans	twin study	Jepson et al 1997
Physician-diagnosed bronchi(oli)tis and pneumonia	0.3 - 0.4	humans	twin study	Boomsma 2000
Cowdria ruminantium resistance	0.49 - 0.85	Guadeloupe goats	experimental inoculation	Matheron et al 1987

continue table 1. Estimated heritabilities of some infectious disease phenotypes

Disease phenotype	Heritability (h^2)	Population	Study design	Ref.
Haemonchus contortus resistance	0.49	Merino lambs	experimental inoculation	Sreter et al 1994
Trypanosoma cruzi seropositivity	0.56	humans, Brazil	pedigree study	Williams-Blangero et al 1997
Several production and disease manifestations following trypanosome infection	0.39 - 0.64	N'Dama cattle	pedigree study	Trail et al 1991
Acquisition of *Helicobacter pylori* infection	0.6 - 0.7	humans	twin studies	Graham et al 1994 Malaty et al 1994

This finding, however, could be underestimated because the cows were selected for high milk production.

A concept related, but not identical to heritability is attributable risk. Attributable risk estimates the contribution of a genetic factor to the burden of disease as if it were possible to remove the factor. In contrast to heritability, estimates of the various components of variance (for example those explained by specific loci or specific mutations) and allele frequencies in the population are required for its calculation (Chakraborty 1984).

Families with increased susceptibility to infection clearly indicate a genetic influence on disease severity and they allow unique opportunities to identify susceptibility loci (Levin et al 1995). Segregation analysis may thus provide evidence that host genetic factors partly determine susceptibility to infectious disease. Without further information on the mode of inheritance and frequencies of the gene variants in the general population, the results of family studies do not provide information on the genetic contribution to disease variability in the general population. An advantage of family studies, however, is that they may also suggest the patterns of inheritance of susceptibility genes using segregation analysis. Segregation analysis explains the inheritance patterns of complex phenotypes in families by testing various models to estimate the number of Mendelian loci influencing the phenotype and the magnitude of specific genetic effects on phenotypic variability. Segregation analysis in families may for example suggest the existence of a major locus and may at least in part define its properties (Gambaro et al 2000). Otherwise segregation analysis may suggest oligogenic, polygenic or complex modes of inheritance.

Segregation analysis in Brazilian families demonstrated that the level of infection with *Schistosoma mansoni* was strongly influenced by a single codominant gene (Abel et al 1991). Later the gene, designated as *SM1*, was mapped by linkage analysis to chromosome 5q31-q33, a region that contains genes encoding proteins that control T cell differentiation, such as the gene for colony stimulating factor-1 receptor, colony stimulating factor-2, IL-3, IL-4, IL-5, IRF1, and IL-13 (Marquet et al 1996). The frequency of the deleterious gene was estimated to be 0.20 to 0.25. As a result about 5 % of the population was predisposed to high infection, 60 % was resistant, and 35 % had an intermediate level of resistance (Abel et al 1991). In Schistosomiasis not only infection level, but also disease phenotype appears under genetic control. Infected individuals differ in the severity of disease characterized by periportal fibrosis and portal blood hypertension. An association study found evidence for a contribution of genetic factors, notably HLA-DQB1*0201, to this disease phenotype (Secor et al 1996).

Adoption studies

A somewhat different approach to estimate the genetic contribution to infectious disease susceptibility is by comparing infectious disease among adoptees and natural children (Sorensen et al 1988). These authors analyzed causes of premature death in 960 families with children adopted early in life. Adoptees with a biological parent who had died before the age of 50 from an infectious (usually respiratory) disease had an increase in relative risk of 5.8 (95 % confidence interval [CI]: 2.5 – 13.7) of dying from an infectious disease. The relative risk of infectious disease mortality among the adoptees was not increased when their adoptive parents had died from infectious disease. (Note that relative risk is the increased chance of acquiring the disease for individuals bearing the risk factor relative to those lacking it.) This genetic influence on infectious disease mortality was greater than the genetic influence on death due to cardiovascular disease and cancer.

Ethnic, racial, and breed differences
Epidemiological studies comparing interpopulation differences in disease prevalence may suggest a role for genetic factors in disease. Because differences in environment, exposure, and strains may occur, they should be interpreted with caution. Moreover patterns of genetic variation between populations may not only involve the action of natural selection, but also simply population movement and random changes in gene frequencies.

Several data suggested that the immune response to syphilitic infections, cholera, plague, infantile diarrheas, smallpox, and *Helicobacter pylori* infection depend on ABO blood groups and could therefore explain differences in ABO in various ethnic groups (Vogel and Motulsky 1986). However, perhaps with the exception of the correlations between *Helicobacter pylori* infection and cholera with blood group O, the data are inconclusive and provide no mechanistic explanation for the presumed relations.

The evolutionary force of infectious disease is sometimes clear in host characteristics of ethnic groups. The influence of the most severe form of malaria, caused by *Plasmodium falciparum*, is obvious from the geographical distribution of the hemoglobinopathies, which give protection against the infection. However, selective forces from the same pathogen and different levels of exposure may also result in subtle differences in populations from different regions and with different genetic make-ups. One study found strong evidence for different susceptibility to malaria of ethnic groups living in the same area in Burkina Faso and thus exposed to a similar parasite load (Modiano et al 1996).

In New Zealand, rates of both meningococcal and pneumoccocal disease are higher in Maori and Pacific Island children than in Caucasian children, suggesting that common, yet unknown mechanisms may explain the high susceptibility to both pathogens. These may include both genetic as well as non-genetic risk factors as overcrowding (Baker et al 2000, Voss et al 1994, Wilson et al 1995).

Evidence of genetically based variation in susceptibility to infection with helminths has also been obtained by examining differences in and between ethnic groups. Large surveys carried out in the southern states of the USA clearly indicated racial differences between black and white people in the prevalence of infection. Hookworm infection prevalences of 22 % for whites and 4 % for blacks were recorded, with the whites harboring worm burdens that were twice as heavy in the black population. The hookworm species involved, *Necator americanus*, is considered to have had its evolutionary origin in Africa and was presumably taken to the USA by slave trade (Keller et al 1937, Wakelin and Blackwell 1988). Subsequently, evidence has been obtained for genetic variation in human susceptibility to schistosomiasis, lymphatic filariasis, onchocerciasis, ascariasis, trichuriasis, and hookworm infection. In addition, detailed surveys have shown that the distribution of parasites is characteristically aggregated. Almost invariably, irrespective of the species of host or parasite, most hosts harbor few parasites and a few harbor large numbers of parasites. These observations have led to the concept of "wormy persons", who are responsible for a major part of the parasite's transmission (Croll and Ghadirian 1981, Wakelin and Blackwell 1988). In addition to differences in genetic susceptibility, it seems likely that dominant mechanisms in the generation of parasite aggregation are differences in host exposure and differences in acquired immunity.

Several veterinary examples of breed differences in susceptibility to infectious disease illustrate the contribution of host genetics. For example, tsetse-transmitted trypanosomiasis is one of the major diseases affecting livestock in Africa. Because of antigenic variation no vaccine is available. However, natural selection has generated cattle, such as the N'Dama and West African Shorthorn breeds of West and Central Africa, that possess a significant degree of innate resistance. The ability to resist the development of anemia after infection appears a key trait of trypanotolerance (Murray et al 1990).

Experimental inoculation of laboratory animals
Numerous workers have reported differences between mouse strains in susceptibility or resistance to infection, which suggest genetically regulated differences in host response. Subsequent infection of mutant strains, as well as recombinant inbred or congenic mutant mice, has clearly provided evidence for a genetic basis for differences in infectious disease susceptibility. A very powerful, though artificial approach to examine the influence of host genes on infectious disease is by disrupting genes in murine models of infectious disease (Newport and Blackwell 1997). Using this approach a plethora of genes, involved for example in macrophage function, the cytokine network, the major histocompatibility complex, T cell function, have been shown to affect disease resistance. Obviously increased disease susceptibility in a knockout mouse model does not necessarily imply that the human counterpart of the disrupted gene is polymorphic and involved in the variability of disease expression in man. Much work (and some luck) is needed to find such a link. However, the *Nramp1* gene, a gene first identified in mice and shown to be responsible for resistance to intracellular bacteria, has set an example. Low-stringency screening of a human cDNA library isolated the human homolog NRAMP1. Although there are some differences, the human and mouse gene products have similar functions (see chapter 8).

Conclusion
A variety of approaches and parameters have been used to estimate the extent of the host genetic contribution to variation in infectious disease susceptibility. Although they clearly provide indications for a contribution of host genetics, they should all be interpreted with caution. Differences in exposure and strain virulence may be serious confounders in these studies. While family studies may indicate a genetic contribution and suggest a mode of inheritance, they do generally not allow extrapolating the findings to the general population. Estimates of heritability, by definition, provide the best parameter for estimating the genetic component of infectious disease. For infectious diseases they differ from nearly zero to 0.7 − 0.8. Association studies examine the contribution of one or more candidate genes in infectious disease. While this approach is powerful for one or more single genes, they do not provide an estimation of the contribution of the total genetic contribution to disease variability.

References

Abel, L., Demenais, F., Prata, A., Souza, A.E., Dessein, A. (1991). Evidence for the segregation of a major gene in human susceptibility/resistance to infection by *Schistosoma mansoni*. Am. J. Hum. Genet. 48, 959-970.

Baker, M., McNicholas, A., Garrett, N., Jones, N., Steart, J., Koberstein, V., Lennon, D. (2000). Household crowding a major risk factor for epidemic meningococcal disease in Auckland children. Pediatr. Infect. Dis. J. 19, 983-990.

Bisset, S.A., Morris, C.A. (1996). Feasibility and implications of breeding sheep for resilience to nematode challenge. Int. J. Parasitol. 26, 857-868.

Boomsma, D. (2000). Personal communication.

Bulmer, M.G. (1970). The biology of twinning in man. Clarendon Press, Oxford.

Casanova, J.L., Newport, M., Fisher, A., Levin, M. In: Primary Immunodeficiencies: A Molecular and Genetic Approach, Ed. by Ochs, H., Puck, J., Smith, C. New York: Oxford University Press; 1997.

Chagnon, N.A., Melancon, T.F. (1980). Reproduction, numbers of kin, and epidemics in tribal populations: a case study. In: Population and biology, Ed. by N. Kreyfitz, Ordina Editions, Liege, pp. 147-168.

Chakraborty, R. (1984). Relative contributions of gene and envrionment: attributable risk and heritability. J. Indian Antrhropol. Soc. 19, 147-152.

Chakravati, M.R., Vogel, F. (1973). A twin study on leprosy. Topic Hum. Genetic. 1, 1-123.

Chan, L., Bundy, D.A.P., Kan, S.P. (1994). Aggregation and predisposition to *Ascaris lumbricoides* and *Trichuris trichiura* at the familial level. Trans. Roy. Soc. Trop. Med. Hyg. 88, 46-48.

Chang, J., Naif, H.M., Li, S., Sullivan, J.S., Randle, C.M., Cunningham, A.L. (1996). Twin studies demonstrate a host cell genetic effect on productive human immunodeficiency virus infection of human monocytes and macrophages *in vitro*. J. Virol. 70, 7792-7803.

Comstock, G. (1978). Tuberculosis in twins: a re-analysis of the Prophit Survey. Am Rev. Respir. Dis. 117, 621-624.

Couvruer, J., Desmonts, G., Girre, J.Y. (1976). Congenital toxoplasmosis in twins. J. Pediatr. 89, 235-240.

Croll, N.A., Ghadirian, E. (1981). Wormy persons: contributions to the nature and patterns of overdispersion with *Ascaris lumbricoides, Ancylostoma duodenale, Necator americanus*, and *Trichuris trichiura*. Tropical and Geographical Medicine 33, 241-248.

Cummins, S.L. (1929). Virgin "soil"- and after. A working conception of tuberculosis in children, adolescents and aborigines. Br. Med. J. 2, 39-41.

Dean, M., Carrington, M., Winkler, C., Huttley, G.A., Smith, M.W., Alikmets, R., Goedert, J.J., Buchbinder, S.P., Vittinghof, E., Gomperts, E., Donfield, S., Vlahov, Kaslow, R., Saah, A., Rinaldo, C., Detels, R., O'Brien, S.J. Genetic restriction of HIV-1 infection and progression to AIDS by a deletion allele of the CKR5 structural gene. (1996). Science 273, 1856-1862.

Dekaban, G.A., Digilio, L., Franchini, G. (1995). The natural history and evolution of human and simian T cell leukemia/lymphotropic viruses. Curr. Opin. Genet. Dev. 5, 807-813.

Detilleux, J.C., Kehrli, M.E., Freeman, A.E., Whetstone, C.A., Kelly, D.H. (1995). Two retroviral infections of periparturient Holstein cattle: a phenotypic and genetic study. J. Dairy Sci. 78, 2294-2298.

Dobyns, H.F. (1966). Estimating aboriginal american population. Current Anthropol. 7, 395-416.

Donald, P.R. (1998). The epidemiology of tuberculosis in South Africa. In: Genetics and tuberculosis. Wiley, Chisester (Novartis Foundation Symposium 217), 24-41.

Douch, P.G., Green, R.S, Morris, C.A., McEewan, J.C., Windon, R.G. (1996). Phenotypic markers for selection of nematode-resistant sheep. Int. J. Parasitol. 26, 899-911.

Fine, P.E.M. (1981). Immunogenetics of susceptibility to leprosy, tuberculosis, and leishmaniasis. An epidemiological perspective. International Journal of Leprosy 49, 437-454.

Garred, P., Madsen, H.O., Balsev, U., Hofman, B., Pedersen, C., Gerstoft, J., Svejgaard, A. (1997). Susceptibility to HIV infection and progression of AIDS in relation to variant alleles of mannose-binding lectin. Lancet 349, 236-240.

Gambaro, G., Anglani, F., D'Angelo, A. (2000). Association studies of genetic polymorphisms and complex disease. Lancet 355, 308-311.

Graham, D.Y., Malaty, Go, M.F. (1994). Are there susceptible hosts to *Helicobacter pylori* infection? Scand. J. Gastroenterol. Suppl. 210S, 6-10.

Haldane, J.B.S. (1949). Disease and evolution. La Ricerca Sci. 19S, 68-76.

Herndon, C.N., Jennings, R.G. (1951). A twin study of susceptibility to poliomyelitis. Am. J. Hum. Genet. 3, 17.

Hill, A.V.S. (1998). The immunogenetics of human infectious diseases. Annu. Rev. Immunol. 16, 593-617.

Holmes, E.C. (1999). Molecular phylogenies and the genetic structure of viral populations. In: Evolution in health and disease. Ed. by S.C. Stearns, Oxford University Press, pp. 173-182.

Irfan, U.M., Dawson, D.V., Bissada, N.F. (1999). Assessment of familial patterns of microbial infection in periodontitis. J. Periodontol. 70, 1406-1418.

Janss, L.L., Bolder, N.M. (2000). Heritabilities of and genetic relationships between *Salmonella* resistance traits in broilers. J. Anim. Sci. 78, 2287-2291.

Jepson, A., Banya, W., Sisay-Joof, F., Hssan-King, M., Nunes, C., Bennet, S., Whittle, H.C. (1997). Quantification of the relative contribution of major histocompatibility complex (MHC) and non-MHC genes to human immune responses to foreign antigens. Infect. Immun. 65, 872-876.

Jouanguy, E., Altare, F., Lamhamedi, S., Revy, P., Emile, J.F., Newport, M., Levin, M., Blanche, S., Seboun, E., Fisher, A., Casanova, J.L. (1996). Interferon-γ-receptor deficiency in an infant with fatal Bacillus Calmette-Guérin infection. N. Engl. J. Med. 335, 1956-1960.

Keller, A.E., Leathers, W.S., Knox, J.C. (1937). The present status of hookworm infestation in North Carolina. American Journal of Hygiene 26, 437-454.

Koets, A.P., Adugna, G., Janss, L.L., van Weering, H.J., Kalis, C.H., Wentink, G.H., Rutten, V.P., Schukken, Y.H. (2000). Genetic variation of susceptibility to *Mycobacterium avium subsp. paratuberculosis* infection in dairy cattle. J. Dairy Sci. 83, 2702-2708.

Lander, E.S., Botstein, D. (1987). Homozygosity mapping: A way to map human recessive traits with the DNA of inbred children. Science 236, 1567-1570.

Levin, M., Newport, M.J., D'Souza, S., Kalabalikis, P., Brown, I., Lenicker, H., Vassallo Agius, P., Davies, E.G., Thrasher, A., Blackwell, J.M. (1995). Familial disseminated atypical mycobacterial infection in early childhood: a human mycobacterial susceptibility gene? Lancet 345, 79-83.

Lin, T.M., Chen, C.J., Wu, M.M., Yang, C.S., Chen, J.S., Lin, C.C. Kwang, T.Y., Hsu,S.T., Lin, S.Y., Hsu, L.C., (1989). Hepatitis B virus markers in Chinese twins. Anticancer Res. 9, 737-741.

Marquet, S., Abel, L., Hillaire, D., Dessein, H., Kalil, J., Feingold, J., Weissenbach, J., Dessein, A.J. (1996). Genetic localization of a locus controlling the intensity of infection by *Schistosoma mansoni* on chromosome 5q31-q33. Nature Genet. 14, 181-184.

Malaty, H.M., Engstrand, L., Pedersen, N.L., Graham, D.Y. (1994). *Helicobacter pylori* infection: genetic and environmental influences, a study of twins. Ann. Intern. Med. 120, 982-986.

Matheron, G., Barre, N., Camus, E., Gogue, J. (1987). Genetic resistance of Guadeloupe native goats to heartwater. Onderstepoort J. Vet. Res. 54, 337-340.

McNeill, W.H. (1976). Plagues and people. Garden City, N.Y.: Anchor Press/Doubleday.

Modiano, D., Petrarca, V., Sirima, B.S., Nebie, I., Diallo, D., Esposito, F., Coluzzi, M. (1996). Different response to *Plasmodium falciparum* malaria in west African sympatric ethnic groups. Proc. Natl. Acad. Sci. USA 93, 13206-13211.

Motulsky, A.G. (1960). Metabolic polymorphisms and the role of infectious diseases in human evolution. Hum. Biol. 32, 28-62.

Murphy, F.A. (1998). Personal communication.

Murray, M., Trail, J.C., D'Ieteren, G.D. (1990). Trypanotolerance in cattle and prospects for the control of trypanosomiasis by selective breeding. Rev. Sci. Tech. 369-386.

Newport, M.J., Huxley, C.M., Huston, S., Hawrylowicz, C.M., Oostra, B.A., Williamson, R., Levin, M. (1996). A mutation in the interferon-γ-receptor gene and susceptibility to mycobacterial infection. N. Engl. J. Med. 335, 1941-1949.

Newport, M.J., Blackwell, J.M. (1997). Genetic susceptibility to tuberculosis. In: Mycobacterial infections. Ed. By McAdam, K.P.W.A., Malin, A. London, Bailliere Tindall.

Pagel, M. (1999). Inferring the historical patterns of biological evolution. Nature 401, 877-884.

Poppel, F. van, Imhoff, E. van (2000). Demographie van de doden. Natuur en Techniek, 68, 40-43.

Rihet, P., Trarore, Y., Abel, L., Aucan, C, Traore-Leroux, T., Fumoux, F. (1998). Malaria in humans: *Plasmodium falciparum* blood infection are levels are linked to chromosome 5q31-33. Am. J. Hum. Genet. 63, 498-505.

Samson, M., Libert, F., Doranz, B.J., Rucker, J., Liesnard, C., Farber, C.M., Saragosti, S., Lapoummerouilie, C., Cogneaux, J., Forceille, C., Muyldermans, G., Verhofstede, C., Burtonboy, G., Georges, M., Imai, T., Rana, S., Yi, Y., Smyth, R.J., Collman, R.G., Doms, R.W., Vassart, G., Parmentier, M. (1996). Resistance to HIV-1 infection in caucasian individuals bearing mutant alleles of the CCR-5 chemokine receptor gene. Nature 382, 722-725.

Secor, W.E., del Corral, H., dos Resi, M.G., Ramos, A.G., Zimon, A.E., Matos, E.P., Reiss, E.A.G., do Carmo, T.M.A., Hirayama, K., David, R.A., David, J.R., Harn, D.A. Jr. (1996). Association of hepatosplenic schistosomiasis with HLA-DQB1*0201. J. Infect. Dis 174, 1131-1135.

Sorensen, T.I., Nielsen, G.G., Andersen, P.K., Taesdale, T.W. (1988). Genetic and environmental influences on premature death in adult adoptees. N. Engl. J. Med. 318, 727-732.

Sreter, T., Kassai, T., Takacs, E. (1994). The heritability and specificity of responsiveness to infection with *Haemonchus contortus* in sheep. Int. J. Parasitol. 24, 871-876.

Strassmann, B.I., Dunbar, R.I.M. (1999). Putting the stone age in perspective. In: Evolution in health and disease. Ed. by S.C. Stearns, Oxford University Press, 91-101.

Todd, J.A., Farral, M., (1996). Panning for gold: genome-wide scanning for linkage in type 1 diabetes. Hum. Mol. Genet, 5, 1443-1448.

Trail, J.M.C., d'Ieteren, G.D.M., Murray, M. (1991). Practical aspects of developing genetic resistance to trypanosomiasis. In: Axford, R.F.E., and Owen, J.B. (Eds): Breeding for disease resistance in farm animals. CAB International, Wallington, pp. 224-234.

Van der Zijpp, A.J. (1987) Genetische aspecten van ziekteresistentie bij landbouwhuisdieren. Tijdschr. Diergeneesk. 112, 927-933.

Vogel, F., Motulsky, A. (1986). Human genetics: problems and approaches, 3rd edn. Springer-Verlag, Heidelberg.

Voss, L., Lennon, D., Okesene-Gafa, K., Ameratunga, S., Martin, D. (1994). Invasive pneumococcal disease in a pediatric population, Auckland, New Zealand. Pediatr. Infect. Dis. J. 13, 873-878.

Wakelin, D., Blackwell, J.M. (1988). Genetic of resistance to bacterial and parasitic infection. Taylor and Francis, London, Philadelphia, New York.

Weeks, D.E., Lathrop, G.M. (1995). Polygenic disease: methods for mapping complex disease traits. Trends in Genet. 11, 513-519.

Williams-Blangero, S., Vandeberg, J.L., Blangero, J., Teixeira, A.R. (1997). Genetic epidemiology of seropositivity for *Trypanosoma cruzi* infection in rural Goias, Brazil. Am. J. Trop. Hyg. 57, 538-543.

Williams-Blangero, S.,Subedi, J., Upadhayay, R.P., Manral, DB., Rai, D.R., Jha, B., Robinson, E.S., Blangero, J. (1999). Genetic analysis of susceptibility to infection with *Ascaris lumbricoides*. Am. J. Trop. Med. Hyg. 60, 921-926.

Wilson, N., Baker, M., Lennon, D., O'Hallahan, J., Jones, N., Wenger, J., Mansoor, O., Thomas, M., Jefferies, C. (1995). Meningococcal disease epidemiology and control in New Zealand. N. Z. Med. J. 108, 437-442.

Zanotto, P.M., Gould, E.A., Gao, G.F. Harvey, P.H., Holmes, E.C. (1996). Population dynamics of flaviviruses revealed by molecular phylogenies. Proc. Natl. Acad. Sci. USA 93, 548-553.

Chapter 2.   Genetic diversity and infectious disease

Mammalian genes clearly differ in their degree of variation. At the one end of the spectrum are monomorphic genes, genes that, as far as we know, do not exhibit variation among the species' members, perhaps with the exception of rare and often lethal alleles. At the other end of the spectrum are polymorphic genes, genes encoding for example blood group and major histocompatibility complex (MHC) antigens and enzymes, which may exhibit a large number of alleles. "Polymorphism" is the term for the situation where multiple allelic forms are maintained at population frequencies too high to be solely attributed to recurrent mutation. A genetic trait is defined as polymorphic if the prevalence of an allelic variant is higher than 1 – 2 % (Vogel and Motulsky 1986). They usually reflect benign sequence variants. Obviously before the completion of the Human Genome Diversity Project, we will not have a comprehensive understanding of genetic variation and its role in disease.

Possible mechanisms for phenotypic variation include polymorphism in the gene itself, its regulation of expression, and polymorphisms in receptor or signaling molecules. Variation in gene expression may not only be dependent on promoter-based regulation, but also on message stability and coordinated regulation by transactivators such as transcription factors. Potential regions of polymorphism therefore include coding, intronic, as well as 3' areas containing regulatory sequences. Any genetic variation may result in an unaltered function, or in loss, gain, or alteration of function. Any relation between molecular variants and disease must therefore be confirmed by assessment of *in vivo* and *in vitro* functional activity and by association in population studies (Rosenwasser 1997).

Many authors have pointed to an advantageous role of genetic diversity in providing resistance to a species against a wide range of pathogens and other environmental stresses. If true, selective pressure should maintain and develop genetic variation. Furthermore, if genetic variation is important in providing resistance against infection, then infections must be and have been a major evolutionary force in providing genetic variation among genes involved in infectious disease resistance. Thus genetic diversity appears to constitute a buffer against changes in the environment and is a key in selection for adaptability. Another, be it a somewhat academic, question is how much infections have contributed to genetic diversity.

The advantageous effect of genetic variability may be obvious at two levels. First, genetic variability in disease susceptibility directly reduces the number and density of susceptibles in a population, which therewith limits the spread of a pathogen (expressed as R_0 or basic reproduction number). Moreover, because the density of susceptibles in a heterogeneous population may be lower (compared to a completely homogeneous susceptible population) some of them may escape infection, a mechanism denoted as genetic herd immunity (Wills 1996). In this model the strongest force driving host-pathogen co-evolution is variation in ease of spread of the pathogen rather than variation in the pathogen's virulence. Second, genetic variation may give a species larger chance of survival when attacked by major and diverse selective pressures. It is therefore likely that the more selective pressures are acting on a gene, the more diversity the gene may exhibit.

Although in theory the advantages of genetic variability and the influence of infections in generating variability are plausible and easy to comprehend, remarkable little observational and experimental data are available to support these views or to estimate their effects in more quantitative terms. For example, Philip Murphy (1993) found, by comparing sequences of genes common to rodents and humans, that proteins involved in host defense have diverged both within and between species three times as quickly as other proteins. The finding suggest also that selection in many species has resulted from exposure to different pathogens. Despite this finding, there appear to be important exceptions. While significant population-based promoter polymorphisms have been revealed within the cytokine genes for IL-3, IL-4, IL-9, and others, such polymorphisms have not been identified for other immune genes. However, as mentioned above, a lack of a functional polymorphism in a cytokine promoter region does not rule out polymorphisms in other regulatory, receptor or signaling functions that are related to the protein's function. For example, toxic nitric oxide (NO) and transforming growth factor-β (TGF-β) have been identified as critical effector mechanisms in the control of infections with *Leishmania* species in mouse. Yet, no functional polymorphisms or allelic variation in the control of NO or TGF-β production or activity have been reported (McLeod et al 1995). (The function of the C to T promoter polymorphism at position -509 in the gene encoding TGF-β has not been elucidated yet.) Differences in susceptibility to *Leishmania* therefore appear to be regulated by genes regulating the Th1 and Th2 dichotomy (see chapter 10). To date, common polymorphisms have also not been identified in the *IL-13RA2* and *IL-16* genes, and the *JAK1, JAK3,* and *TYK2* genes, genes that are involved in the signaling pathway of cytokines (Shirakawa et al 2000). Initial reports mentioned a lack of polymorphism in the IL-13 gene promoter, a gene possibly involved in the generation of asthma (Anderson et al 1999, Wills-Karp et al 1998, Wills-Karp and Rosenwasser 1999). Recently, however, several single nucleotide polymorphisms (SNPs) have been identified in the promoter, coding, and 3' untranslated regions of the IL-13 gene (Graves et al 2000, Laundy et al 2000, Liu et al 2000). The Arg130Gln polymorphism in the IL-13 gene, or others in close linkage with it, appears strongly associated with the development of elevated serum IgE levels.

In populations showing genetic diversity in susceptibility to infectious disease mechanisms must operate to maintain it. In contrast, it is likely that in the absence of infection, individuals who are resistant to infection have a reduced fitness. In the presence of infection, additional mechanisms must thus operate to maintain diversity. The most important of these are heterozygous advantage (syn. overdominance) and frequency- and density-dependent selection. Other mechanisms that can maintain balanced polymorphisms include migration, sex linkage, unequal selection on the different sexes, and fluctuating selection pressures due to intermittent epidemics.
Heterozygous advantage occurs when the heterozygote has a selective advantage over both homozygous individuals. The best-documented evidence for infectious diseases favoring heterozygous advantage is malaria. For example, individuals who are homozygous for the sickle cell gene (or other inherited hemoglobinopathies) have a reduced viability due to anemia, while normal individuals have a reduced viability due to malaria. In endemically infected areas the heterozygotes have an advantage over both and thus maintain the polymorphism (Allison 1964, Anderson 1988). This and similar relations make that wherever there is, or has been, malaria there are hemoglobinopathies, often at very high frequencies. Likewise, when there is no malaria, these disorders are absent or rare. Exceptions to this rule are rare and can usually

be explained by gene flow, the action of drift on small populations, and other mechanisms that do not contradict the selective influence of malaria (Weatherall 1996).

One form of balancing selection is frequency- and density-dependent selection, which occurs when selection depends on the relative frequencies of the various host genotypes and on the total number of hosts. Because temporal and geographical variation in the protective effect of host genetic polymorphisms might often occur, this form of selection is typically exerted by infectious disease. While selective effects of pathogens require exposure to the pathogens, common susceptibility alleles will suffer more when their frequency is high. In contrast, individuals with rare alleles may have the greatest resistance to infectious diseases. Such frequency- and density-dependent selections in infectious disease have been suggested for blood group antigens, MHC class I genes, and parasite receptors with complex immunological functions, such as CD36 and ICAM-1 (Craig et al 2001). This mode of selection is necessarily transient: When the selected allele increases in frequency, the pathogen likely adapts itself. This form of adaptation is therefore known as the Red Queen hypothesis (see Alice in Wonderland). For example, in human populations where the HLA-A11 allele is present but not common, the prevalent Epstein-Barr virus strains have a conserved epitope that is presented by HLA-A11 to relevant cytotoxic T cells. In contrast, in populations where HLA-A11 is common, the prevalent virus strains harbor an amino acid substitution in this same epitope that prevents binding to HLA-A11. Thus this particular virus strain has likely escaped presentation by HLA-A11, because this common MHC haplotype exerts strong selection pressure for variant viruses (Apanius et et al 1997).

Because of their nature, infectious disease will only exert a selective pressure if the density of susceptible hosts is above a critical threshold value (Anderson 1988). Below the critical host density the pathogen is unable to persist within the population and will have no selective impact. Furthermore, geographical strain differences of pathogens may produce different selection pressures in particular regions. Moreover particular MHC types may show different associations with major infectious diseases in different geographic regions. Likewise, selection may fluctuate in time due to repetitive epidemics (Gillespie 1985). Thus the simple occurrence of epidemics of fatal diseases may lead to significant variation in selection pressures over time and space. Complex interactions may therefore occur that relate to the local circumstances of an infectious disease with regard to its incidence, antigenic composition and transmissibility.

Another proposed mechanism for generating and maintaining diversity is sexual selection. Sexual selection has been hypothesized to be influenced by pathogens, in the sense that individuals in good health and vigor are preferred in mate choice because they probably have heritable resistance to the predominant pathogens. Preferring healthy mates should yield more resistant progeny and result in populations of hosts less susceptible to the local pathogens and in a lower average level of virulence (Hamilton and Zuk 1982, Wedekind 1999).

Because many co-evolutionary paths are possible, the final outcome of the selective forces is very complex. Moreover, insufficient data from field studies are available to permit assessment of which of these types of selection have been most important in generating diversity at particular loci. Some examples follow to illustrate this point. The unknown long-term genetic effects of the bubonic plague caused by *Yersinia pestis* may underline our ignorance of the precise role of infections in generating genetic diversity. Although this disease is estimated to have caused 200 million human deaths, sometimes nearly wiping out whole cities and villages, we do not know the effects of this disease on our genetic make-up. Its selective effect may however have been very large as almost everywhere the first wave of plague killed between twenty to fifty percent of the people in the affected communities (Wills 1996). Because the reasons for the disappearance of plague in Europe (for example in London around 1665) are unknown, it is tempting to hypothesize a contribution of genetic resistance. The same is true for leprosy, a disease that caused many deaths in the 12th and 13th centuries in Europe, and for influenza that caused 20 million deaths worldwide during the epidemic of 1919. Malaria, on the other hand, which still exerts selective pressure, appears to give more information on its role in maintaining certain polymorphisms (see chapter 21).

In conclusion, it seems likely that many delicate and complex mechanisms that provide and maintain genetic variability are unknown. A full understanding of the role of infectious diseases in generating and maintaining genetic diversity is therefore impossible. Not only have many known and unknown pathogens evolved to lesser virulence or disappeared, but there is likely also a smaller or greater role of many less virulent or (nowadays) avirulent saprophytic micro-organisms.

Balanced and transient polymorphisms

In many parts of the world where malaria is endemic there are thalassemias with completely different patterns of α or β globin gene mutations (Weatherall 1996). This suggests that these mutations have arisen locally and have reached a high frequency by selection. However, the mechanisms of selection appear to differ. The high frequency of the β-thalassemias, like that of the sickle cell trait, appears to result from heterozygote protection against severe malaria. For example, heterozygosity for hemoglobin S (HbS) provides 80-95% protection against severe forms of *P. falciparum* infection, notably severe anemia and cerebral involvement (Hill et al 1991). The mortality of Hb SS is 100 % in childhood in Africa. These traits thus represent balanced polymorphisms. From the frequency of the gene for HbS, it is possible to estimate the mortality from malaria by using the Hardy-Weinberg equation. In malaria-endemic areas, Hb SA heterozygotes have a survival advantage compared with Hb AA children. The mortality from malaria must have been as high as 25 % to account for the HbS gene frequencies that are present today (Miller 1994).

In contrast, the mechanism for selection for the mild form of α-thalassemia may be more complex. There are two α globin genes per haploid genome ($\alpha\alpha$). The commonest forms of α-thalassemias result from deletions of either one or both of these genes. Thalassemia due to a single deleted α gene, of which there are several different forms, is extremely common in many tropical populations. However, it appears that only homozygotes are resistant to *P. falciparum* malaria to a level of about 25 %, which is considerably less than sickle-cell heterozygotes. Because this very frequent genotype does not appear to have deleterious effects, it is likely that these forms of α-thalassemias represent transient polymorphisms, rather than balanced polymorphisms, that would have gone to fixation if the selective pressure of malaria had continued (Weatherall 1996).

Mechanisms of maintaining diversity of major histocompatibility complex (MHC) loci

Accumulation of point mutations, gene duplications, deletions, intra- and intergenic recombination events, and selection are basic molecular mechanisms that have resulted in genetic diversity at MHC loci, the most diverse antigenic system of mammals that has a function in antigen presentation (see chapter 9). It consists of a large number of genes, in humans on the short arm of chromosome 6. Interestingly, the range of MHC allele frequencies at some loci is very large with some alleles quite common and many others rare, while at other loci the range of allele frequencies is far narrower. The question which evolutionary mechanisms have maintained polymorphism is particularly relevant for the MHC. Most researchers agree that the diversity at MHC loci is maintained by selective pressures produced by fatal infectious diseases among the ancestors of modern species (Parham and Ohta 1996, Hughes and Yeager 1998). The observation that different allelic lineages at the MHC locus have remained polymorphic within humans since their evolution from a common ancestor with Old World monkeys is strong evidence for a polymorphic-maintaining (or balancing) selection at this locus. Indeed MHC alleles and their frequencies often have long persistence times, in some cases tens of millions of years. MHC diversity apparently predates speciation by millions of years and is largely inherited as part of the genetic diversity of the founders of a new species. However, because the allelic composition of the MHC polymorphism does change over evolutionary time, the MHC system must be capable of accommodating new alleles with similar properties without destructing the equilibria that permit the maintenance of the older alleles. In the absence of selection, most alleles should have been lost through genetic drift. Thus the question is which selective forces have raised and lowered the frequencies of specific alleles. Because very strong associations of the MHC (including the human leukocyte antigen complex [HLA]) with infectious diseases have not been identified, the role of infectious diseases in maintaining the unique diversity of the MHC polymorphism has been questioned. However, relatively weak associations, with odds ratios close to 1.0, would appear sufficient to maintain polymorphism over long periods of time (Hill 1996). In addition, the very old age of MHC diversity may argue that old parasites that have co-evolved with their hosts for millions of years in the past have had more influence on MHC diversity than the more recently evolved parasites that we know today (Klein and O'Huigin 1994). This argument would also explain why it has been so difficult to find convincing association between MHC alleles and resistance to infectious disease.

The role of infectious diseases co-evolving with their hosts in shaping and maintaining MHC polymorphism is further underlined by their mode of action in selecting determinants for T cellular immunity. In fact, the only known function of classical MHC molecules is to present peptides to T lymphocytes. The capacity to select a wide range of T cell determinants likely results in heterozygote advantage (or overdominant selection) as one explanation for the extraordinary polymorphisms found at MHC loci. Because MHC expression is codominant, heterozygous advantage for MHC loci is explained by the increased repertoire of epitopes presented to the immune system upon infection. Indeed, individuals heterozygous at MHC genes may themselves have increased disease resistance. For example, in HIV-infected individuals maximum *HLA* heterozygosity of class I loci (A, B, and C) delayed AIDS onset, whereas individuals who were homozygous for one or more loci progressed rapidly to AIDS and death (Carrington et al 1999). As indicated, this heterozygous advantage in AIDS is probably explained by an increased repertoire of viral epitopes presented to CD8+ cytotoxic T cells (Fig. 2). But in addition to heterozygote advantage, frequency- and density-dependent selection has likely also contributed to maintaining MHC polymorphism.

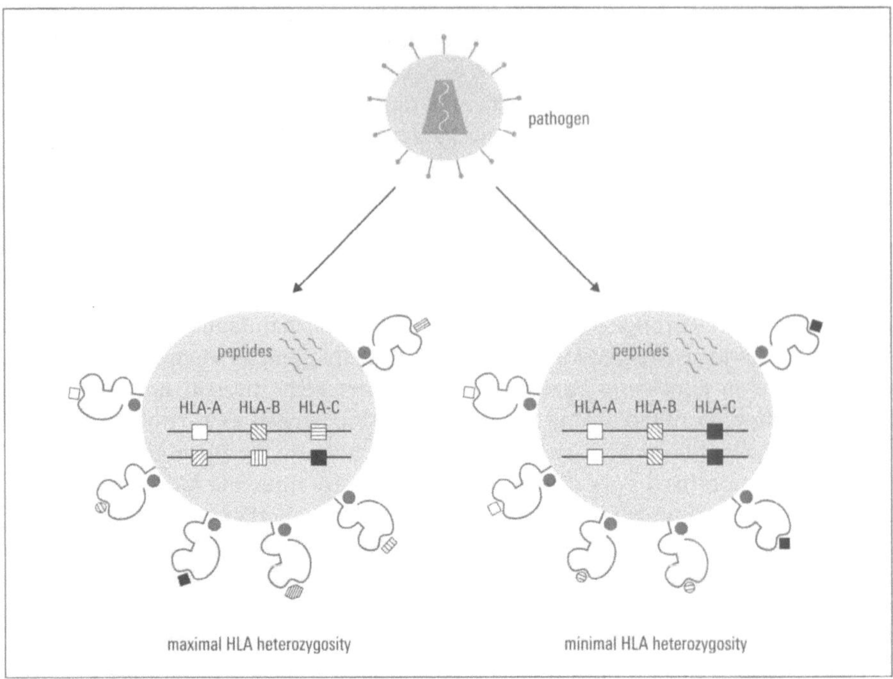

Figure 2. Maximal HLA heterozygosity may protect against infectious disease. People heterozygous for HLA alleles at different loci (left) express a more diverse range of peptide-presenting molecules on the surface of their antigen-presenting cells. Hence a stronger and broader immune response may follow in these persons than in those that only express one type of antigen-presenting molecule at each locus. The figure further illustrates that selection on the entire HLA family is weaker than on a single HLA gene.

Other evidence for selective forces acting at maintaining MHC polymorphism is the finding that mutations that produce amino acid changes cluster around the peptide-binding cleft of the HMC molecules, indicating that selection must have been acting on this function during evolution. Also introns have been homogenized relative to exons over evolutionary time, suggesting that balancing selection acts to maintain diversity in the latter, in contrast to the former (Hughes and Yeager 1998). Other adjunctive evolutionary mechanisms that have been suggested, but not proven, to contribute to the maintenance of the MHC polymorphism are disassortative mating and fetal-maternal interactions (Clarke and Kirby 1991, Hill 1992, Potts et al 1991, Wills 1991). Fetal-maternal interactions might lead to increased rates of fetal loss in mothers who share relatively more MHC types with their offspring.

The MHC system is not just very polymorphic, but allele frequencies vary also greatly between populations. For example, HLA-A2 is present in all 87 human populations tested, except in New Guinea, with an average gene frequency of around 25 %. In contrast, HLA-B70 is not found at all in 30 populations, has an average frequency of 1.7 % and reaches frequencies of 19 % and 17 % in South African Blacks and Khoi respectively (Bodmer 1996). This observation agrees with the assumption that modest selective forces, in combination with a significant production rate of new alleles by gene conversion or recombination, could give rise to marked changes in gene frequency distributions within a few hundred generations.

References

Allison, A.C. (1964). Polymorphism and natural selection in human populations. Cold Spring Harbor Symposium on Quantitative Biology 29, 137-149.

Anderson, K.L., Mathieson, P.W., Gillespie, K.M. (1999). Polymorphisms not found in the IL-13 gene promoter. Science 284, 5419.

Anderson, R.M. (1988). The population biology and genetics of resistance to infection. In: Genetics of resistance to bacterial and parasitic infection. Ed. by D. Wakelin and J.M. Blackwell, Taylor and Francis, London, Philadelphia, New York.

Apanius, V., Penn, D., Slev, P.R., Ruff, L.R., Potts, W.K. (1997). The nature of selection on the major histocompatibility complex. Crit. Rev. Immunol. 17, 179-224.

Bodmer, J. (1996). World distribution of HLA alleles and implications for disease. In: Variation in the human genome. Wiley, Chichester (Ciba Foundation Symposium 197), 233-258.

Carrington, M., Nelson, G.W., Martin, M.P., Kissner, T., Vlahov, D., Goedert, J.J., Kaslow, R., Buchbinder, S., Hoots, K., O'Brien, S.J. (1999). *HLA* and HIV-a: Heterozygote advantage and *B*35-Cw*04* disadvantage. Science 282, 1748.

Clarke, B., Kirby, D.R. (1996). Maintenance of histocompatibility polymorphisms. Nature 211, 999-1000.

Craig, A., Hastings, I., Pain, A., Roberts, D.J (2001). Genetics and malaria-more questions than answers. Trends in Parasitol. 17, 55-56.

Gillespie, J.H. (1985). The interaction of genetic drift and mutation with selection in a fluctuating environment. Theor. Populat. Biol. 27, 222-237.

Graves, P.E., Kabesch, M., Halonen, M., Holberg, C.J., Baldinini, M., Fritzsch, C., Weiland, S.K., Erickson, R.P., von Mutius, E., Martinez, F.D. (2000). A cluster of seven tightly linked polymorphisms in the IL-13 gene is associated with total serum IgE levels in three populations of white children. J. Allergy Clin. Immunol. 105, 506 – 513.

Hamilton, W.D., Zuk, M. (1982). Heritable true fitness and bright birds: a role for parasites? Science 218, 384-387.

Hill, A.V.S., Allsopp, C.E.M., Kwiatkowski, D., Anstey, N.M., Twumasi, P., Rowe, P.A., Bennett, S., Brewster, D., McMichael, A.J., Greenwood, B.M. (1991). Common west African HLA antigens are associated with protection from severe malaria. Nature 352, 595-600.

Hill, A.V.S. (1996). Genetic susceptibility to malaria and other infectious diseases: from the MHC to the whole genome. Parasitology 112 (Suppl.), S75-S84.

Hughes, A., Yeager, M. (1998). Natural selection at major histocompatibility loci of vertebrates. Annu. Rev. Genet. 32, 415-435.

Klein, J., O'Huigin, C. (1994). MHC polymorphism and parasites. Phil. Trans. R. Soc. Lond. B 346, 351-358.

Laundy, G.J., Spink, C.F., Keen, L.J., Wood, N.A. Bidwell, J.L. (2000). A novel polymorphism in the human interleukin-13 (IL-13) promoter. Eur. J. Immunogenet. 27, 53-54.

Liu, X., Nickel, R., Beyer, K., Wahn, U., Ehrlich, E., Freidhoff, L.R., Bjorksten, B., Beaty, T.H., Huang, S.K. (2000). An IL13 coding region variant is associated with a high total serum IgE level and atopic dermatitis in the German multicenter atopy study (MAS-90). J. Allergy Clin. Immunol. 106, 167 – 170.

McLeod, R., Buschman, E., Arbuckle, L.D., Skamene, E. (1995). Immunogenetics in the analysis of resistance to intracellular pathogens. Current Opinion Immunol. 7, 539-552.

Miller, L.H., (1994). Impact of malaria on genetic polymorphisms and genetic diseases in Africans and African Americans. Proc. Natl. Acad. Sci USA 91, 2415-2419.

Murphy, P.M. (1993). Molecular mimicry and the generation of host defense protein diversity. Cell 72, 823-826.

Parham, P., Ohta, T. (1996). Population biology of antigen presentation by MHC class I molecules. Science 272, 67-74.

Potts, W.K., Manning, C.J., Wakeland, E.K. (1991). Mating patterns in seminatural populations of mice influenced by MHC genotype. Nature 352, 619-621.

Rosenwasser, L.J. (1997). Interleukin-4 and the genetics of atopy. New Engl. J. Med. 337, 1766-1767.

Shirakawa, T., Deichmann, K.A., Izuhara, K., Mao, X.-Q., Adra, C.N., Hopkin, J.M. (2000). Atopy and asthma: genetic variants of IL-4 and IL-13 signalling. Immunology today 21, 60-64.

Vogel, F., Motulsky, A. (1986). Human genetics: problems and approaches, 3rd edn. Springer-Verlag, Heidelberg.

Weatherall, D.J. (1996). Host genetics and infectious disease. Parasitology 112, S23-S29.

Wedekind, C. (1999). Pathogen-driven sexual selection and the evolution of health. In: Evolution in health and disease. Ed. by S.C. Stearns, Oxford University Press, pp. 102-107.

Wills, C. (1991). Maintenance of multiallelic polymorphism at the MHC region. Immunological Reviews 124, 165-220.

Wills, C. (1996). Plagues. Harper Collins Publishers, London.

Wills-Karp, M., Luyimbazi, J., Xu, X., Schofield, B., Neben, T.Y., Karp, C.L., Donaldson, D.D. Interleukin-13: central mediator of allergic asthma. Science 282, 2258-2263.

Wills-Karp, M., Rosenwasser, L.J. (1999). Polymorphisms not found in the IL-13 gene promoter. Science 284, 5419.

Chapter 3.   Co-evolutionary forces and balanced pathogenicity

Balanced pathogenicity

Numerous examples exist of micro-organisms that maintain themselves in their host population without causing many harms. This more or less peaceful co-existence is explained partly by specific immunity (which may be transferred by mothers to the next generation) and partly by genetic adaptation of the pathogen to the host and vice versa. One example (of many) is the co-existence of *Yersinia pestis*, the cause of plague, and native mice or voles (*Peromyscus, Onychomys*, and *Microtus* species) in western parts of the United States. These species appear to be partially resistant to plague and are presumed to serve, along with their fleas, as reservoirs of *Y. pestis* during interepizootic periods (Gage et al 2000, Thomas et al 1988). Interestingly, in certain geographic areas both plague-susceptible and plague-resistant mice and rat strains, which appear to coexist, can be discriminated. Both intraspecies and interspecies differences in susceptibility to plague thus have been observed. These differences in susceptibility appear to be large and are controlled by multiple genes, among others genes controlling the antimicrobial activity of phagocytic cells and genes affecting the nutritional status of the bacterium (see below) (Chen and Meyer 1974, Hubert and Goldenberg 1970, Isin and Suleimenov 1987).

In case of peaceful co-existence, micro-organisms and their hosts have thus arrived at a state of balanced pathogenicity (Mims 1995), which optimally guarantees survival of the host and transmission of the pathogen. In this context it is important to note that virulence (pathogenicity) not only depends on the properties of the pathogen. Virulence is only expressed after interaction of molecules from pathogen and host. Hence virulence is a property of the host – pathogen interaction.

Several interactions, including host and pathogen adaptation, may be involved to reach such a state of balanced pathogenicity. A high rate of mutation in pathogens, especially rapidly evolving RNA viruses, leads to the generation of a highly variable mixture of closely related genomes, referred to as a quasispecies that persists and continuously evolves in infected individuals or populations (Eigen and Schuster 1979). This is especially the case when RNA viruses, such as hepatitis C or foot-and-mouth disease virus, establish persistent infections. Viral evolution is further affected by phenomena as genetic bottlenecks (strong reductions in numbers of populations resulting in a loss of genetic diversity), Muller's ratchet (mutations cumulatively and irreversibly eroding fitness in ratcheted fashion), random drift, and perhaps even punctuated equilibrium (Murphy 1998). Reassortment of segmented genomes (in influenza, arena and bunya viruses), and homologous and non-homologous recombination events further add to genetic diversity. Such events are also referred to as modular evolution. Bacteria also use several ways of increasing genetic variation, including reassortment/recombination, exchange of DNA plasmids that carry genes for virulence-associated characteristics, the acquisition of novel genes, site-specific sequences of repetitive DNA (contingency loci), and interaction with the immune system. Particular strains display high mutation rates ("mutator phenotypes") due to mutations in DNA repair enzymes or the development of enzymes ("DNA mutases") designed to generate mutations (Radman 1999, Rainey and Moxon 2000). Evidently such evolutionary adaptations may lead to drastic changes in the pathogen's properties.

As illustrated by the myxomatosis case (see below), modification of disease severity occurs through mechanisms that differ in speed. A host population may rapidly acquire specific immunity that may be transferred to offspring as maternal immunity. Genetic adaptation of pathogens follows immunologic adaptation of hosts. Especially RNA viruses with a mutation frequency of 1 in 10^4 nucleotides, a small genome, and rapid replication may rapidly adapt to new hosts. DNA genomes of pathogens and hosts naturally evolve more slowly.

Though it explains many peculiarities of host-pathogen interactions, the weakness of the balanced pathogenicity theory is that it emphasizes evolution towards decreased virulence rather than stable virulence or evolution towards increased virulence or fluctuations in virulence. It has however often been observed that the level of virulence increases when an infection is introduced into densely clustered groups, such as schools, hospitals, military barracks, or families. In such circumstances transmission of the pathogen may be so easy, that its transmission is not hampered by its virulence. So an increase in virulence may ensue. For example the virulence of measles transmitted within households in West Africa increased three- to fourfold between the primary and secondary cases, while the virulence did not increase in the general population (Aaby 1988). The explanation for this phenomenon is not clear, but may involve both genetic (within host competition and adaptation to the host genotype) and non-genetic factors (higher inoculating dose) (Read et al 1999). Hence substantial virulence may be maintained provided that this is not associated with a reduction in transmissibility. Alternatively, virulent strains may coexist with more transmissible mild strains, provided that there is a low degree of cross-immunity between mild and virulent strains (Gupta and Hill 1995).

Many researchers have thus questioned the function of virulence. Virulence may during some stages in the evolution be important for a micro-organism if it is useful for its transmission and survival. Tissue damage may be necessary or unavoidable for effective shedding. Falciparum malaria is still virulent despite its coexistence with its human host for approximately 7,000 years. However, only strains that produce large numbers of parasites in their host's blood are likely to be transmitted through mosquitoes to new hosts, thus providing evidence for a direct relation between the level of parasitemia (and probably virulence) and transmission. Indeed severe malaria is associated with high levels of parasitemia. A high rate of antigen switching and other immune evasion strategies may help in maintaining virulence and transmissibility of *P. falciparum*.

The most pathogenic strains of *Yersinia,* those that cause bubonic or pneumonic plague in animals and humans, also appear only successful in maintaining themselves, because their increased pathogenicity also increases their ability to spread. These strains are unable to invade the cells lining the gut and thus pass out through the feces. However, to ensure their transmission, they must multiply extracellularly in huge numbers that allow them to be transmitted by fleas through the air. This appears to be associated by a high mortality of their hosts due to toxic shock syndrome. Thus increased pathogenicity of such strains increases their ability to spread (Wills and Green 1995). Likewise, yellow fever virus has seemingly remained constant in its capacity to kill humans over the last 300 years. Rabies virus evolution, in contrast, appears to favor lethality in reservoir hosts, which hosts may increase breeding rates and refill niches quickly with susceptible young hosts to favor the perpetuation of the virus. Furthermore, micro-organisms may not evolve towards a less virulent form in a species when that species is irrelevant for the survival of the micro-organism. Otherwise virulence may be an "accident" (for example in a non-natural host) or an indication that the pathogen had not yet had the time to reach the state of balanced pathogenicity.

Another example of a (still) very virulent pathogen is HIV, which may be or remain very virulent to humans because its pathogenic effect, notably the destruction of CD4+ T helper cells, usually takes place after the infected individual has been able, during the symptom-free stages of infection, to transmit the virus to others. However, an indication that HIV may continue to evolve to a lesser virulent pathogen may be the behavior of simian immunodeficiency virus (SIV), which causes symptomless infections in wild African monkeys.

Thus at particular times and places other co-evolutionary paths than balanced pathogenicity alone appear possible. Micro-organisms further developed ways of dealing with competing microbes. For example, both *Helicobacter pylori* and HIV produce anti-microbials that inhibit other infections, and the host's immune response to one pathogen, for example influenza virus, may inhibit the growth of others.

Despite exceptions and unexplained observations, many examples exist of host-pathogen interactions illustrating that host and pathogen must often have co-evolved towards a situation where the pathogen does no or only little harm to its host. For example, several cattle breeds indigenous in West and Central Africa are resistant to sleeping sickness caused by trypanosomes (see chapter 22). Such breeds have been estimated to be in Africa for 5,000 to 7,000 years (Murray et al 1984). Another example is the peaceful coexistence of many mammalian species and herpesviruses. Many mammalian species harbor one or a few herpesviruses, which cause little or no disease. Except in very young (herpes simplex virus in young children, Aujeszky's disease virus [pseudorabies virus] in young pigs) or immunocompromised (cytomegalovirus infections in transplant recipients) individuals, the disease is usually is very mild. Herpesviruses have even evolved mechanisms that allow them to establish inapparent or latent infections in their host, usually in the central nervous system. During latency the viral genome is not even expressed (except the latency-associated transcripts), thus allowing survival of the virus for decades in an infected, healthy host. Following reactivation, the virus starts to replicate again, is transported to mucosal surfaces, and may be excreted and transmitted to new susceptible individuals. In contrast to this situation of balanced pathogenicity, very severe infections may occur when individuals are accidentally infected with a herpesvirus from another host. Examples are lethal herpes simplex virus (a natural pathogen of humans) infections in rodents, lethal herpes B virus (a natural pathogen in *Macaca* species) in humans, and lethal Aujeszky's disease virus (a mild pathogen for pigs) infections in cats, dogs, cattle, sheep, and horses. It appears needless to note that the virus is not transmitted in such accidentally infected "new" hosts.

The best-studied example of co-evolution is –beyond doubt- the recent history of myxomatosis in wild rabbits in Australia (Fenner 1983). After introduction of a highly virulent strain of the myxoma virus in 1950, genetic changes in both the virus and the host were carefully monitored in order to assess the relation between virulence and transmissibility of the virus and susceptibility of the host. Following introduction successively less virulent strains rapidly evolved. The original strain killed 99 % of the infected animals leading to a substantial decrease in the rabbit population in the years immediately after the introduction of the virus. In the year following the introduction less virulent strains already circulated and they predominated 3 to 4 years later. Because less virulent strains allow a longer life expectancy of the rabbits, they increase the period during which the rabbits are infectious. However virulence was not completely lost. In a period of approximately 10 years a more or less steady polymorphic equilibrium developed in which the predominant strains have intermediate grades of virulence. The relation between rapid viral multiplication and virulence probably explains this. Rabbits infected with more pathogenic strains experience more open skin lesions which facilitate viral transmission via the intermediate mosquito vector. Rabbit resistance has evolved more slowly and appeared to vary greatly in degree, both in space and time, depending on the predominant virulence grade of the virus. Within a period of seven years rabbits developed in which mortality upon inoculation with virulent strains was only 25 % compared to the 90 – 99 % mortality previously. Herd immunity probably contributed to the decrease in transmissibility of the virus and its development to less virulence. Hence the interaction ends with the maintenance of polymorphic populations of both host and pathogen (Anderson 1988, Fenner 1983).

There thus appears little doubt that genetic traits of hosts and pathogens are the result of co-evolutionary forces. A trait of one species may have evolved in response to a trait of the other, and may have been followed by a subsequent reciprocal response of the first species. This may especially be true for immune responses and immune evasion strategies, including polymorphisms in epitopes. This concept of "gene for gene" interactions has found support after detailed studies of plant-pathogen interactions. Indeed there are examples of plant resistance inherited at a single locus in a dominant fashion, and pathogen virulence also inherited at a single locus in a recessive fashion, and mutual genetic interaction between the loci (Anderson 1988). Field data for mammalian infections are evidently more sparse. In African children with malaria, evidence has been obtained that human leukocyte antigen (HLA) class I molecules influence the variants of parasites that cause infections. These variants differ in an antigenic region of the parasite's circumsporozoite protein. Cohabiting parasite strains also appear to have evolved strategies to downregulate cellular immune responses that are restricted by such HLA molecules (Gilbert et al 1998). However, it is obvious that most interactions involve more complex and multiple gene interactions with a range of pathogens, making the "gene for gene" concept too simplistic. In addition, other genes that act as suppressors, enhancers, and modifiers may influence the phenotypic expression of allelic variation in a gene. Gene redundancy, compensatory pathways, and robust properties of molecular pathways may buffer allelic variation (Hartman et al 2001). Demographic and behavioral changes likely influence evolutionary adaptations. Evidently new epidemics of an infectious disease may also be triggered by the evolution of types or strains with antigenic properties that allow to escape the acquired immunity within the host population that is created by its predecessors, a process called immune selection. Subsequent models therefore indicated that mutation rates in host and pathogens are influenced by environmental factors and may evolve into a complex host-parasite arm race involving reciprocal genetic interactions and coadaptation (Haraguchi and Sasaki 1996).

The evolution of micro-organisms has –beyond doubt- been directed towards long-term survival strategy. They may therefore become less virulent, only to improve their host-to-host transmission. Improved sanitary measures and other factors that hamper transmission may support such evolution. For example, *Vibrio cholerae* and *Salmonella typhi* strains are replaced by less virulent strains after cleaning up water supplies (Ewald 1994, Wills 1996). This is one often-neglected by-product of improved hygienic and other measures to hamper pathogen transmission. On the other hand, pathogens may spread easily in dense populations (Bouma et al 1995), therewith allowing the emergence of more virulent pathogen strains. As mentioned increased virulence has for example been observed during measles virus epidemics that could easily spread in naive populations.

In conclusion, while the concepts of balanced pathogenicity and mutual co-evolutionary forces may generally be true, there are interesting examples of infectious diseases that both support these concepts, as well as illustrate the exceptions to the rule and that point to subtleties we do not understand yet. A further complicating factor influencing virulence may be competition between pathogens.

Molecular mechanisms of virulence control
The genetic adaptations that have led to balanced pathogenicity of a specific micro-organism are mostly unknown. Obviously the many adaptive and specific immune mechanisms of higher vertebrates are a compromise between mechanisms acting against a wide variety of pathogens. Genetic mechanisms developed by higher vertebrates to resist infection are therefore numerous and include modulation of receptor expression, innate cellular immunity, antimicrobial proteins, and several others. For example, several variants of the receptors on the intestinal epithelium of pigs binding to enterotoxigenic *Escherichia coli* may or may not be expressed. Varying receptor expression results in different susceptibility to diarrhea induced by this organism (see chapter 7) (Francis et al 1998). People in sub-Saharan Africa often do not express the Duffy blood group antigen, a chemokine receptor, on their red blood cells. Because the Duffy blood group antigen is the receptor for the merozoite of *Plasmodium vivax* on red blood cells, these people are resistant for malaria. Interestingly, the Duffy blood group antigen is still expressed on other cells, thus enabling its function as a chemokine receptor on these cells (see chapter 7) (Daniels 1997). As a result of this selective expression, the Duffy mutation is protective against *P. vivax* without deleterious consequences. Evidently this work has opened the possibility to examine whether chemokine analogs can be developed as anti-malarial therapy through receptor blockade and the possibility to develop a vaccine against the Duffy-binding ligand. Another mechanism developed by higher vertebrates is constituted by the *Mx* proteins, proteins that are induced by interferon-α/β and that protect cells of some species against influenza virus (see chapter 18). *Mx* proteins act during early stages of infection by inhibiting different steps of the multiplication cycle of the virus (Pavlovic et al 1992).

Several micro-organisms, on the other hand, have evolved finely regulated mechanisms enabling them to partly or completely evade their hosts' immune response. Among these pathogens are members of the poxvirus, herpesvirus, retrovirus and adenovirus families, but also enteric bacteria as *Salmonella* spp. Herpesvirus latency has already been mentioned. Another mechanism used by herpesviruses is their ability to downregulate MHC class I expression on infected cells, thus hampering the lysis of such cells by CD8+ cytotoxic T cells (Mellencamp et al 1991, Ploegh 1998). Several poxviruses encode proteins, presumably pirated from their hosts, that are secreted from infected cells and that are able to modulate or block their host's cytokine and inflammatory responses. Among these molecules (also known as cytokine response modifying proteins) are proteins that bind IL-1β, IL-18, tumor necrosis factor-α (TNF-α), GM-CSF, type I interferons, and CC chemokines. IL-18-binding protein, for example, functions as a decoy receptor inhibiting interferon-γ production and hence the activation of natural killer (NK) and T lymphocyte helper type 1 cells. The parapoxvirus orf virus encodes a homolog of IL-10, which is likely important in immune evasion, since IL-10 inhibits nonspecific immunity and Th1 effector cells functions. Other poxvirus virulence gene products interfere with interferon signalling within cells and prevent the cleavage of biologically active IL-1β from its precursor protein. Again other intracellular poxvirus-encoded proteins attempt to retard antiviral responses such as apoptosis (Haig 1998, Nash et al 1999, Smith et al 1999, Smith et al 2000, Xiang and Moss 1999).

Non pathogenic *Salmonella* are able to abrogate synthesis of inflammatory cytokines by gut epithelial cells. They do so by preventing ubiquitination and hence degradation of IκB, an inhibitor of the master transcription factor NF-κB. NF-κB itself is a major inducer of the inflammatory response of the gut. The pathogenic plague bacillus *Yersinia enterocolitica* uses a similar mechanism. One of its Yop proteins directly interferes with the activity of IKKβ, which is an activator of the NF-κB pathway, in epithelial cells (Xavier and Podolsky 2000). Other Yop proteins from *Yersinia* spp. are able to induce apoptosis of macrophages. *Shigella* strains are also able to kill macrophages by inducing apoptosis. They also stimulate the release of large quantities of the proinflammatory cytokine interleukin-1 (IL-1). The IpaB protein is responsible for these effects (Donnenberg 2000).

Virulence-moderating genes also appear to act in protozoan parasites as shown by the recent identification of a gene that modulates virulence in *Leishmania*. When the gene encoding pterine reductase 1 (PTR1), which converts biopterin into a biologically active form, is inactivated, the parasite replicates 50-fold more in inoculated mice and the severity of pathologic lesions is enhanced (Pennisi 2000).

Specialization and host specificity

A result of the co-evolution between host and micro-organism is the specialization of the micro-organism, i.e. the property to survive in one host species or a subpopulation of them only (Wills 1996). While the pathogen may be completely avirulent for its own host to which it is fully adapted, it may maintain full virulence for other host species. I find this nicely illustrated by the herpesviruses. Pigs, cattle, horses, and humans all have their "own" herpesvirus, which (very young animals excepted) causes only mild or inapparent infections. After infection, the virus may establish latent infection in nervous tissues to survive for long times. After reactivation of the latent virus, which is often symptomless, the virus may be shed again and spread to other hosts.

While Aujeszky's disease virus, the herpesvirus of pigs, is only mildly pathogenic for pigs, it is highly lethal for almost any other animal species, except horses and humans (Kimman et al 1991). Specialization of this virus therefore implies that it has been completely adapted to pigs in which it is only mildly virulent and has evolved numerous mechanisms to establish latency from which reactivation is possible. Latency of herpesviruses appears possible in one species only. The poxviruses have similarly evolved towards a situation that mammalian species harbor their own pathogenic poxviruses, with myxoma virus infecting rabbits, ectromelia virus infecting mouse, orf virus infecting sheep, and molluscum contagiosum virus and variola virus infecting humans.

The molecular mechanisms of specialization and host specificity have been elucidated in some instances. Host specificity of poliovirus to humans appears largely explained by the presence of its receptor in this species and its absence in others. In contrast, the low pathogenicity of some *Yersinia pestis* strains for guinea pigs (while they are fully virulent for mice) appears to be explained by a host genetic influence on the nutritional adequacy of the bacterium. Guinea pigs have a circulating aspariginase that limits the availability of asparigine on which *Y. pestis* strains are partly dependent for growth (Burrows and Gillett 1971).

New diseases caused by old pathogens
Following the concept of balanced pathogenicity a micro-organism can only cause severe disease in a host which is irrelevant for its survival and transmission, or in a "new" host, for example after the micro-organism has crossed an ecosystem barrier or has jumped to a new species. In the latter case, virulence is only a temporal phenomenon, which occurs before complex genetic adjustments between host and micro-organisms have taken place. Cholera for example may still be virulent because it has only recently, around 1800, crossed species boundaries. Notably, it appears to evolve to lesser virulence. The first recorded cholera outbreak was in 1817 in the river deltas of Bengal near Calcutta (Wills 1996). From then pandemic waves of cholera spread from the Ganges River delta to elsewhere in the world. Cholera was attested in Europe for the first time in Europe in 1832, and continued to be a problem until the beginning of the 20th century, when water sanitation became widespread. The organism's natural habitat is estuarine and coastal, where it lives free and is associated with invertebrates. Most strains lack the transmissible genetic elements that encode the toxin and adhesin virulence factors that are necessary for causing severe disease in humans.

Another example is *Legionella pneumophila*, the cause of Legionnaires disease, which may cause severe and lethal lung disease in humans, because they are not essential for the survival of this pathogen. This bacterium normally lives in amoebas. Other examples of severe infectious diseases in this species that are irrelevant for the survival of the causative micro-organisms are rabies and Lassa fever. Examples of pathogens invading new host populations and causing severe epidemics are shown in Table 2.

Table 2. Species jumps of pathogens leading to severe " first-contact" disease and high mortality in the newly invaded species

Pathogen	Source species	Newly invaded species	Location	Time	Ref.
Vibrio cholerae	invertebrates	humans	India	1800	Wills 1996
Rinderpest virus	Asian bovines	buffaloes	Africa	1889	Daszak et al 2000
Human T-cell lymphotropic virus (HTLV)-I and II	chimpanzees,	humans mandrills	Africa		Poiesz 1980, Koralnik 1994
Human immunodeficiency virus types 1 and 2 (HIV-1, HIV-2)	monkeys (chimpanzees and sooty mangabey monkeys respectively)	humans	Africa	1950	Myers et al 1992
Marburg and Ebola virus (filoviruses)	insectivorous or fruit bat and rodent reservoirs suspected	captive and wild nonhuman primates, humans	several (Sub-Saharan Africa, Indonesia, Philippines, Germany, USA)	1950 – 2000 (several events)	Daszak et al 2000
Canine parvovirus	cats	dogs	world-wide	1978	Parrish 1999
Aquatic morbilliviruses	seals, whales, dogs	seals, dolphins, porpoises	North sea, Mediterranean sea, Lake Baikal	1988 (several events)	Barret et al 1995
Canine distemper virus	domestic dogs	African wild dogs, lions	Serengeti	1991	Daszak et al 2000
Nipah and Hendra viruses	fruit bats	horses, swine, dogs, humans	Australia, Malaysia and Singapore	1994, 1998	Chua et al 2000

Evolutionary trade-offs

Evolution of higher vertebrates towards resistance against infectious diseases may have led to unfavorable consequences or unwanted responses to other environmental factors, for example leading to enhanced susceptibility to chronic degenerative diseases such as atherosclerosis. Macrophages are involved in atherogenesis because they are the major sources of lipid-laden foam cells in early "fatty streak" stages of atherogenesis. Knockout mice homozygous for null mutations in the gene encoding apolipoprotein E develop marked hyperlipidemia and atherosclerosis. When these mice also have mutations in the macrophage type-I and type-II class-A (MRS-A) scavenger receptor gene, they develop less atherosclerosis, but become highly susceptible to infection with *Listeria monocytogenes* and herpes simplex virus type-1. MRS-A, which can bind an extraordinarily range of ligands including bacterial pathogens, thus appears to play a part in host defense against pathogens. In addition, atherosclerosis may be a trade-off of the evolution of macrophage function towards handling infectious pathogens (Suzuki et al 1997).

Sometimes, simple observations have suggested selective importance of severe disorders for resistance to infectious disease. For example, genetic and epidemiological studies have suggested that the high frequency of Tay Sachs disease in certain Jewish populations reflects heterozygous resistance to tuberculosis in east European ghetto's (O'Brien 1991). As these populations are estimated to originate from as few as several thousand individuals, also founder effects likely contributed to the high carrier frequency. Tay Sachs disease is an autosomal recessive disease that is characterized by an early childhood onset, neurodegenerative disease caused by mutations in the gene encoding the hexosaminidase A enzyme. Patients usually die before puberty. In any case, the presence of at least two different Tay Sachs alleles of independent evolutionary origin supports the notion that carriers of these alleles did have a fitness advantage.

Another example has been proposed by Miller (1994), who suggested that the high frequency of hypertension in black populations associated with high sodium and low potassium content of red blood cells may be a genetic response to malaria. Miller also suggested that hemochromatosis in sub-Saharan Africa might have arisen by selection of heterozygotes against malaria. This suggestion is based on the malarial parasite's ability to detoxify intracellular heme-derived iron and its requirement for extracellular iron for its development. The high frequency of hemochromatosis in Caucasians may also have arisen by providing some selective advantage (against infectious agents or not) through increased iron absorption.

In conclusion, these and other observations have led to the understanding that the genetic constitution that protected us against epidemics of infectious diseases in the past may not be optimally suited to protect us against the many non-infectious diseases of today's western society and developing world. Selection in the past has likely be directed towards resistance to infectious disease and towards a "thrifty genotype", a genotype which had been selected to deal with sporadic availability of food (Neel 1962, Weiss et al 1984, Zimmet et al 1990). In today's society we face other hygienic conditions and an excessive diet. Some gene variants that protected against infections or malnutrition in the past might therefore now be responsible for susceptibility to autoimmune disease, allergy, or metabolic diseases as insulin-resistant diabetes and obesity.

References

Aaby, P. (1988). Malnutrition and overcrowding-exposure in severe measles infection. A review of community studies. Rev. Infect. Dis. 10, 478-491.

Anderson, R.M. (1988). The population biology and genetics of resistance to infection. In: Genetics of resistance to bacterial and parasitic infection. Ed. by D. Wakelin and J.M. Blackwell, Taylor and Francis, London, Philadelphia, New York.

Barret, T., Blixenkrone-Moller, M., Di Guardo, G., Domingo, M., Duigan, P., Hall, A., Mamaev, L., Osterhaus, A.D. (1995). Morbilliviruses in aquatic mammals: report on a round table discussion. Vet. Microbiol. 44, 261-265.

Bouma A., Jong, M.C.M. de, Kimman, T.G. (1995). Transmission of pseudorabies virus within pig populations is independent of the size of the population. Preventive Veterinary Medicine, 23, 163-172.

Burows, T.W., Gillet, W.A. (1971). Host specificity of Brazilian strains of *Pasteurella pestis*. Nature 229, 51-52.

Chen, T.H., Meyer, K.F. (1974). Susceptibility and antibody response of *Rattus* species to experimental plague. J. Infect. Dis. 129, S62-S71.

Chua, K.B., Bellini, W.J., Rota, P.A., Harcourt, B.H., Tamin, A., et al. (2000). Nipah virus: A recently emergent deadly paramyxovirus. Science, 288, 1432-1435.

Daniels, G. (1997). Blood group polymorphisms: molecular approach and biological significance. Transfus. Clin. Biol. 4, 383-390.

Daszak, P., Cunningham, A.A., Hyatt, A.D. (2000). Emerging infectious diseases of wildlife-Threats to biodiversity and human health. Science 287, 443-449.

Donnenberg, M.S. (2000). Pathogenic strategies of enteric bacteria. Nature 406, 768-774.

Eigen, M., Schuster, P. (1979). The hypercycle: a principle of natural self-organisation. Springer Verlag, Berlin.

Ewald, P.W. (1994). Evolution of infectious disease. Oxford University Press, Oxford, New York.

Francis, D.H., Grange, P.A., Zeman, D.H., Baker, D.R., Sun, R., Erickson, A.K. (1998). Expression of mucin-type glycoprotein K88 receptors strongly correlates with piglet susceptibility to K88(+) enterotoxigenic *Escherichia coli*, but adhesion of this bacterium to brush borders does not. Infect. Immun. 66, 4050-4055.

Fenner, F. (1983) Biological control, as exemplified by smallpox eradication and myxomatosis. Proceedings of the Royal Society of London B 218, 259-283.

Gage, K.L., Dennis, D.T., Orloski, K.A., Ettestad, P., Brown, T.L., Reynolds, P.J., Pape, W.J., Fritz, C.L., Carter, L.G., Stein, J.D. (2000). Cases of cat-associated human plague in the western United States. Clin. Infect. Dis. 30, 893-900.

Gilbert, S.C., Plebanski, M., Gupta, S., Morris, J., Cox, M., Aidoo, M., Kwiatkowski, D., Greenwood, B.M., Whittle, H.C., Hill, A.V.S. (1998). Association of malaria parasite population structure, HLA, and immunological antagonism. Science 279, 1173-1177.

Gupta,. S., Hill, A.V.S. (1995). Dynamic interactions in malaria: host heterogeneity meets parasite polymorphism. Proc. Roy. Soc. Lond. B. 261, 271-277.

Haig, D.M. (1998). Poxvirus interference with the host cytokine response. Vet. Immunol. Immunopathol. 63, 149-156.

Haraguchi, Y., Sasaki, A. (1996). Host-parasite arms race in mutation modifications: indefinite escalation despite a heavy load. J. Theor. Biol. 183, 121-137.

Hartman, J.L., Garvik, B., Hartwell, L. (2001). Principles for the buffering of genetic variation. Science 291, 1001-1004.

Hubbert, W.T. and Goldenberg, M.I. (1970). Natural resistance to plague: genetic basis in the vole (*Microtus californicus*). Am. J. Trop. Med. Hyg. 19, 1015-1019.

Isin, Zh, M., Suleimenov, B.M. (1987). Comparative investigation of PMN leucocyte antimicrobial potential in animals of different species susceptibility to the plague agent. J. Hyg. Epidemiol. Immunol. 3, 307-312.

Kimman T.G., Brinkhorst, G.J., Ingh, T.S.G.A.M. van den, Pol, J.M.A., Gielkens, A.L.J., Roelvink, M.E. (1991). Aujeszky's disease in horses fulfils Koch's postulates. Veterinary Record 128, 103-106.

Koralnik, I.J., (1994). A wide spectrum of simian T-cell leukemia/lymphotropic virus type I variants in nature: Evidence for interspecies transmission in Equatorial Africa. J. Virol. 68, 2693-2707.

Mellencamp, M.W., O'Brien, P.C.M., Stevenson, J.R. (1991). Pseudorabies virus-induced suppression of major histocompatibility complex class I antigen expression. J. Virol. 65, 3365-3368.

Miller, L.H. (1994). Impact of malaria on genetic polymorphisms and genetic diseases in Africans and African Americans. Proc. Natl. Acad. Sci. USA 91, 2415-2419.

Mims, C.A., Dimmock, N.J., Nash, A., Stephen, J. (1995). Mims' pathogenesis of infectious disease, 4th ed. Academic Press.

Murphy, F.A. (1998). Personal communication.

Murray, M., Trail, J.C., Davis, C.E., Black, S.J. (1984). Genetic resistance to African trypanosomiasis. J. Infect. Dis. 149, 311-319.

Myers, G., MacInnes, K., Korber, B. (1992). The emergence of simian/human immunodeficiency viruses. AIDS Res. Hum. Retrovir. 8, 373-386.

Nash, P., Barret, J., Cao, J.X., Hota-Mitchell, S., Lalani, A.S., Everett, H., Xu, X.M., Robichaud, J., Hnatiuk, S., Ainslie, C., Seet, B.T., McFadden, G. (1999). Immunomodulation by viruses: the myxoma virus story. Immunol. Rev. 168, 103-120.

Neel, J.V. (1962). A "thrifty" genotype rendered detrimental by progress? Am. J. Hum. Genet. 14, 353-361.

O'Brien, S.J. (1991). Ghetto legacy. Molecular Evolution 1, 209-211.

Parrish, C.R. (1999). Host range relationships and the evolution of canine parvovirus. Vet. Microbiol. 69, 29-40.

Pavlovic, J., Haller, O., Staeheli, P. (1992). Human and mouse *Mx* proteins inhibit different steps of the influenza virus multiplication cycle. J. Virol. 66, 2564-2569.

Pennisi, E. (2000). Parasitology: Close encounters: Good, bad and ugly. Science 290, 1491-1493.

Ploegh, H. (1998). Viral strategies of immune evasion. Science 280, 248-253.

Poiesz, B.J. (1980). Detection of type C retrovirus particles from fresh and cultured lymphocytes of a patient with T cell lymphoma. Proc. Natl. Acad. Sci. USA 77, 7415-7419.

Radman, M. (1999). Enzymes of evolutionary change. Nature 401, 866-869.

Rainey, P.B., Moxon, E.R. (2000). When being hyper keeps you fit. Science 288, 1186-1187.

Read, A.F., Aaby, P., Antia, R., Ebert, D., Ewald, P.W., Gupta, S., Holmes, E.C., Sasaki, A., Shields, D.C., Taddei, F., Moxon, E.R. (1999). What can evolutionary biology contribute to understanding virulence? In: Evolution in health and disease. Ed. by S.C. Stearns, Oxford University Press, pp. 205-215.

Smith, G.L., Symons, J.A., Alcami, A. (1999). Immune modulation by proteins secreted from cells infected by vaccinia virus. Arch. Virol. Suppl. 15, 111-129.

Smith, V.P., Bryant, N.A., Alcami, A. (2000). Ectromelia, vaccinia and cowpox viruses encode secreted interleukin-18-binding proteins. J. Gen. Virol. 81, 1223-1230.

Suzuki, H., Kurihara, Y., Takeya, M., Kamada, N., Kkataoka, M., Jishage, K., Ueda, O., Sakaguchi, H., Higashi, T., Suzuki, T., Takashima, Y., Kawabe, Y., Cynshi, O., Wada, Y., Honda, M., Kurihara, H., Aburatani, H., Doi, T., Matsumoto, A., Azuma, S., Nosa, T., Toyoda, Y., Itakura, H., Yazaki, Y., Kodama, T. (1997). A role for macrophage scavenger receptors in atherosclerosis and susceptibility to infection. Nature 386, 292-296.

Thomas, R.E., Barnes, A.M., Quan, T.J., Beard, M.L., Carter, L.G. (1988). Susceptibility to *Yersinia pestis* in the northern grasshopper mouse (*Onychomys leucogaster*). J. Wildlife Dis. 24, 327-333.

Weiss, K.M., Ferrel, R.E., Harris, C.L. (1984). A New World syndrome of metabolic diseases with a genetic and evolutionary basis. Yearbook of Physical Anthropology 27, 153-178.

Wills, C., Green, D.R. (1995). A genetic herd-immunity model for the maintenance of MHC polymorphism. Immunological Reviews 143, 263-292.

Wills, C. 1996. Plagues. Harper Collins Publishers, London.

Xavier, R.J., Podolsky, D.K. (2000). How to get along-Friendly microbes in a hostile world. Science 289, 1483-1484.

Xinag, Y., Moss, B. (1999). IL-18 binding and inhibition of interferon γ induction by human poxvirus encoded proteins Proc. Natl. Acad. Sci. USA 96, 11537-11542.

Zimmet, P., Dowse, G., Finch, C., Serjeantson, S., King, H. (1990). The epidemiology and natural history of NIDDM; lessons from the South Pacific. Diabetes and Metabolic Reviews 6, 91-124.

Chapter 4.   Mapping susceptibility loci

Several approaches (see Fig. 3) have been developed to localize disease genes. Genetic approaches for mapping susceptibility genes have been more frequently applied to infections with a protracted course, such as HIV, leprosy and tuberculosis, than to very acute infections, as the former diseases usually permit to recruit larger number of cases in a given period. Nonetheless the principles of studying chronic and acute diseases are similar.

Family-based linkage and association-based case-control studies have both proven their value in identifying genes involved in infectious disease. These two study designs both have advantages and disadvantages, and they are therefore complementary. For example, because association is only detectable over short genetic distances (generally < 1 cM), this approach can be used in fine mapping to localize the disease susceptibility gene in a region, which was initially identified by linkage analysis to segregate with the disease (Bellamy and Hill 1998, Feder et al 1996). (Note that one

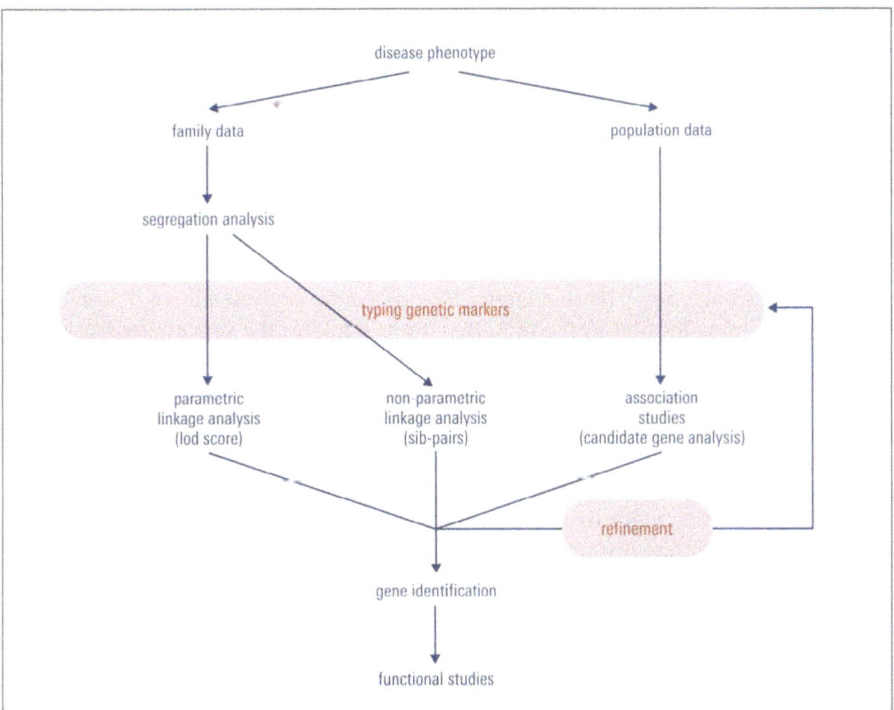

Figure 3. Schematic representation of steps that may be involved in the identification of genes involved in infectious diseases. When evidence for linkage disequilibrium between genetic markers and disease phenotype is obtained, fine genetic and physical gene mapping may allow the identification of candidate genes. Polymorphisms are identified and tested for association with disease. Association studies test for association between a disease phenotype and a given marker allele. These may directly identify a susceptibility gene when the marker allele is the susceptibility gene itself, or they may identify a susceptibility gene when the tested marker allele is in strong linkage disequilibrium with the disease allele. Functional studies, for example in transgenic or knockout mouse models, are usually required to establish whether candidate genes are involved in causing the disease.

Morgan (M) represents a distance in which on average one genetic crossover occurs each time that a gamete is formed. The total genetic map length in humans is 26 Morgans or 2600 cM. On average 1cM contains approximately 1 million base pairs.)

Genetic markers

Linkage studies require genetic markers that make use of variation within a species. Marker variants (polymorphisms, alleles) are inherited in a Mendelian way and may help in elucidating the genetics of disease susceptibility. Evidently, to be useful in mapping, the markers must be polymorphic, that is they must differ among individuals. If the inheritance of a disease phenotype correlates with the inheritance of a marker, for instance in pedigrees, the marker is genetically linked to the gene responsible for that phenotype. This co-segregation indicates that the marker and the gene are proximate on the same chromosome.

The most often used molecular genetic markers are the multi-allelic microsatellites or short tandem repeats (STRs), for example from the Généthon collection. Microsatellites are di-, tri-, or tetranucleotide tandem repeats of a very simple sequence (such as CA or GATA) in which the number of repeats varies at a given locus throughout the population. They are often less than 0.1 kb and may be found in up to 200.000 positions in the genome. Some microsatellites may affect enhancer/promoter function, but a common biological function has up till now not been ascribed to microsatellites. Their length variation is probably created through slippage of DNA polymerase I during replication. This accounts for a very high mutation rate that is 10,000 times higher than elsewhere in the genome. Several thousands of microsatellites in the human genome are now known and used in genome-wide screening studies. They are densely distributed with an average interval between markers of 1.6 cM (Dib et al 1996). They are determined by polymerase chain reaction (PCR) and analyzed by software such as Genotyper™ (Perkin-Elmer). The detection of these alleles is straightforward. PCR primers located in the unique regions flanking the repeat amplify a DNA fragment containing the microsatellite, and length variants are determined on polyacrylamide gels (Fig. 4). Other examples of markers are minisatellites. These are intermediate size (0.1 kb – 20 kb) arrays of short tandemly repeated DNA sequences. Differences in their copy number form the basis of polymorphism in the variable number of tandem repeats (VNTR polymorphism). In addition to the microsatellite markers, other comprehensive maps have been assembled on the basis of a collection of more than tens of thousands distinct transcribed sequences, including known genes and gene fragments, or expressed sequence tags (ESTs) (Adams et al 1993, Schuler et al 1996). They thus identify human cDNAs. Using these markers comprehensive physical maps of the human genome are being assembled in which the relative location of the markers is defined by the actual length along the chromosome rather than by recombination events (Chumakov et al 1995, Hudson et al 1995, Quereshi et al 1999). Thus the transcript map identifies the chromosomal locations of expressed genes.

Other markers are restriction fragment length polymorphisms (RFLP), based on point mutations in restriction sites, and, finally, single nucleotide polymorphisms (SNP's) that are based on detailed sequence information. SNP's are the most frequent sequence variants that occur about once in every 1,250 bases on average of the 3 billion bases in the human genome. Therewith they constitute a map representative of diversity across the genome. SNPs, however, are not uniformly distributed over the entire

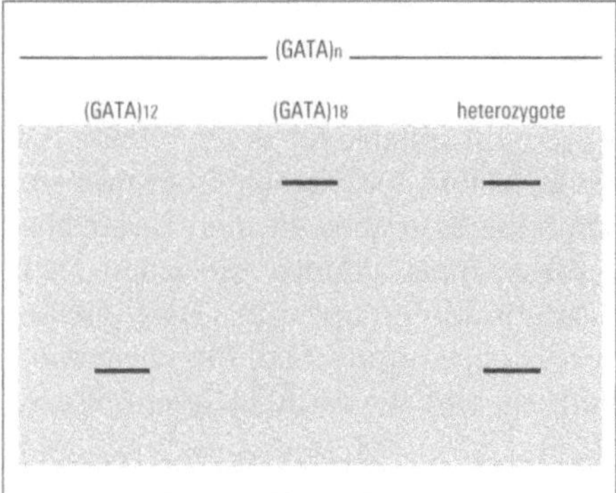

Figure 4. Result of a microsatellite marker analysis. The (GATA)n repeat sequence is amplified by polymerase chain reaction (PCR) and analyzed by gel electrophoresis. The length of the repeat varies between 12 and 18. Because the larger microsatellite migrates more slowly in the gel than the smaller repeat, individuals homozygous for one or the other allele of the marker, and individuals heterozygous for both alleles can be discriminated.

genome and only 1 % of them affect protein function. The identification of SNPs proceeds at rapid pace. Approximately 2.3 million of them have now been mapped to their positions on the genome and they have been made available at the internet (http://snp.cshl.org, http://www.ncbi.nlm.nih.gov/SNP, http://hgbase.crg.ki.se). A great advantage of these markers is that SNPs can be screened by high throughput hybridization methods, such as DNA chips. Although these markers may be very helpful in tracking down the location of susceptibility genes by whole-genome scans, one problem is that many of the common SNPs (as other markers) will have no role in disease. The number (density) of SNPs required for mapping therefore depends on the degree of linkage disequilibrium on specific chromosomal regions in the population. Occasionally SNPs in genes or control regions may causally affect susceptibility to disease. Using SNPs in coding and promoter regions with potential functional significance rather than random SNPs may therefore enhance the power of this approach (Risch 2000). The number of SNPs needed for mapping may also depends on the ethnic group studied. Only approximately one-third of the SNPs appear widely applicable in all populations. Therefore more SNPs may be needed to map susceptibility genes in a particular ethnic group. Integrated genetic and physical maps together with nucleotide and protein sequence databases are available through sites on the World Wide Web (for example at http://www.ncbi.nlm.nih.gov/SCIENCE96/).

Modes of inheritance: monogenic and polygenic disease
Evidently mapping studies are arguably more effective when focused on diseases with a clear genetic basis. A few examples exist in which a single major gene affects the outcome of infectious disease and in which the gene is transmitted via a simple monogenic Mendelian pattern of inheritance. However the severity of most infectious disease is influenced by multiple host genes interacting with each other and with microbial genes, and with environmental factors, thus creating a spectrum of disease expression. Loci responsible for such quantitative variation are sometimes referred to as QTLs (for quantitative trait loci). The degree and type of epistasis, or interlocus interactions between genes, strongly influence the chances of detecting them via a linkage analysis. Interaction may be multiplicative or additive. Another consequence of epistasis is that effects of single genes likely vary from population to population according to the prevalence of alleles at other loci. Several approaches to maximize power in QTL studies, such as examining siblings at opposite ends of a risk factor distribution (discordant sibling-pairs), have been described (Risch and Zhang 1995).

However, even if there is no epistasis (or "modifier genes"), detecting susceptibility loci may be difficult when several distinct loci independently influence the disease (Weeks and Lathrop 1995).

An important question is which phenotypes are most suitable for study. If there is a strong indication for the involvement of genetic factors, for example from twin studies, it appears a useful approach in case control studies to study the more severe phenotype rather than mild cases of infection. For example, positive associations with tuberculosis and malaria were seen most clearly in the groups with severe disease. This finding may result from the frequent occurrence of infection, the uncommon occurrence of clinical disease among the infected, and the rarity of severe disease (Brahmajoth et al 1991, Hill 1996).

Segregation analysis

Segregation analysis may be a first step in determining from family data how a given phenotype is inherited, and to discriminate between shared environment and genetic relationships. A genetic effect can be expressed in terms of relative risk, odds ratio (for binary clinical phenotypes), or in the heritability due to the gene (or proportion of the phenotypic variance explained by the major gene for a quantitative phenotype). Segregation analysis uses maximum likelihood methods to test for a major gene and to estimate the measurements for the phenotype/genotype model. Using these estimates, parametric linkage analysis by the LOD score method may subsequently confirm and locate the gene.

Linkage studies

Linkage analysis tests whether in families a phenotype of interest is transmitted with genetic markers of known chromosomal location. Genetic linkage analyses therewith localize genes contributing to disease to approximate (candidate) chromosomal regions. Linkage is due to the close proximity of two genes or markers along a chromosome and is inferred to exist when co-segregation between a genetic marker and a gene affecting the phenotype of interest in families is detected more often than would be expected by chance (Ellsworth and Manolio 1999). Demonstrating linkage therewith gives statistical proof that a disease is influenced by a genetic mechanism. Linkage analysis needs the typing of genetic markers. The most commonly used statistical test for this purpose is the LOD (log of the odds) score method. The LOD score method is a likelihood ratio, testing the hypothesis of linkage (against the hypothesis of no linkage) for different genetic distances (or recombination fractions) between the phenotype locus and the marker locus. A LOD score ≥ 3 is most commonly used as critical value for establishing linkage, while a LOD score ≤ 2 argues against linkage. Linkage can be tested marker by marker (two-point analysis) or by a set of linked markers (multipoint analysis) (Abel and Dessein 1998). Because the traditional LOD score analysis, which uses parametric methods, assumes the presence of a single major disease locus with a specific mode of inheritance that accounts for the majority of the genetic variation, it is seldom powerful for the analysis of genetic susceptibility to infectious disease. Nonetheless, this approach may be used when a clear major gene model can be inferred and when reliable epidemiological information for each family member is available (which has for example been the case in schistosomiasis or in recessive autosomal immune disorders) (Abel and Dessein 1998).

Parametric, or model-based, methods (segregation analysis and linkage analysis by the LOD score method) require to define the model and to specify the relationship between the phenotype and other factors, for example environmental factors or exposure levels, that may influence the expression of the phenotype. As a result, the genotype/phenotype model includes, in addition to the frequency of the susceptibility allele, all the parameters that describe and quantify the relationship between susceptibility and the relevant genetic and environmental factors. This relationship can be mathematically expressed in several ways, most often using regression methods that define model parameters in terms of regression coefficients. Regression methods may be used to analyze both binary as well as quantitative phenotypes. However, serious errors may occur under an incorrectly specified model or if parameters, such as allele frequencies or penetrance, have not been estimated accurately. An alternative strategy is to conduct segregation and linkage analysis simultaneously. This may increase power and generate less biased estimates of inheritance models and linkage parameters. In addition, such a combined segregation and linkage analysis may increase the efficiency for detecting gene-environment interactions (Ellsworth and Manolio 1999). Model-free nonparametric methods are ideal for searching for susceptibility loci (including QTLs) involved in genetically complex diseases. Nonparametric methods test whether the alleles of a given marker are distributed at random in persons with the same phenotype. These methods are free from assumptions that the data follow defined probability distributions. They thus examine the genetic factors influencing a phenotype without specifying the phenotype/genotype mode. These studies are strongly recommended when little is known about the relationship between the phenotype and a putative gene. This may be the case in the study of complex traits, including infectious diseases, when no segregation analysis has been performed, or when segregation analysis failed to infer a clear major gene.

One of these methods is the affected-sib-pair method, which tests for increased marker similarity in affected sib pairs and which thus requires no assumptions about the mode of inheritance of the disease. For any chromosomal region containing a major susceptibility gene, affected sib-pairs inherit the same parental alleles more often than expected by chance.

The affected sib-pair method is frequently used in the analysis of complex diseases such as infectious diseases. Knowledge about penetrance of genes, mode of inheritance, or number of genes involved is not required. The sib-pair method is also advantageous with regard to ease of collection and statistical simplicity. The method is most powerful when identity-by-descent (IBD) of marker alleles can be determined. IBD refers to inheritance of the same allele from a given parent. The method therefore preferably requires two affected siblings and preferably also their parents, who are preferably unaffected. The inheritance of polymorphic markers from parents to offspring is determined to calculate the number of alleles, which are shared by the sibs and which are identical by descent. At any marker zero, one or two alleles may be shared by the sibs. In a large number of families the expected frequencies of sharing 2:1:0 alleles are 25 %: 50 %: 25 % when there is no linkage between the disease and the marker. If however the marker is linked to a susceptibility locus then there is an increase in families where the affected sibs share two alleles identical by descent and a reduction in the number sharing zero alleles. The power of sib-pair methods decreases when they have to rely on identity-by-state (IBS). In this case IBD cannot be established because parents are not included in the study. Alleles are considered to be identical-by-state if they are not known to be derived from a common ancestral allele, but seem to be identical.

Usually, at least 100 sib pair families are examined in a genome screen. Approximately 300 or more evenly spaced microsatellite markers are typed. However, because many microsatellites are typed, false positive linkages to regions that do not contain susceptibility genes could occur by chance. It is important therefore to confirm loci identified in a first genome screen (Weeks and Lathrop 1995). A solution to that problem is to perform a genome screen in two stages (Bellamy and Hill 1998). First, an entire genome screen is made using the affected sib pair method. Subsequently the markers that demonstrate linkage are genotyped in a second set of families (see chapter 16).

A limitation of the affected sib pair method may be the limited availability of siblings affected by the disease, which in particular may be the case for many acute infectious diseases. Evidently the power of this and similar methods to detect a disease suscepti-bility locus depends on the contribution the locus makes to the genetic variation of the trait (the risk attributable to the locus, expressed as the risk ratio or λ_s for that locus [see chapter 1]), the number of genes involved, the way they interact with each other, and the recombination fraction between the marker and the locus. As indica-ted, interaction may be multiplicative or additive with regard to the penetrance of the disease phenotype, where penetrance is defined as the probability of being affec-ted given a certain genotype (Weeks and Lathrop 1995).

A strategy to find loci involved in a polygenic mode of inheritance is known as staged searching. An initial genomic scan is carried out at a low marker density (for example 20 cM), and regions of interest are subsequently investigated in subsequent stages by typing more markers and preferably also in more sib pairs (Brown et al 1994). The power of this method can be improved by using a highly polymorphic and multimar-ker analysis, and by adopting a candidate gene approach. Other approaches to detect linkage include the affected-pedigree-member method, which measures marker simi-larity between affected individuals of extended families in terms of identity-by-state, and linkage-disequilibrium-based approaches (see below).

In these approaches, excess allele sharing can be tested by a simple chi-square (in par-ticular when all parental marker data are known), or by maximum likelihood methods, such as the maximum likelihood score and the maximum likelihood bino-mial approach (Abel and Dessein 1998). This method studies affected relatives in a pedigree to show how often a particular copy of a chromosomal region is inherited from a common ancestor compared with that expected under random Mendelian segregation. Popular software packages, such as MAPMAKER/SIBS, further allow the testing of quantitative phenotypic responses by regression methods, as well as multi-point analysis of sib-pair data (Kruglyak and Lander 1995).

Linkage disequilibrium analysis

Linkage disequilibrium describes an association at the population level of a particular marker allele with a disease and can be used to help positionally clone genes. Linkage disequilibrium occurs when haplotype combinations of alleles at different loci occur more frequently than would be expected from random association. This can arise from various causes, including recent admixture of different populations with diffe-rent allele frequencies, selection for cooperation in function, genetic drift, population bottlenecks (strong reductions in numbers of populations resulting in a loss of genetic diversity), and new mutations. Disequilibrium because of shared ancestors is usually seen only with markers within 1 Mb (or approximately 1 cM) or less of a disease locus. Sometimes it is restricted to very close markers (Strachan and Read 1996). Thus linkage disequilibrium between a marker and a trait locus may lead to the identification of a susceptibility gene in the near vicinity of the marker. Mapping by linkage disequilibrium works best with data from genetically isolated populations with a limited number of founders, which may provide a relatively simple genetic background with an increased frequency of founder effects (Jorde 1995). In these studies geneticists study the genetic make up of families affected by a particular disease, looking for marker sequences that inherit with the condition. Eventually this may result in a search for the disease gene itself. Unfortunately, when genes contribute only little to the condition, linkage disequilibrium analysis may not be very powerful (Risch and Merikangas1996).

In conclusion, linkage analysis test for co-segregation within families of a disease phenotype with a random marker locus. Linkage studies may also screen the entire genome to locate regions with susceptibility genes (Davies et al 1994). Although useful for localizing genes responsible for Mendelian disorders, they are less appropriate for complex diseases when different genes contribute significantly to the disease in different families (genetic heterogeneity). Because the power of this approach is rather low compared with whole-genome association studies, it will only detect chromosomal areas with a major effect (or high heritability) on disease susceptibility (see Risch and Merikangas [1996] for a discussion on sample size). In fact, genome-wide linkage scans designed to localize disease genes have yielded few significant results. Furthermore, they can only implicate relatively large areas of a chromosome and do not identify the gene responsible for linkage. However, they may implicate regions in which multiple genes have an effect on disease. This will often be the case when linkage studies suggest the involvement of the MHC, which may involve class I genes, such as HLA-B, class II genes, such as HLA-DR, or to class III genes, such as TNF-α (Hill 1996).

The advantage of the linkage approach is that unknown genes may be mapped and identified. In addition, in family linkage studies, there is no requirement to identify a control group. After demonstrating linkage between microsatellite markers and disease susceptibility, candidate gene(s) may subsequently be identified by demonstrating association between a small genomic region and disease (see below). Such studies depend on high-resolution genetic and physical maps. Moreover, association studies are necessary in identifying a susceptibility gene once its chromosomal position has been determined by linkage. In addition to linkage analysis, a number of experimental techniques have been reported to help localize genes, including *in situ* hybridization, somatic cell hybridization, homology mapping, deletion mapping, and detection of chromosomal aberrations.

Association studies
As indicated above, most infectious disease may be considered, from the point of view of the geneticist, as complex traits, where a number of genetic and environmental factors determine the outcome of disease. A number of genes may either increase or decrease susceptibility, but these susceptibility genes alone are neither necessary nor sufficient to cause the trait. Segregation analysis and linkage analysis face difficulties

with such non-Mendelian conditions. Indeed segregation analysis is very sensitive to biases, whereas parametric linkage analysis requires the specification of a precise genetic model, including disease gene frequency, and the penetrance of each genotype (Gambaro et al 2000). The power of linkage analysis is therefore reduced, and this may explain why disease association studies have been widely and successfully used to identify such susceptibility genes. Note that linkage is a relationship between loci, while association is a relationship between alleles. A powerful approach to finding susceptibility loci is therefore to examine statistical association between the occurrence of disease (or a disease manifestation) and a marker genotype in a population-based case-control study. Alternatively, cross-sectional and longitudinal study designs may be used. Association studies thus provide a complementary approach for detection of susceptibility genes and are usually essential in the identification of a susceptibility gene once its chromosomal position has been determined by linkage (Weeks and Lathrop 1995).

Association can arise if the marker is causally implicated in the disease, or if it is in linkage disequilibrium with a susceptibility locus. Linkage disequilibrium (or population allelic association) is the situation where closely linked genes on a chromosome tend to remain associated rather than undergo genetic randomization. (However, it is important to note that linkage disequilibrium does not necessarily imply linkage between the loci considered, i.e. their physical proximity on a single chromosome.) The strength of this disequilibrium for a combination of alleles is measured by the observed frequency of an haplotype compared with the expected frequency, based on the frequency of the individual alleles. As a result of linkage disequilibrium, the frequency of a pair of alleles occurring together is greater than the product of the individual gene frequencies. Linkage disequilibrium thus creates difficulties in establishing which genes are actually responsible for the phenotype, or which are only secondarily associated because of linkage disequilibrium. As outlined above, linkage disequilibrium may, among others, result from natural selection promoting a particular haplotype (a combination of alleles at neighboring genes or markers that occur close together on the same chromosome), or from insufficient time elapsing since the first appearance of closely located alleles to allow them to acquire a random distribution in the population. Therefore the possible bias of population stratification or ethnicity, as a cause of confounding in association studies, should be excluded. Indeed, population stratification due to recent admixture or inappropriate matching of patients and controls is the major confounding factor in association studies, thus establishing the need for confirmatory studies that use ethnically distinct populations (Ewans and Spielman 1995, Gambaro et al 2000, Weeks and Lathrop 1995). Linkage disequilibrium testing may also be used to help positionally clone genes (see above). Because association studies can be regarded as hypothesis generating, further molecular studies (positional cloning, search for mutations, *in vitro* functional studies, and transgenic animal studies) should then elucidate how the relation between a candidate gene and the disease is accomplished. (Or, to speak with Blaise Pascal: proof follows the experiment, not the reasoning.)

The strength of an association is usually expressed as relative risk (RR) (also called risk ratio) or odds ratio (OR). The relative risk reflects the ratio of the cumulative incidence among persons with the genetic risk factor over that among persons without the risk factor. However, this measure of association would require often long-lasting and expensive cohort studies. Disease frequencies are therefore frequently ascertained in case-control studies and hence expressed as the odds of disease. (Because of the sampling designs a direct measure of relative risk cannot be obtained in case-control studies.) The odds ratio is the ratio of the odds of disease among persons with the (genetic) risk factor over the odds of disease among persons without the risk factor. Odds ratios have become popular for measuring effects of genetic and other risk factors, because they give a good approximation of the (the more relevant) relative risk for rare diseases, and because they can easily be obtained from cohort, case-control, as well as from cross-sectional studies. Disease associated with a marker is therefore usually examined by comparing allele frequencies at the marker locus in random samples of unrelated patients and controls in a case-control design. Cases and controls are carefully matched for ethnicity and other factors. As indicated, another more laborious and expensive approach to defining the role of candidate genes is to undertake a prospective cohort study rather than a hospital-based case-control study (Williams et al 1996). The advantage of such an approach is that it may easy circumvent experimental bias and may easy be combined with functional or clinical studies.

From the risk or odds ratio the magnitude of the protection or susceptibility afforded by particular alleles can be estimated. Alternatively when there is quantitative variation in disease expression caused by variation in several or many genes of smaller effect, alleles will appear to be associated with different levels of disease expression. Such genes are thus typically quantitative trait loci or QTL. This association can be detected by statistical techniques such as regression or maximum likelihood. Genetic interactions can be recognized by testing for simultaneous association between multiple loci and the disease phenotype, but this may give rise to the problem of multiple testing (see below). This means that the likelihood that a specific association exceeds a nominal or pointwise significance level increases according to the number being tested.

Relative risk and odds ratio do not provide an estimation for the overall contribution of a genetic risk factor to a disease in a particular population. The term attributable fraction is therefore designed to reflect the excess fraction of cases of a disease in a population that would not have occurred within a certain time period had the risk factor been absent. This measure is important in quantifying the role of a specific genetic factor in disease etiology and in terms of public health impact of the factor. Thus the attributable fraction estimates the proportion of disease that could be prevented if the factor were absent. Several methods for calculating the attributable fraction have been described. Therefore allele frequencies and the relationship of alleles to disease in terms of relative risk or odds ratio must be known (Khoury et al 1993).

Population association studies are more sensitive than linkage studies especially when susceptibility loci are common and have modest effects or when there is linkage heterogeneity. In addition to the candidate gene approach, it is also possible to use association studies to localize a disease susceptibility gene by typing many closely linked markers. The region showing a peak of association with disease should represent the location of the disease susceptibility gene (Feder et al 1996). Subsequent studies should then confirm that a putative gene in that region is really responsible for susceptibility or in close linkage disequilibrium with a gene responsible for susceptibility. Genome-wide association studies may use data from unrelated families with only a single affected case, or they may use unrelated cases. While approximately 300 microsatellite markers may be sufficient to scan the genome in families, far more (30.000 or more) are needed to scan the genome of unrelated cases. To avoid false-positivity due to the problem of multiple testing, high thresholds for significance, for example LOD-scores of 3.3 – 3.8, should be used. An alternative method to eliminate associations significant only by chance (not only caused by testing multiple markers) is by a Bonferroni correction. First the number of independent tests n has to be determined. If the required overall level of significance is 0.05, then an association only becomes significant if p exceeds $0.05/n$ (Craig et al 2001).

Thus, compared to linkage studies the power of association studies is generally high. Moreover, because population association studies do not require family studies, they are also relatively simple. The power to detect association using a marker depends on several factors, including the strength of the linkage disequilibrium between the marker and disease, the frequency of the mutation or polymorphism, the recombination fraction between the disease and the marker, the increase in risk attributable to the susceptibility locus, and the penetrances of the different disease locus genotypes (Schaid and Sommer 1993, Weeks and Lathrop 1995). Because it is usually unknown in advance to predict the increase in risk attributable to the susceptibility locus, many studies employ a sample size of at least 200, which may give a reasonable power and attainable sample size and genotyping effort. When reasonable assumptions on allele frequencies and odds ratios can be made, required sample sizes can easily be calculated using for example the Epi Info program (Dean et al 1995). Whether genetically isolated populations are advantageous for association studies, due to a relatively high level of linkage disequilibrium, is in debate (Abbott 2000). However, data from the general population may be advantageous, because they are more representative of the population as a whole. One should further keep in mind that failure to find an association does not imply that a gene is unimportant in the susceptibility of a disease. The gene may simply not be polymorphic in the studied population.

While appealing because of their simplicity and high power, there are several drawbacks to population associations.
- When we examine loci that may cause susceptibility to infectious disease, the cases and controls must be exposed to the infectious agent in similar ways and in equal quantities. This requires, at least, rigorous matching for age, sex, location and time of exposure, population subsets, and strain differences.
- Population admixture and population history and structure notably disturb association studies and may give rise to artifacts.
- Allelic association may identify a candidate gene directly responsible for susceptibility, or it may point to a candidate gene in close linkage disequilibrium to the marker. Stratified analysis of linked genes may not be able to attribute the highest risk to a specific allele, but rather to the extended or ancestral haplotype. This does not negate the value of studying genetic variants at a specific locus if they are part of an extended haplotype, because such studies may point to the site of the functional locus. In any case, it may be misleading to look at genetic variants of single loci in isolation.
- Association between a disease and a marker may be coincidental, especially when the controls have not been selected carefully. Therefore further *in vivo* and *in vitro* functional studies are required to provide additional and mechanistic proof of a causal relationship that is biologically plausible. Such studies must elucidate that the polymorphisms are functional and thus lead to altered function. Alternatively, when the polymorphisms are in non-coding regulatory regions, it must be shown

that they affect the amount of gene product that is being produced.

- Statistical analysis should be very rigorous and correct for the number of hypotheses tested (Strachan and Read 1996). Large sample sizes are generally required, especially to examine rare alleles or multi-allelic genes.

Association studies with internal controls

To overcome the problem that cases and controls are from genetically different subpopulations (population stratification), association study designs have been developed that use internal controls. These study designs, notably the haplotype relative risk (HRR) test and the transmission disequilibrium test (TDT), use parent/child trios and are therefore insensitive to population stratification. These tests appear very sensitive and detect linkage between a genetic marker and disease susceptibility only if population association is also present. Thus for each proband two parents have to be tested, which may be difficult if the disease occurs at older age. These tests therefore appear very suited for common childhood infections, such as infections with respiratory syncytial virus (RSV). The general idea of these tests is to examine whether a marker, an allele possibly associated with the disease, is more frequently transmitted to the affected child than would occur by chance. The "controls" are therefore hypothetical persons carrying the two parental alleles that were not transmitted to the affected child. In the TDT test, in contrast to the HRR test, at least one of the parents has to be heterozygous for the marker. Although typing parents of affected individuals is more expensive than typing of patients and controls only, the approach ensures that the control and case populations are well matched. Significance of the associations is tested by chi-square (Bink et al 2000, Evans et al 1995, Khoury 1994, Lander and Schork 1994, Spielman et al 1993). While family-based linkage analyses have a power to detect linkage between marker loci in candidate regions and the putative disease susceptibility over a distance of 10-20 cM, the TDT test may detect allelic associations over 1 cM (Blackwell et al 1997). As always when multiple comparisons are performed for association between many loci and disease susceptibility, there is the real chance of getting false-positive results. One may therefore restrict the association tests to a limited set of candidate genes. Otherwise a two-stage approach may be used. For example, in the first stage one applies the HRR or the TDT test. In the second stage, potential association encountered in the previous stage can be confirmed by examining a separate set of 200 unrelated cases along with 200 ethnically and age-matched controls (Schaid and Sommer 1993, Sobell et al 1992). Other methods that have been developed make corrections, based on the genome itself, for population stratification in association studies (Bacanu et al 2000).

References

Abel, L., Dessein, A.J. (1998). Genetic epidemiology of infectious diseases in humans: Design of population-based studies. Emerg. Infect. Dis. 4, 593-603.

Abbott, A. (2000). Manhattan versus Reykjavik. Nature 406, 340-342.

Adams, M.D., Kerlavage, A.R., Fields, C., Venter, J.C. (1993). 3,4000 new expressed sequence tags identify diversity of transcripts in human brain. Nature Genet. 4, 256-267.

Bacanu, S.A., Devlin, B., Roeder, K. (2000). The power of genomic control. Am. J. Hum. Genet. 66, 1933-1944.

Bellamy, R.J., Hill, A.V.S. (1998). Host genetic susceptibility to human tuberculosis. In: Genetics and tuberculosis, Wiley, Chichester (Novartis Foundation Symposium 217), 3-23.

Backwell, J.M., Black, G.F., Peacock, C.S., Miller, E.N., Sibthorpe, D., Gnananandha, D., Shaw, J.J., Silveira, F., Lins-Lainson, Z., Ramos, F., Collins, A., Shaw, M.A. (1997). Immunogenetics of leishmanial and mycobacterial infections: the Belem Familiy Study. Philos. Trans R. Soc. Lond. B. Biol. Sci. 352, 1331-1345.

Bink, M.C., Te Pas, M.F., Harders, F.L., Janss, L.L. (2000). A transmission/disequilibrium test approach to screen for quantitative trait loci in two selected lines of large white pigs. Genet. Res. 75, 115-121.

Brahmajoth, V., Pitchappan, R.M., Kakkanaiah, V.N., Sashidar, M., Rajaram, K., Ramu, S., Palanimurugan, K., Paramasivan, C.N., Prabhakar, R. (1991). Association of pulmonary tuberculosis and HLA in south India. Tubercle 72, 123-132.

Brown, D.L., Gorin, M.B., Weeks, D.E. (1994). Efficient strategies for genomic searching using the affected-pedigree-member method of linkage analysis. Am. J. Hum. Genet. 54, 242-253.

Chumakov, I.M., Rigault, P., Le Gall, I., Bellane-Chantelot, C., Billault, A., Guillou, S., Soularue, P., Guasconi, G., Poullier, E., Gros, I. et al. (1995). A YAC contig map of the human genome. Nature 377 (6546 Suppl.), 175-297.

Craig, A., Hastings, I., Pain, A., Roberts, D.J (2001). Genetics and malaria-more questions than answers. Trends in Parasitol. 17, 55-56.

Davies, J.L., Kawaguchi, Y., Bennet, S.T., Copeman, J.B., Cordell, H.J., Pritchard, L.E., Reed, P.W., Gough, S.C., Jenkins, S.C., Palmer, S.M. et al. (1994). A genome-wide search for human type 1 diabetes susceptibility genes. Nature 371, 130-136.

Dean, A.G. et al (1995). Epi Info, Version 6: A word-processing, database and statistics program for Public Health on IBM-compatible microcomputers. Centers for Disease Control and Prevention, Atlanta, Georgia, USA.

Dib, C., Fauré, S., Fizames, C., Samson, D., Druout, N., Vignal, A., Milasseau, P., Marc, S., Hazan, J., Seboun, E., Lathrop, M., Gyapay, G., Morissette, J., Weissenbach, J. (1996). A comprehensive genetic map of the human genome based on 5,264 microsatellites. Nature 380, 152-154.

Ellsworth, D.L., Maniolo, T.A. (1999). The emerging importance of genetics in epidemiologic research III. Bioinformatics and statistical genetic methods. Ann. Epidemiol. 9, 207-224.

Ewens, W.J., Spielman, R.S. (1995). The transmission/disequilibrium test: History, subdivision, and admixture. Am. J. Hum Genet. 57, 455-464.

Feder, J.N., Gnirke, A., Thomas, W., Tsuchihashi, Z., Ruddy, D.A., Basava, A., Dormishian, F., Domingo, R. Jr., Ellis, M.C., Fullan, Hinton, L.M., Jones, N.L., Kimmel, B.E., Kronmal, G.S., Lauer, P., Lee, V.K., Loeb, D.B., Mapa, F.A., McClelland, E., Meyer, N.C., Mintier, G.A., Moeller, N., Moore, T., Morikang, E., Wolff, R.K. et al. (1996). A novel MHC class I-like gene is mutated in patients with hereditary haemochromatosis. Nat. Genet. 13, 399-408.

Gambaro, G., Anglani, F., D'Angelo, A. (2000). Association studies of genetic polymorphisms and complex disease. Lancet 3555, 308-311.

Hill, A.V.S. (1996). Genetic susceptibility to malaria and other infectious disease: from the MHC to the whole genome. Parasitology 112, S75-S84.

Hudson, T.J., Stein, L.D., Gerety, S.S., Ma, J., Castle, A.B., Silva, J., Slonim, D.K., Baptista, R. Kruglyak, L., Xu, S.H. et al (1995). An STS-based map of the human genome. Science 270, 1945-1954.

Jorde, L.B. (1995). Linkage disequilibrium as a gene-mapping tool. Am. J. Hum. Genet. 56, 11-14.

Khoury, M.J., Beaty, T.H., Cohen, B.H. Fundamentals of genetic epidemiology. Monographs in epidemiology and biostatistics, volume 22, Oxford University Press, New York, Oxford, 1993.

Khoury, M.J. (1994). Case-parental control method in the search for disease-susceptibility genes. Am. J. Hum. Genet. 55, 414-415.

Kruglyak, L., Lander, S. (1995). Complete multipoint sib-pair analysis of qualitative and quantitative traits. Am. J. Hum. Genet. 57, 439-454.

Lander, E.S., Schork, N.J. (1994). Genetic dissection of complex traits. Science 265, 2037-2048.

Qureshi, S.T., Skamene, E., Malo, D. (1999). Comparative genomics and host resistance against infectious diseases. Emerging Inf. Dis. 5, 36-47.

Risch, N. (2000). Searching for genetic determinants in the new millennium. Nature 405, 847-856.

Risch, N., Zhang, H. (1995). Extreme discordant sib pairs for mapping quantitative trait loci in humans. Science 268, 1584-1589.

Risch, N., Merikangas, K. (1996). The future of genetic studies of complex human diseases. Science 221, 1181-1183.

Schaid, D.J., Sommer, S.S. (1993). Genotype relative risks: methods for design and analysis of candidate-gene association studies. Am. J. Hum. Genet. 53, 1114-1126.

Schuler, G.D., Boguski, M.S., Stewart, E.A., Stein, L.D., Gyapay, G., Rice, K., White, R.E., Rodriguez-Tome, P., Aggarwal, A., Bajorek, E., Bentolila, S., Birren, B.B., Butler, A., Castle, A.B. et al (1996). A gene map of the human genome. Science 274, 540-546.

Sobell, J.L., Heston, L.L., Sommer, S.S. (1992). Delination of genetic predisposition to multifactorial disease: a general approach on the threshold of feasibility. Genomics 12, 1-6.

Spielman, R.S., McGinnis, R.E., Ewens, W.J. (1993). Transmission test for linkage disequilibrium: The insulin gene region and insulin-dependent diabetes mellitus (IDDM). Am. J. Hum. Genet. 52, 506-516.

Strachan, T., Read, A.P. (1996). Human molecular genetics. BIOS scientific publishers Ltd. Williams, T.N., Maitland, K., Bennett, S., Ganczakowski, M., Peto, T.E., Newbold, C.I., Bowden, D.K., Weatherall, D.J., Clegg, J.B. (1996). High incidence of malaria in alpha-thalassaemic children. Nature 383, 522-525.

Weeks, D.E., Lathrop, G.M. (1995). Polygenic disease: methods for mapping complex disease traits. Trends in Genet. 11, 513-519.

Chapter 5.   The use of animal models and comparative genomics

Several genes determining susceptibility for human infectious disease have been found by first studying laboratory animal models. Because susceptibility to infection is often a complex genetic trait, researchers try to improve their insight for example by reducing the complexity into monogenic phenotypes. While this is difficult to do in humans, the inheritance and mechanisms of disease resistance have been analyzed in animals. Especially strains of laboratory mice have been very useful. Mice are easy to breed, several well-characterized inbred strains are available, their genome contains a high density of genetic markers, genetic manipulation is possible, and there is extensive (> 85 %) homology with humans, thus allowing comparative gene mapping (Malo and Skamene 1994). In addition, environmental factors can be experimentally controlled to reduce their effects on phenotypic variance or to allow the study of gene-environment interactions.

Experimental crosses between strains with contrasting phenotype have the advantage that a single set of genetic factors segregate in each cross. In addition, much of the phenotypic variation is attributed to genetic factors (i.e. the genetic heritability is high), and experimental inoculation with pathogens can easily be standardized. For inbred animal models, the statistical analysis of quantitative trait analysis is relatively easy using analysis of variance or related procedures. These can also be applied if the inheritance is polygenic (Hilbert et al 1991, Weeks and Lathrop 1995). Additional statistical approaches have been described for QTL mapping when phenotype values are not normally distributed (Boyartchuk et al 2001).

To identify host-resistance genes in mice a step-wise approach has proven its feasibility. First, experimental inoculation should identify inbred strains that differ in their susceptibility to a pathogen. These experiments likely give information on the mode of inheritance. When there is a continuous variation in disease susceptibility in a large panel of inbred strains, there is likely a complex trait controlled by multiple genes. In contrast, when there is a pattern of discontinuous variation, with clearly distinct susceptible and resistant strains, the trait is likely controlled by a single locus with two alternative alleles.

Second, to further determine the mode of inheritance, resistant and susceptible strains are crossed. Correlation of the resistance or susceptibility phenotype with a specific chromosomal region is then performed by linkage analysis. The localizations obtained for genes affecting complex traits mapped by linkage in experimental crosses span at least 15-20 cM. This would allow finding at least one candidate gene in most regions of linkage. Evidently the power of linkage analysis in animal model systems is much higher than in humans, which is explained by the use of inbred strains that differ in a limited number of loci and the large number of informative animals that can be used (Lathrop 1996, Risch 2000).

Third, the underlying mechanisms of the resistance or susceptibility phenotype should be elucidated by evaluating known genes and by positionally cloning of new genes in the identified genetic interval.

Finally, however, the relationship between a particular gene and susceptibility to infection should be proven through modification of the phenotype after transfer of a disease-associated variant into a control strain. Alternative approaches could include the analysis of mice bearing a gene knockout or a gene multiplication.

Recombinant inbred mouse strains and experimental crosses
Especially when a single major locus is involved, recombinant inbred strains are easily used to map the resistance/susceptibility gene. Recombinant inbred strains are created from resistant and susceptible parental strains, for example from reciprocal crosses between strains A/J and C57BL/6J, which are subsequently maintained by brother-sister matings for 20 generations. These strains have been very useful in finding susceptibility loci, because they differ in susceptibility to a large number of infectious and non-infectious diseases (Malo and Skamene 1994, Marshall et al 1992). This approach has identified several loci controlling susceptibility to a wide range of infections. For example, the *Mx* and *Nramp* genes have first been identified in mice before human homologs were characterized.

Also when different loci are likely involved, the availability of different inbred strains showing continuous variation in their susceptibility to a pathogen allows the identification of some of the loci involved. This approach has for example proven its value in delineating the mechanism of resistance to infections with *Salmonella typhimurium*. Different inbred strains of mice show different susceptibility to experimental inoculation with *S. typhimurium*. The A/J strain is fully resistant, the BALB/cJ and C3H/HeJ strains are very susceptible, and the DBA/2J and CBA/N strains are intermediate. Crosses between resistant and susceptible strains identified several distinct loci operating via different mechanisms: *Ity* (later shown to be identical with *Bcg*, *Lsh* and *Nramp*), *Lps*, *xid*, and several H-2 linked loci. *Ity* and *Lps* are genes involved in macrophage function, while *xid* modulates humoral immunity (Malo and Skamene 1994, Plant and Glynn 1979, O'Brien et al 1980, O'Brien et al 1981, Gros et al 1983, Scher 1982).

In a similar approach, modifiers of *Listeria monocytogenes* susceptibility were mapped to chromosomes 5 and 13. *L. monocytogenes* injected intravenously causes death of BALB/cByJ animals within 72 hours, whereas all C57BL/6ByJ mice survive indefinitely. The F2 cross revealed a complex phenotypic segregation pattern, with a wide range of survival times. Influx of innate immune cells was likely involved in this effect. Although two loci with major effects were identified, additional, yet undetected QTL likely affected the outcome of *L. monocytogenes* in this model (Boyartchuk et al 2001).

This approach has further been followed in a number of studies examining the course of infection with several eukaryotic parasites, bacteria, and viruses. The designations of the identified loci usually refer to the organism studied, and include *Ack*, *Anth2*, *Bcga*, *Lsr1*, *Pscr1*, *Pscr2*, *Ric*, *Cms*, *crz*, *Hc*, *Ld*, *Mal1*, *mph1*, *Rsm-1*, *Sc11*, *Sc12*, *Ts1-4*, *Av1*, *Av2*, *Cmv1*, *Fhe*, *Flv*, *Fv1-6*, *Hv1*, *Hv2*, *If-x*, *Ptv1*, *Rmc1*, *Rmcf*, *Rmp1*, *Rmp2*, *Rvil1*, *Tmevd1*, *Tmevd2*, *Rmp3*, *Rvf1*, *Rvf2*, *Rmv1*, *Rmv2*, *Rrs*, *Rrv1*, and several other H-2 linked loci (reviewed by McLeod et al 1995). Another approach to dissect polygenic inheritance is the use of recombinant congenic strains of mice.

Recombinant congenic mouse strains
When susceptibility is controlled by a limited number of genes, a strategy to identify them is to separate them by random breeding so that they can be studied either individually or in a setting in which the number and complexity of interactions is largely reduced. For that purposes inbred recombinant congenic strains have been produced

(Demant and Hart 1986, Groot et al 1992). Strains are congenic when they differ only at a single or several genetic loci as a result of backcrossing the differential locus or loci to a standard inbred strain. They make it possible to study the biological effects of different alleles at the differential locus (loci) free from noise and interactions with other loci. Initially two inbred strains, a donor and a recipient strain, are crossed. Backcrossing, without selection, with the recipient strain produces the next two generations. The offspring is crossed by strict brother-sister mating to produce a set of homozygous inbred lines, which thus have mosaic genomes. Each of the recombinant congenic inbred lines contains on average a random 12.5 % fraction of the donor genome. The remaining 87.5 % is from the common background inbred strain. Because the fractions donor genome overlap, 20 – 25 recombinant congenic lines are required to represent 95 % of the donor genome in the recombinant congenic lines. Using these strains epistatic effects of other genes that affect susceptibility are minimized by separating the parts from the donor genome (Frankel and Schork 1996, van Wezel et al 1996). Genotyping using microsatellite markers indicated that the strains differ in the proportion and fragment sizes of the donor genome. These strains not only simplify the tasks of identifying genes involved in the multigenic control of infectious disease, but they also allow to dissect complex aspects of the pathogenesis. Mapping genes that affect resistance to infectious disease is done by using F2 hybrids or backcrosses between a suitable recombinant congenic strain and its "background" strain. Subsequent fine mapping is done by producing a number of recombinant mice, each carrying different recombinations, i.e. different fragments of the originally identified gene segment. The progeny of each recombinant is tested for presence of the tested fragment and the effect of the susceptibility gene. If the shorter fragment does contain the novel susceptibility gene, there will be a phenotypic difference between the progeny which did and which did not inherit the tested fragment. If the relevant susceptibility gene is located outside the shorter fragment, there will be no difference between the mice, which did and did not inherit it.

Three series of recombinant congenic lines were produced using the background and donor strain pairs: BALB/cHeA and STS/A (the CcS/Dem series), C3H/DiSnA and C57/BL/10ScSnA (the HcB/Dem series), and O20A and B10.O20/Dem (the OcB/Dem series)(Groot et al 1992).

The HcB/Dem series were tested for susceptibility to infection with *M. tuberculosis* and the strains that were most resistant or susceptible were genotyped using microsatellite markers (Kramnik et al 1998). Survival times showed a continuous distribution pattern indicative of multigenic control. In addition, genotyping using microsatellite markers at intervals of 5-10cM indicated that no donor chromosomal segment was common to all resistant or susceptible mice, and that no identical or overlapping patterns for the recombinant congenic lines, resistant or susceptible to *M. tuberculosis*, were observed. Some B10 donor genome fragments appeared clearly associated with enhanced susceptibility upon interaction with the C3H background genome. These results thus confirm the multigenic control of susceptibility to tuberculosis. Further mapping of the susceptibility genes and phenotypic analysis is in progress (Kramnik et al 1998).

In our laboratory we started to investigate the genetic control of susceptibility to infection with *Bordetella pertussis* using the CcS/Dem and HcB/Dem recombinant congenic lines. Mice were therefore inoculated intranasally with 5×10^8 colony-forming units of *B. pertussis*. The initial results indicate a genetic control of the number of viable bacteria in the lung at day 7 postinoculation. The CcS4 strain appeared the most resistant strain of the CcS hybrids, and the HcB28 strain appeared the most resistant strain of the HcB hybrids. F2 hybrids from these strains and their recipient BALB/c and C3H strains are therefore used to map the gene(s) involved in their phenotype.

Comparative genomic analysis: from mouse to man
When susceptibility genes are identified in mice, knowledge of the genomic organization of human and mouse facilitates direct localization and identification of their human orthologs. Orthologs are chromosomal segments or DNA sequences with a close similarity between species (Strachan and Read 1996). Such genes can then be evaluated in humans by polymorphism analysis. Comprehensive high-resolution genetic maps, based on genes or microsatellites and spanning the genomes of mouse and human, have been created recently for that purpose. They are essential tools for developing mouse models of human disease and for identifying the function of genes (Anderson et al 1996, Eppig and Nadeau 1995, Copeland et al 1993, Jordan and Collins 1996, Qureshi et al 1999). Genetic mapping studies are further underway in many species, including rats (Ratmap), Sheep (SheepBase), and pigs (PigBase) (Anderson et al 1996). Databases for species-specific or comparative mapping studies are available (Anderson et al 1996, Nadeau et al 1995, Wakefield and Graves 1996). Evidently, the comparative map of the mouse and human genomes is the best developed and has defined a large number of conserved linkage groups or syntenic relationships (DeBry and Seldin 1996). Unfortunately however, polymorphisms in one species will usually not be carried through to polymorphisms in other species, although there are some exceptions to this rule. (For example, some MHC alleles are found in different species and must therefore have been arisen millions of years ago before speciation, suggesting that selection has been directed at maintaining the polymorphisms [Figueroa et al 1988, Lawlor et al 1988].) Despite such drawbacks, animal models may be very useful to identify important molecular pathways, pathogenic mechanisms, and candidate genes for further studies in humans. Thus, even without functional variation in human orthologs of genes mapped in animal models, these studies may provide important understanding of infectious disease pathogenesis.

Caenorhabditis elegans and Drosophila melanogaster as animal models?
While genetic approaches to study host-pathogen interactions may be more feasible in rodents than in man, this may even be more the case in model organisms with small, well characterized genomes, such as the nematode *Caenorhabditis elegans* and the fruitfly *Drosophila melanogaster*. While the benefits of this approach must be awaited, it has been shown that some pathogens, notably *Salmonella typhimurium* and *Pseudomans aeruginosa,* can lethally infect *C. elegans.*

References

Anderson, L., Archibald, A., Ashburner, M., Audun, S., Barendse, W., Bitgood, J. et al. (1996). Comparative genome organization of vertebrates. Mamm. Genome 7, 717-734.

Boyartchuk, V.L., Broman, K.W., Mosher, R.E., D'Orazio, S.E.F., Starnbach, M.N., Dietrich, W.F. (2001). Multigenic control of *Listeria monocytogenes* susceptibility in mice. Nature Genet. 27, 259-260.

Copeland, N.G., Jenkins, N.A., Gilbert, D.J., Eppig, J.T., Maltais, L.J., Miller, J.C., Dietrich, W.F., Weaver, A., Lincoln, S.E., Steen, R.G. et al. (1993). A genetic linkage map of the mouse: current applications and future prospects. Science 262, 57-66.

DeBry, R.W., Seldin, M.F. (1996). Human/mouse homology relationships. Genomics 33, 337-351.

Demant, P., Hart, A.A.M. (1986). Recombinant congenic strains: a new tool for analyzing genetic traits determined by more than one gene. Immunogenetics 24, 416-422.

Eppig, J.T., Nadeau, J.H. (1995). Comparative maps: the mammalian jigsaw puzzle. Curr. Opin. Genet. Devel. 5, 709-716.

Figueroa, F., Günther, E., Klein, L. (1988). MHC polymorphism pre-dating speciation. Nature 335, 265-267.

Frankerl, W., Schork, N.J. (1996). Who's afraid of epistasis? Nat. Genet. 14, 371-373.

Groot, P.C., Moen, C.J., Dietrich, W., Stoye, J.P., Lander, E.S., Demant, P. (1992). The recombinant congenic strains for analysis of multigenic traits: genetic composition. FASEB J. 6, 2826-2835.

Gros, P., Skamene, E., Forget, A. (1983). Cellular mechanisms of genetically controlled host resistance to *Mycobacterium bovis* (BCG). J. Immunol. 131, 1966-1972.

Hilbert, P., Lindpainter, K., Beckmann, J.S., Serikawa, T., Soubrier, F., Dubay, C., Cartwright, P., De Gouyon, B., Julier, C., Takahasi, S. et al. (1991). Chromosomal mapping of two genetic loci associated with blood-pressure regulation in hereditary hypertensive rats. Nature 353, 521-529.

Jordan, E., Collins, F.S. (1996). A march of genetic maps. Nature 380, 111-112.

Kramnik, I., Demant, P., Bloom, B.B. Susceptibility to tuberculosis as a complex genetic trait: analysis using recombinant congenic strains of mice. In: Genetics and tuberculosis, Wiley, Chichester (Novartis Foundation Symposium 217), 120-137.

Lathrop, G.M. (1996). Quantitative phenotype analysis for localization and identification of disease-related genes in a complex genetic background. In: Variation in the human genome. Wiley, Chichester (Ciba Foundation Symposium 197), 284-299.

Lawlor, D.A., Ward, F.E., Ennis, P.D., Jackson, A.P., Parham, P. (1988). HLA-A and B polymorphism predate the divergence of humans and chimpanzees. Nature 335, 268-271.

Malo, D., Skamene, E. (1994). Genetic control of host resistance to infection. Trends in Genet. 10, 365-371.

Marshall, J.D., Mu, J.L., Cheah, Y.C., Nesbitt, M.N., Frankel, W.N., Paigen, B. (1992). The AXB and BXA set of recombinant inbred mouse strains. Mamm. Genome 3, 669-680.

McLeod, R., Buschman, E., Arbuckle, L.D., Skamene, E. (1995). Immunogenetics in the analysis of resistance to intracellular pathogens. Current Opinion Immunol. 7, 539-552.

Nadeau, J.H., Grant, P.L., Mankala, S., Reiner, A.H., Richardson, J.E., Eppig, J.T. (1995). A Rosetta Stone of mammalian genetics. Nature 373, 363-365.

O'Brien, A.D., Rosenstreich, D.L., Scher, I., Campbell, G.H., MacDermott, R.P., Formal, S.B. (1980). Genetic control of susceptibility to *Salmonella typhimurium* in mice: role of the LPS gene. J. Immunol. 124, 20-24.

O'Brien, A.D., Scher, I., Metcalf, E.S. (1981). Genetically conferred defect in anti-*Salmonella* antibody formation renders CBA/N mice innately susceptible to *Salmonella typhimurium* infection. J. Immunol. 126, 1368-1372.

Plant, J., Glynn, A.A. (1979). Locating *Salmonella* resistance gene on mouse chromosome 1. Clin. Exp. Immunol. 37, 1-6.

Qureshi, S.T., Skamene, E., Malo, D. (1999). Comparative genomics and host resistance against infectious diseases. Emerging Inf. Dis. 5, 36-47.

Risch, N. (2000). Searching for genetic determinants in the new millennium. Nature 405, 847-856.

Scher, I. (1982). The CBA/N mouse strain: an experimental model illustrating the influence of the X-chromosome on immunity. Adv. Immunol. 33, 1-71.

Strachan, T., Read, A.P. (1996). Human molecular genetics. BIOS scientific publishers Ltd.

Van Wezel, T., Stassen, A., Moen, C., Hart, A.A.M., van der Valk, M.A., Demant, P. (1996). Gene interaction and single gene effects in colon tumour susceptibility in mice. Nat. Genet. 14, 468-470.

Weeks, D.E., Lathrop, G.M. (1995). Polygenic disease: methods for mapping complex disease traits. Trends in Genet. 11, 513-519.

Wakefield, M.J. Graves, J.A.M. (1996). Comparative maps of vertebrates. Mamm. Genome 7, 715-716.

Chapter 6.   How to identify relevant candidate genes?

Several approaches may be used to identify candidate host genes that may have relevant functions in the pathogenesis of infectious disease, and that deserve to be examined in further studies for association with resistance or susceptibility. They can be
summarized as follows (see also Fig. 5):

1. Genetic approach: from genetic variation to protein and function.

To search for relevant candidate genes involved in a specific disorder, a successful
approach may be to start with a familial linkage analysis to find chromosomal locations containing one or more candidate genes that nonrandomly segregate with the
trait ("positional cloning"). As discussed, this approach requires families in which
affected and non-affected family members can easily be discriminated by a reliable
identification of the disease phenotype, as well as informative polymorphic DNA markers, for example from the Généthon collection (Dib et al 1996). One may choose to
type a few markers in a limited number of chromosomal regions (for example because the region may contain genes that are related to the disease under study), or one
may choose a random search along the whole genome. Once evidence for linkage is
obtained, fine genetic and physical mapping is performed to narrow down the genetic intervals. Within the chromosomal intervals that are significantly linked to one or
more markers, one subsequently searches for genes that may affect the pathogenesis
of the disease. These genes are subsequently examined for polymorphisms or muta-

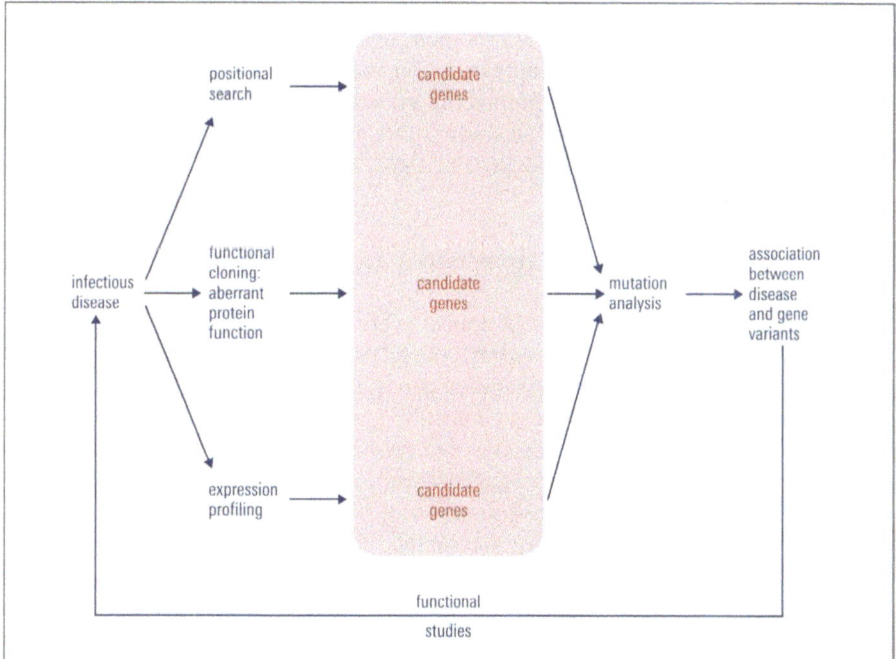

*Figure 5. Several pathways of searching for candidate genes for association and subsequent functional
studies.*

tions. When the tested polymorphism is in strong linkage disequilibrium with the disease allele or is the disease allele itself, subsequent association studies may directly identify the disease gene and confirm its involvement in the disease. This has to be confirmed, for example by functional studies of the gene variants.

This approach has been successful in finding a candidate region in schistosomiasis. In a Brazilian population, segregation analysis showed that a major gene, designated *SM1*, controlled the intensity of infection by *Schistosoma mansoni*. The gene could account for 66% of the infection intensity variance after correction for other factors. A genomewide scan, using the model estimated from the segregation analysis in the parametric linkage analysis, mapped *SM1* to human chromosome 5q31-q33, a genetic region that contains several genes controlling T cell differentiation, IL-5 levels, IgE responses, eosinophilia, and bronchial hyperresponsiveness in asthma. Because resistant homozygotes mount a Th2 response upon *Schistosoma* infection, characterized, among others, by strong IL-5 responses, polymorphisms controlling IL-5 production could well be responsible for susceptibility/resistance to schistosomiasis (Abel et al 1991, Marquet at al 1996).

2. Pathophysiological approach: from protein function to genetic variation.

This approach, also referred as "functional cloning", starts with the identification of an important biochemical or immunological function, a biochemical defect, or a variation in protein function. Subsequent steps are a functional characterization of the variable protein, the identification of the gene, and a search for gene variants. Finally the association of gene variants with susceptibility to disease is examined.

This approach has been followed to identify one of the genes implicated in malaria resistance. Pathologic studies showed high TNF-α blood levels in cerebral malaria. Because a variant of the TNF-α gene promoter, denoted as TNF2, is a much stronger transcriptional activator than the more common allele TNF1 and indeed affects TNF-α expression, a subsequent study examined the association between TNF2 with cerebral malaria. Persons homozygous for TNF2 were subsequently found to have an increased risk for cerebral malaria (Abel and Dessein 1998, McGuire et al 1994, Wilson et al 1997).

3. Gene expression approach: from mRNA to protein function and genetic variation.

This method is still in its infancy, but appears very promising. Using cDNA microarrays containing clones of complementary DNA (cDNA) for a large number (for example > 15,000) of human or laboratory animal genes, it is possible to systematically analyze gene expression in response to infection. In contrast to more conventional hybridization techniques, such analyses can now be performed in a miniaturized, massively parallel, "genome-wide" hybridization format, also designated as DNA or genome chips. This approach therewith offers an "expression signature" of a disease. Comparisons of mRNA levels in multiple samples are accomplished with differential fluorescence labeling (Iyer et al 1999, Schena et al 1998). Although expensive at the moment, advances in DNA microarray fabrication techniques promise to reduce the cost of arrays (Johnston and Fields 2000). Microbial pathogens, beyond doubt, induce a complex repertoire of host gene expression profiles that reflect different pathogenic mechanisms and that may point to interesting pathways and candidate genes influencing susceptibility to disease (Alizadeh et al 2000, Brown and Hartwell 1998, Eisen et al 1998, Relman 1998). Expressed genes may be further studied with regard to their function, tissue- and disease stage-specific expression, contribution to susceptibility or resistance, interaction with other genes, and polymorphisms. Evidently gene expression profiling may yield interesting findings upon experimental inoculation of susceptible and resistant strains of laboratory animals, but it may be a difficult task to dissect which expressed genes are really causally affecting the variation in disease phenotypes. Results of expression studies might even point to genotypic variation at a locus. Thus this initial holistic approach may generate hypotheses that subsequently require a reductionist approach to gain further insight into gene function. The micro-array technology can further be used in polymorphism detection and genotyping on a genomic scale.

References

Abel, L., Demenais, F., Prata, A., Souza, A.E., Dessein, A. (1991). Evidence for the segregation of a major gene in human susceptibility/resistance to infection by *Schistosoma mansoni*. Am. J. Hum. Genet. 48, 959-970.

Abel, L., Dessein, A.J. (1998). Genetic epidemiology of infectious diseases in humans: Design of population-based studies. Emerg. Infect. Dis. 4, 593-603.

Alizadeh, A.A., Eisen, M.B., Davis, R.E. et al. (2000). Distinct types of diffuse large B-cell lymphoma identified by gene expression profiling. Nature 403, 503-511.

Brown, P.O., Hartwell, L. (1998). Genomics and human disease-variations on variation. Nat. Genet. 18, 91-93.

Dib, C., Fauré, Flzames, C., Samson, D., Drouot, N., Vignal, A., et al. (1996). A comprehensive genetic map of the human genome based on 5,264 microsatellites. Nature 380, 152-154.

Eisen, M.B., Spellman, P.T., Brown, B.O., Botstein, D. (1998). Cluster analysis and display of genome-wide expression patterns. Proc. Natl. Acad. Sci. USA 95, 14863-14868.

Iyer, V.R., Eisen, M.B., Ross, D.T., Schuler, G., Moore, T., Lee, J.C.F., Trent, J.M., Staudt, L.M., Hudson, J.Jr., Boguski, M.S., Lashkari, D., Shalon, D., Botstein, D., Brown, P.O. (1999). The transcriptional program in the response of human fibroblasts to serum. Science 283, 83-87.

Johnston, M., Fields, S. (2000). Grass-roots genomics. Nature Genetics, 24, 5-6.

Marquet, S., Able, L., Hillaire, D., Dessein, H., Kalil, J., Feingold, J., et al (1996). Genetic localization of a locus controlling the intensitiy of infection by *Schistosoma mansoni* on chromosome 5q31-q33. Nat. Genet. 14, 181-184.

McGuire, W., Hill, A.V.S., Allsopp, C.E.M., Greenwood, B.M., Kwiatkoski, D. (1994). Variation in the TNF-α promoter region associated with susceptibility to cerebral malaria. Nature 371, 508-511.

Relman, D.A. (1998). Detection and identification of previously unrecognized microbial pathogens. Emerg. Infect. Dis. 4, 382-389.

Schena, M., Heller, R.A., Theriault, T.P., Konrad, K., Lachenmeier, E., Davis, R.W. (1998). Microarrays: biotechnology's discovery platform for functional genomics. Trends Biotechnol. 16, 301-306.

Wilson, A.C., Symons, J.A., McDoqwell, T.L., McDevitt, H.O., Duff, G.W. (1997). Effects of a polymorphism in the human tumor necrosis factor α promoter on transcriptional activation. Proc. Natl. Acad. Sci. USA 94, 3195-3199.

Part II   Genes and gene functions influencing susceptibility to infectious disease

Chapter 7.   Genes involved in attachment and entry of micro-organisms into the body

The following chapters deal with functional mechanisms that may be responsible for resistance or susceptibility to infectious disease. However, throughout this text the terms susceptibility, insusceptibility, tolerance, resistance and resilience are used somewhat loosely and they usually do not exactly refer to the physiological mechanisms involved. However, these terms may reflect different situations upon contact of a micro-organism with its host. For example the micro-organism may fail to establish infection, often caused by the lack of a specific receptor and leading to a failure in attachment or entry. This situation is also termed natural resistance. It may depend on a single gene and may result in complete resistance. Otherwise, the micro-organism may establish infection, which then fails to completely develop. This incomplete development may be caused by innate mechanisms of resistance,   including acute inflammatory mechanisms, and is sometimes called non-specific resistance. The micro-organism may also establish and develop infection, which may then be quickly controlled or eliminated by the host's adaptive immune response. Variations in such innate or specific immune responses, usually involving many genes, may thus lead to variations in the clinical outcome of the infection, in time to recovery, or in variable mortality. Genes responsible for these responses are sometimes referred to as severity genes. Finally, the capacity to tolerate the physiological disturbances caused by infection, or to maintain health and productivity, is sometimes referred to as resilience or tolerance (Stear and Wakelin 1998).

Most micro-organisms establish infection via the mucosae of the alimentary, respiratory, and urogenital tracts. A first step in colonization and sometimes penetration is usually the specific attachment of the micro-organism to one or more specific receptors on the eukaryotic cell or the extracellular matrix. Some micro-organisms do not invade the body via healthy mucosae or the skin, but enter the body via damaged tissues, or they may be introduced by biting arthropods or contaminated needles. After entry into the body, some micro-organisms may further attach to specific receptors on certain cell types, such as erythrocytes (*Plasmodium* spp., *Babesia*), macrophages (*Bordetella pertussis*), or T helper cells (human immunodeficiency virus). The location of the micro-organism in the body is therewith determined to a large extent by these ligand/receptor interactions. The nature of a number of these receptor/ligand interactions, and the influence of some specific receptor mutations or polymorphisms, has been characterized in detail (see Table 3). Receptors usually contain carbohydrate moieties, most often galactose, mannose, and sialic acid, of surface glycoproteins. Among the eukaryotic receptors are members of the integrin family, a family of eukaryotic glycoprotein adhesion molecules that mediate cell-cell and cell-extracellular matrix interactions. Integrins bind to proteins expressing the triplet motif Arg-Gly-Asp. The nature of the interaction between host and micro-organism may be more complex when pathogens use more than one adhesin system in a particular sequen-

Table 3. Receptor ligand interactions between micro-organisms and host cells and the influence of receptor mutations

Micro-organism	Microbial ligand(s)	Cellular (co)receptor(s)	Functional mutation/phenotype	Ref.
Enterotoxigenic *E. coli* (ETEC)	K88ab, K88ac, K88ad on fimbriae (pigs)	various forms of mucin-type glycoprotein receptors bcd on intestinal epithelium	lack of receptor results in resistance to ETEC-induced diarrhea	Francis et al 1998
Uropathogenic E. coli	P fimbriae	blood group P substance on uroepithelial cells	higher susceptibility in persons with high receptor density	Svanborg Eden et al 1983
Salmonella typhi	type I fimbriae	cystic fibrosis transmembrane conductance regulator	ΔF508 mutation results in less bacterial uptake *in vitro* (resistance)	Pier et al 1998
Pseudomonas aeruginosa		cystic fibrosis transmembrane conductance regulator	ΔF508 mutation results in reduced binding, endocytosis, and clearance of bacteria (susceptibility)	Piet et al 1997
Bordetella pertussis	filamentous haemagglutinin (FHA), pertactin, pertussigen	CR3 and others on respiratory epithelium and macrophages		
	fimbriae	very late antigen-5 (VLA-5) on monocytes and macrophages		Hazenbos et al 1993
Mycobacterium leprae		laminin-2/α-dystroglycan complex on Schwann cells		Spear 1998

continue table 3. Receptor ligand interactions between micro-organisms and host cells and the influence of receptor mutations

Micro-organism	Microbial ligand(s)	Cellular (co)receptor(s)	Functional mutation/phenotype	Ref.
HIV	gp120	CD4, CCR5, CXCR4 on T cells and macrophages; CCR 3 and CCR5 on microglia	homozygous 32 bp deletion in CCR5 results in resistance	Cohen et al 1998 He et al 1997
Arena viruses (Lassa fever virus, lymphocytic choriomeningitis virus)	GP-1	α-dystroglycan on a wide range of cells		Spear 1998
Coxsackievirus A13, A18, A21	viral capsid	intercellular adhsion molecule-1 (ICAM-1) (Ig superfamily)		
Coxsackievirus A21, B1, B3, B5, echoviruses, enterovirus 7 (and others)	viral capsid	CD55 (decay-accelerating factor [DAF])		
rhinoviruses (major receptor group rhinoviruses)	viral capsid	ICAM-1 on respiratory epithelium		Rossman et al 2000
poliovirus	viral capsid	CD155 (poliovirus receptor, member of Ig superfamily)		
alphaherpesviruses (herpes simplex virus and others)	glycoprotein D	herpesvirus entry mediator C (HveC) on epithelial and neuronal cells		Geraghty et al 1998
Epstein-Barr virus	viral envelope protein	CR2 (CD21) on B cells		

continue table 3. Receptor ligand interactions between micro-organisms and host cells and the influence of receptor mutations

Micro-organism	Microbial ligand(s)	Cellular (co)receptor(s)	Functional mutation/phenotype	Ref.
parvovirus B19		globoside (blood group P) on erythroid precursor cells and endothelial cells	lack of P antigen results in resistance	Brown et al 1994
measles virus (Edmonston strain)	haemagglutinin	CD46		
measles virus (all strains)	haemagglutinin	CDw150 (SLAM [signalling lymphocyte activation) molecule] on activated B and CD45RO^{high} memory T cells		Tatsuo et al 2000
rabies virus	viral envelope protein	acetylcholine receptor at neuromuscular junction		
Plasmodium vivax		Duffy glycoprotein (chemokine) receptor on erythrocytes	lack of transcriptional activation on erythrocytes confers resistance	Horuk et al 1993
Plasmodium falciparum		glycophorin A, B on erythrocytes		
P. falciparum-infected red blood cells		CD36 on endothelial cells	CD36 deficiency results in enhanced susceptibility to severe malaria	Aitman et al 2000
		ICAM-1 on endothelial cells	point mutation in ICAM-1 is associated with increased risk of cerebral malaria	Fernandez-Reyes et al 1997
Leishmania mexicana	surface glycoprotein gp63	CR3 on macrophages		

tial combination. In addition, opsonized bacteria, including *Legionella pneumophila* and *Mycobacerium tuberculosis*, may enter cells such as macrophages through the interaction of adsorbed C3bi with the complement receptor 3 (CR3) on these cells.

While lack of a receptor usually results in lack of susceptibility or resistance, there may be exceptions to this rule. For example, lack of CD36, a major receptor for *P. falciparum*-infected red blood cells on endothelial cells and monocytes, results in enhanced susceptibility to severe malaria, particularly cerebral malaria. This may perhaps be explained, because binding of infected red blood cells to CD36 contributes to sequestering of the infected red blood cells and inhibiting the immune response to the parasite (see chapter 21) (Aitman et al 2000).

Diarrhea-resistant pigs

Enterotoxigenic *Escherichia coli* (ETEC) strains produce life-threatening diarrhea in young piglets and are therefore of major economic importance for the swine industry. However, not all pigs are susceptible to ETEC. The susceptibility of pigs to these organisms correlates with the capacity of the bacteria to adhere to the pig's small intestine. *E.coli* adheres to small intestinal brush border cells via the adhesin K88 (syn. F4) located on the bacteria's fimbriae. Receptors for K88 are expressed on the small intestine. Three variants of K88 fimbrial adhesins are distinguished (K88ab, K88ac, and k88ad). Pigs clearly differ in their capacity to bind the variant receptors. Three receptors have been recognized: 1) receptor bcd, binding al three K88 variants, 2) receptor bc, binding K88ab and K88ac, and 3) receptor d, binding K88ad (Billey et al 1998, Rapacz and Hasler-Rapacz 1986). Receptor bc is a pair of mucin-type sialoglyco-proteins of 210 and 240 kDa. Segregation analysis and adhesion affinity data further suggested that there are two adhesion affinity receptors for K88ad: a high and a low affinity receptor (Hu et al 1993). Pigs may express none, one, or more of these receptors. The non-adhering phenotype seems to be inherited as a recessive trait (Rapacz and Hasler-Rapcaz 1986), but the exact mode of inheritance has not yet been determined. By linkage in 38 pig families the K88ab- and ac-binding receptor has been localized on chromosome 13 (Guerin et al 1993). In addition, flanking genetic markers that may be used in marker-assisted breeding programs have been identified. Subsequently, an intestinal mucin-type glycoprotein (IMTGP), located in the small intestinal brush border, has been shown to function as receptor for K88ab and K88ac. Its expression was highly correlated with susceptibility to K88(+) ETEC-induced diarrhea. Of the 12 pigs that produced IMTGP, 11 developed severe diarrhea upon experimental inoculation. Colonizing bacteria were observed histologically in the small intestines of the pigs that expressed IMPTG. In contrast, only 2 of 18 pigs that did not produce IMPTG developed severe diarrhea. The bacterial concentration in the jejuna and ilea of pigs expressing IMPTG was significantly greater than that in pigs not expressing IMTGP (Francis et al 1998). Subsequent work showed that the capacity to develop an immune response to fimbriae was dependent on expression of the receptor for the fimbriae (Van den Broeck et al 1999). Similar observations on the requirement for a pathogen receptor for the induction of specific immunity to the pathogen have been reported for other infection models as well. For example, the poliovirus receptor is necessary for the development of a mucosal immune response upon infection with this pathogen, as shown by mice either transgenic or not for this receptor (Buisman et al 2000).

AIDS-resistance genes

AIDS has become the principal cause of death among those 24 to 44 years old. Many more than those killed harbor the causative virus. Already early after detection of the virus, researchers wondered why some people escape HIV-1 infection despite high exposure to the virus and why some people who contract the virus progress very slowly to AIDS. For example, 10 – 25 percent of the hemophiliacs who received tainted blood evaded the virus in the era before donated blood was screened for HIV. Many researchers focussed on genetic characteristics of the virus and on other factors, such as infections with other pathogens. Stephen O'Brien and Michael Dean started a search for AIDS-resistance genes, well knowing that in mice, specific alleles of more than 30 genes confer resistance to retroviruses (O'Brien and Dean 1997) (see chapter 19). In the 90's, the gene encoding the CC chemokine receptor 5 (CCR5) was isolated, and subsequently it was shown that CCR5 is the cell surface receptor for the chemokines MIP-1α and MIP-1β, RANTES, as well as the second receptor for macrophage-tropic (M-tropic) strains of HIV. CCR5 is a member of the seven-transmembrane, G-protein coupled receptor family, and normally functions in the chemotaxis of leukocytes towards sites of inflammation. HIV normally enters macrophages via gp120 binding to CD4 and CCR5 (Feng et al 1996). However, experiments have shown that the CD4 molecule is necessary, but not sufficient for HIV entry into cells.

The CXCR4 chemokine receptor is the ligand for another chemokine, stromal cell derived factor-1α (SDF-1α) and functions as co-receptor for T lymphocyte tropic strains of HIV. The absence of major mutations in CXCR4 is probably related to its essential role in hematopoiesis, cardiogenesis, and vascular and cerebellar development, because of which CXCR4-lacking individuals would die early during embryogenesis (Blaak 1999). Two different point mutations, one of which was a silent mutation, were found in the CXCR4 gene. The mutated proteins functioned comparably to wild-type controls in an HIV-envelope-dependent fusion assay and did not appear to play a dominant mechanistic role in susceptibility to HIV/AIDS (Cohen et al 1998).

In contrast, a well-described major polymorphism in CCR5 is the 32-base-pair deletion that results in a premature stop codon at amino acid 185 and lack of cell surface expression. Homozygosity for the deletion mutant appeared strongly protective against HIV (Liu et al 1996, Dean et al 1996, Samson et al 1996). Homozygosity for the deletion mutant (occurring in approximately 1 percent of Caucasians) does not affect immunological function, and homozygous individuals are completely healthy. As with many other immunological functions, the CCR5 activity therefore appears redundant.

The resistance conferred by the ΔCCR5 mutation appears not to be total, since isolated cases of HIV-positive CCR5Δ32 deletion homozygotes have been found. This may be explained by the transmission of variants that use other coreceptors. In a least three such cases, the invading virus appeared to be a T lymphocyte tropic strain, or a strain having a V3 loop sequence indicating a CXCR4 selectivity. Indeed, one of the HIV-1-infected CCR5 Δ32/Δ32 individuals only harbored CXCR4-using viruses (Biti et al 1997, Michael et al 1998, O'Brien et al 1997, Theodorou et al 1997).

Heterozygosity for the deletion mutant (occurring in approximately 15 - 20 percent of Caucasian-Americans) appears to confer partial resistance, but is mainly associated with a delay in disease onset. Healthy individuals who are heterozygous for the Δ32 mutation showed decreased numbers of CCR5-expressing PBMC and decreased mean CCR5 expression levels on PBMC, as compared with individuals with a homozygous

CCR 5 wild-type genotype (Wu et al 1997). HIV-1 infected individuals who are hetero-zygous for the Δ32 mutation have slower rates of CD4 decline, lower viral loads, and survive longer than individuals with the wild genotype (De Roda Husman et al 1997, Huang et al 1996). The rate of disease progression in heterozygous individuals is typi-cally 2 - 4 years. *In vitro*, CCR5Δ32 heterozygous PBMC are permissive to CCR5-depen-dent HIV-1 strains (Liu et al 1996, Samson et al 1996).

The CCR5Δ32 deletion mutant allele is absent in African, eastern Asian, and in native Americans, but is rather prevalent among Caucasians. The frequency of the mutant allele varies in most European and America Caucasian countries from 5 to 16 %, with southern European populations showing lower frequencies than northern Europeans. An exceptional high allele frequency of nearly 21 % is recorded in Ashkenazi Jews, a population of ancient Israeli and east European descent known to be very endoga-mous. In populations from Saudi Arabia, India and Pakistan, the frequency of the mutant CCR5Δ32 allele is 2-5 % (Martinson et al 1997). The lack of the mutant in Afri-cans may indicate that it originated after humans left Africa, estimated to have occur-red 130,000 to 200,000 years ago. Because the size of the deletion is identical in all populations, it seems likely that the mutation has been produced only once. However, additional data on flanking sequences are needed to definitely resolve this question. By use of coalescence theory to interpret haplotype genealogy, the origin of the CCR5Δ32-resistance allele has been estimated approximately 700 years ago (Stephens et al 1998). However other authors, basing their estimates on the geographic distribu-tion of allele frequencies and intrahaplotypic variation determined by flanking micro-satellites, came to a date of origin approximately 4,000 years ago (Carrington et al 1997, Libert et al 1998). Because the genotypes are in Hardy-Weinberg equilibrium there appears no selective advantage or disadvantage in the absence of HIV-1 (Mart-inson et al 1997). Furthermore, because it is clear that HIV-1 has not yet had the time to be the selective agent, the findings have led to speculations of previous epidemics of pathogens (including *Yersinia pestis* as the cause of the "black death" in Northern Europe from 1347-1350) using CCR5 as (co-)receptor, and giving a strong selection advantage to individuals bearing the ΔCCR5 mutant (Altschuler, E.L., 2000, O'Brien and Dean 1997, Stephens et al 1998). Thus the relationship between the ΔCCR5 mutant and resistance to HIV-1 is a warning that relationships between polymorp-hisms and specific selective agents may not always be clear. Selecting pathogens may give resistance to related or to unrelated pathogens, and the selecting agent itself could have disappeared, thus obscuring its selective influence on the genetic make-up of the host.

Because HIV-1 will spread more slowly in populations with 20 % of the individuals partially or completely resistant to infection than in completely susceptible popula-tions, the ΔCCR5 allele frequency may be an important factor in predicting AIDS endemicity in public health forecast studies. For example, in the United States, a gre-ater risk for HIV infection has been clearly documented among African-Americans than among Caucasians, when known risk factors such as social class were controlled. As mentioned, the prevalence of the ΔCCR5 allele is lower in African-Americans than in Caucasians (Easterbrook et al 1993).

A variation in the CCR5 promoter, CCR5P1, is associated with high promotor activity and appears not associated with a reduced or increased rate of infection, but with an increased rate of progression to AIDS. The other variant of the promotor polymorp-hism appears associated with a 45 % lower promoter activity *in vitro* and with slow progresion to AIDS (Martin, et al 1998, McDermott et al 1998). An estimateed 10 to 17 percent of patients who develop AIDS within 3.5 years of HIV-1 infection do so because they are homozygous for CCR5P1/P1, and 7 to 13 percent of all people carry this susceptible genotype. This finding corroborates the notion that strong expression of CCR5 on CD4+ T cells correlates with rapid disease progression. Another CCR5 mutation associated with HIV resistance has been identified in a high-risk homosexual man. PBMC of this person were totally resistant to infection by CCR5-dependent viruses. This person carried the ΔCCR5 allele in combination with a single T to A point mutation at position 303 on the other allele. The m303 mutation introduces a premature stop codon and prevents the expression of a functional coreceptor. The CCR5 m303 allele was subsequently detected in 3 heterozygous individuals out of 500 healthy blood donors (Quillent et al 1998). Several other CCR5 mutations have been described, but because of their very low frequency it is impossible to evaluate their association with the HIV resistance phenotype (Ansari-Lari et al 1997, Carrington et al 1997).

The CCR5 deletion is beyond doubt the strongest genetic factor contributing resistance to HIV. However, in a group of uninfected persons that had high exposure to HIV, only 20 percent were homozygous for the CCR5 deletion mutant. Thus other genetic and non-genetic factors still not identified may explain the lack of infection in the others (see chapter 19). Evidently the discovery of CCR5 as co-receptor for HIV and determinant of infection susceptibility stimulated research directed at blocking its function, including the delivery of chemokines, synthetic chemokine-derivatives, specific CCR5-antagonizing molecules, and vaccination against CCR5 (Arenza-Seisdedos et al 1996, Bai et al 2001).

The malarial parasite receptor
Plasmodium vivax and *P. falciparum* are the major causes of human malaria. Several resistance genes have been isolated, each functioning in one of the many steps in the disease pathogenesis. People in sub-Saharan Africa who lack the Duffy blood group antigen/chemokine receptor (DARC) appear to resist malaria caused by *P. vivax*. The Duffy blood group antigen is the erythrocyte receptor for the merozoites of *P. vivax* and the related monkey malaria, *P. knowlesi*. Cells lacking the Duffy blood group antigen are not invaded by the parasite. Like CCR5, the Duffy blood group antigen functions as a promiscuous chemokine receptor. It functions as the erythrocyte receptor for a variety of chemokines, including interleukin-8 (IL-8) and melanoma growth stimulatory activity (MGSA) (Horuk et al 1993, Horuk 1994, Montoya et al 1994). A point mutation within an erythroid-specific transcription factor-binding site at the GATA box in the promoter of the FY (or DARC) gene is common in African people and rare in other groups and results in the absence of the Duffy glycoprotein from red cells and their resistance to *P. vivax* infection. The Duffy glycoprotein is still expressed in other tissues, thus limiting possible negative effects of its absence to erythroid cells (Daniels 1997). In certain areas of West Africa the human population contains a high proportion of Duffy-negative individuals and the prevalence of *P. vivax* is correspondingly low.

Cystic fibrosis and adherence of Salmonella typhi
Cystic fibrosis (CF) is the most common life-shortening autosomal recessive disorder in Europeans. It is caused by mutations in the cystic fibrosis transmembrane conductan-

ce regulator (CFTR) gene on chromosome 7 and results in chronic sinopulmonary disease, characterized by persistent infection with typical CF pathogens, gastrointestinal and nutritional abnormalities, salt loss syndromes, and male infertility. The CFTR gene encodes a protein of 1,480 amino acids that functions as a cAMP-regulated chloride channel in the apical membrane of epithelial cells (Rosenstein and Zeitlin 1998). Defective Cl⁻ transport across epithelia is therefore the hallmark of the disease. CF occurs in approximately 1 of 2,500 Europeans. Over 600 different CF mutations have been identified, of which the ΔF508 mutation is the most frequent, representing 70 % of the mutant alleles. Several immunologic malfunctions in CF patients result in chronic colonization with several bacterial species, of which mucoid producing strains of *Pseudomonas aeruginosa* are the most common (see chapter 13).

Using the variability of three microsatellite loci close to the ΔF508 mutation, and of the genealogy of microsatellites bearing the mutation, the date of the mutation event has been dated around 50,000 years ago. Such an old mutation could only have survived in human populations if it conferred an advantage to carriers. The high levels of mutant CFTR alleles have therefore been explained by increased resistance to infectious diseases in heterozygotes (Bertranpetit and Calafell 1996). This hypothesis has been corroborated by the finding that *Salmonella typhi*, but not the related murine pathogen *S. typhimurium*, uses CFTR for entry into epithelial cells (Pier et al 1998). *In vitro*, cells expressing the wild-type CFTR internalized more *S. typhi* than cells expressing the ΔF508 mutation. Also transgenic mice expressing the ΔF508 mutation translocated fewer *S. typhi* into their gastrointestinal mucosa than mice expressing wild-type CFTR did. In addition there was a clear gene dosage effect, as mice expressing zero, one or two mutant CFTR genes translocated decreasing numbers of *S. typhi* bacteria. A monoclonal antibody directed towards the first extracellular domain of human CFTR inhibited uptake of the bacteria, whereas antibodies directed against either the fourth extracellular domain or a cytoplasmic domain did not. Thus CFTR is a major epithelial cell-receptor for internalization of *S. typhi*.

Whether there is a real causal epidemiological link between typhoid fever and CF or not, the presumed relationship may illustrate several aspects of the genetics of infectious disease resistance. It illustrates how a severe life-shortening mutation can be maintained in a population by giving a selective advantage to heterozygotes by means of increased resistance to infection. The ΔCFTR allele may be maintained in populations at a frequency of 4-5 % not only by giving heterozygote advantage for resisting typhoid fever, but also for resisting other pathogens that cause diarrhea, such as cholera. (Probably however the period that cholera was a problem in Europe was too short to be the only selective force maintaining mutant CFTR alleles.) Moreover it illustrates how a single, but essential step in the pathogenesis of infectious disease (in this case bacterial adherence) can be characterized and more fully understood by careful study of mutations in the host genome. Because *S. typhi* and *S. typhimurium* appear to use different epithelial cell receptors for translocation into the gastrointestinal tract, the example further illustrates how different host ranges and disease symptoms may be explained by different receptor usage.

References

Aitman, T.J., Cooper, L.D., Norsworthy, P.J., Wahid, F.N., Gray, J.K., Curtis, B.R., McKeigue, P.M., Kwiatkowski, D., Greenwood, B.M., Snow, R.W., Hill, A.V., Scott, J. (2000). Malaria susceptibility and CD36 mutation. Nature 405, 1015-1016.

Altschuler, E.L. (2000). Plague as HIV vaccine adjuvant. Med. Hypoth. 54, 1003-1004.

Ansari-Lari, M.A., Liu, X.M., Metzker, M.L., Rut, A.R., Gibbs, R.A. (1997). The extent of genetic variation in the CCR5 gene. Nat. Genet. 16, 221-222.

Arenzana-Seisdedos, F., Virelizier, J.-L., Rousset, D., Clark-Lewis, I., Loetscher, P., Moser, B., Baggiolini, M. (1996). HIV blocked by chemokine antagonists. Nature 383, 400.

Bai, J., Rossi, J., Akkina, R. (2001). Multivalent anti-ccr5 ribozymes for stem cell-based hiv type 1 gene therapy. AIDS Res. Hum. Retroviruses 20, 385-399.

Bertranpetit, J., Calafell, F. (1996). In: Variation in the Human Genome (eds. Chadwick, D., Cardew, G.), Wiley, Chichester, 97-114.

Billey, L.O., Erickson, A.K., Francis, D.H. (1998). Multiple receptors on porcine intestinal epithelial cells for the three variants of Escherichia coli K88 fimbrial adhesin. Vet. Microbiol. 59, 203-212.

Biti, R., Frech, R., Young, J, Bennets, B., Steart, G. (1997). HIV-infection in an individual homozygous for the CCR5 deletion allele. Nature Med.3, 252-253.

Blaak, H. (1999). Virus and host determinants of HIV-1 infection and AIDS pathogenesis. Thesis, University of Amsterdam.

Brown, K.E., Hibbs, J.R., Gallinella, G., Anderson, S.M., Lehman, E.D., McCarthy, P., Young, N.S. (1994). Resistance to parvovirus B19 infection due to lack of virus receptor (erythrocyte P antigen). New Engl. J. Med. 330, 1192-1196.

Buisman, A.M., Sonsma, J.A., Kimman, T.G., Koopmans, M.P. (2000). Mucosal and systemic immunity against poliovirus in mice transgenic for the poliovirus receptor: The poliovirus receptor is necessary for a virus-specific mucosal IgA response. J. Infect. Dis. 181, 815-823.

Carrington, M., Kissner, T., Gerrard, B., Ivanov, S., O'Brien, S.J., Dean, M (1997). Novel alleles of the chemokine-receptor gene CCR5. Am. J. Hum. Genet. 61, 1261-1267.

Cohen, O.J., Paolocci, S., Bende, S.M., Daucher, M., Moriuchi, H., Moriuchi, M., Cicala, C., Davey, R.T. Jr., Baird, B., Fauci, A.S. (1998). CXCR4 and CCR5 genetic polymorphisms in long-term nonprogressive human immunodeficiency virus infection: lack of association with mutations other than CCR5-Delta32. J. Virol. 72, 6215-6217.

Daniels, G. (1997). Blood group polymorphisms: molecular approach and biological significance. Transfus. Clin. Biol. 4, 383-390.

Dean, M., Carrington, M., Winkler, C., Huttley, G.A., Smith, M.W., Alikmets, R., Goedert, J.J., Buchbinder, S.P., Vittinghof, E., Gomperts, E., Donfield, S., Vlahov, Kaslow, R., Saah, A., Rinaldo, C., Detels, R., O'Brien, S.J. (1996). Genetic restriction of HIV-1 infection and progression to AIDS by a deletion allele of the CKR5 structural gene. Science 273, 1856-1862.

De Roda Husman, A.M., Koot, M., Cornelissen, M., et al. (1997). Association between CCR5 genotype and the clinical course of HIV-1 infection. Ann. Intern. Med. 127, 882-890.

Easterbrook, P.J., Chmiel, J.S., Hoover, D.R., Saah, A.J., Kaslow, R.A., Kingsley, L.A., et al. (1993). Racial and ethnic differences in human immunodeficiency virus type 1 (HIV-1) seroprevalence among homosexual and bisexual men. The multicenter AIDS cohort study. Am. J. Epidemiol. 138, 415-429.

Feng, Y., Broder, C.C., Kennedy, P.E., Berger, E.A. (1996). HIV-1 entry cofactor: functional cDNA cloning of seven transmembrane, G protein-coupled receptors. Science, 272, 872-877.

Fernandez-Reyes, D., Craig, A.G., Kyes, S.A., Peshu, N., Snow, R.W., Berendt, A.R., Marsh, K., Newbold, C.I. (1997). A high frequency African coding polymorphism in the N-terminal domain of ICAM-1 predisposes to cerebral malaria in Kenya. Hum. Mol. Genet. 112, 1357-1360.

Francis, D.H., Grange, P.A., Zeman, D.H., Baker, D.R., Sun, R., Erickson, A.K. (1998). Expression of mucin-type glycoprotein K88 receptors strongly correlates with piglet susceptibility to K88(+) enterotoxigenic Escherichia coli, but adhesion of this bacterium to brush borders does not. Infect. Immun. 66, 4050-4055.

Geraghty, R.J., Krummenacher, C., Cohen, G.H., Eisenberg, R.J., Spear, P.G. (1998). Entry of alphaherpesviruses mediated by poliovirus receptor-related protein 1 and poliovirus receptor. Science 280, 1618-1620.

Guerin, G., Duval-Iflah, Y., Bonneau, M., Bertaud, M., Guillaume, PP., Ollivier, L. (1993). Evidence for linkage between K88ab, K88ac intestinal receptors to Escherichia coli and transferrin loci in pigs. Anim. Genet. 24, 393-396.

Hazenbos, W.L., van den Berg, B.M., and van Furth, R. (1993). Very late antigen 5 and complement receptor type 3 cooperatively mediate the interaction between Bordetella pertussis and human monocytes. J. Immunol. 151, 6274-6282.

He, J., Chen, Y., Farzan, M., Choe, H., Ohagen, A., Gartner, S., Busciglio, J., Yang, X., Hofmann, W., Newman, W., Mackay, C.R., Sodroski, J., Gabuzda, D. (1997). CCR3 and CCR5 are coreceptors for HIV-1 infection of microglia. Nature 385, 645-649.

Horuk, R., Chitnis, C.E., Darbonne, W.C., Colby, T.J., Rybicki, A., Hadley, T.J., Miller, L.H. (1993). A receptor for the malarial parasite Plasmodium vivax: the erythrocyte chemokine receptor. Science 261, 1182-1184.

Horuk, R. (1994). The interleukin-8-receptor family: from chemokines to malaria. Immunol. Today 15, 169-174.

Hu, Z.L, Hasler-Rapacz, J., Huang, S.C., Rapacz, J. (1993). Studies in swine on inheritance and variation in expression of small intestinal receptors mediating adhesion of the K88 enteropathogenic Escherichia coli variants. J. Hered. 84, 157-165.

Huang, Y., Paxton, W.A., Wolinsky, S.M., et al (1996). The role of a mutant CCR5 allele in HIV-1 transmission and disease progression. Nature Medicine 2, 1240-1243.

Libert, F., Cochaux, P., Beckman, G., Samson, M., Aksenova, M., Cao, A. et al (1998). The delta CCR5 mutation conferring protection against HIV-1 in Caucasian populations has a single and recent origin in northeastern Europe. Human Molecular Genetics 7, 399-406.

Liu, R., Paxton, W.A., Choe, S., Ceradini, D., Martin, S.R., Horuk, R., MacDonald, M.E., Stuhlmann, H., Koup, R.A., Landau, N.R. (1996). Homozygous defect in HIV-1 coreceptor accounts for resistance of some multiply-exposed individuals to HIV-1 infection. Cell 86, 367-377.

Martin, M.P., Dean, M., Smith, M.W., et al. (1998). Genetic accelaration of AIDS progresssion by a promoter variant of CCR5. Science 282, 1907-1911.

Martinson, J.J., Chapman, N.H., Rees, D., Liu, Y.-Y, Clegg, J..B. (1997). Global distribution of the CCR5 gene 32-basepair deletion. Nature genet. 16, 100-103.

McDermott, D.H., Zimmerman, P.A., Guignard, F., Kleeberger, C.A., Leitman, S.F., the Multicenter AIDS Cohort Study (MACS), and Murphy, P.M. (1998). (1998). CCR5 promoter polymorphism and HIV-1 disease progression. The Lancet 352, 866-870.

Michael, N.L., Nelson, J.A.E., KewalRamani, V.N., Chang, G., O'Brien, S.J., Mascola, J.R., Volsky, B., Louder, M., White, G.C.2nd, Littman, D.R., Swanstrom, R., O'Brien, T.R. (1998). Exclusive and persistent use of the entry coreceptor CxCR4 by human immunodeficiency virus type 1 from a subject homozygous for CCR5 D32. J. Virol. 72, 6040-6047.

Montoya, F., Restrepo, M., Montoya, A.E., Rojas, W. (1994). Blood groups and malaria. Rev. Int. Med. Trop. Sao Paulo 36, 33-38.

O'Brien, S.J., Dean, M. In search of AIDS-resistance genes. Scientific American, September 1997, 28-35.

O'Brien, T.R., Winkler, C., Dean, M., Nelson, J.A., Carrington, M., Michael, N.L., White, G.C. (1997). HIV-1 infection in a man homozygous for CCR5 delta 32. The Lancet 349, 1219.

Pier, G.B., Grout, M., Zaidi, T., Meluleni, G., Muesenborn, S.S., Banting, G., Ratcliff, R., Evans, M.J. Colledge, W.H. (1998). Salmonella typhi uses CFTR to enter intestinal epithelial cells. Nature 393, 79-82.

Pier, G.B., Grout, M., Zaidi, T.S. (1997). Cystic fibrosis transmembrane conductance regulator is an epithelial cell receptor for clearance of Pseudomonas aeruginosa from the lung. Proc. Natl. Acad. Sci. USA 94, 12088-12093.

Quillent, C., Oberlin, E., Braun, J., Rousset, D., Gonzalez-Canali, G., Metais, P., Montagnier, L., Virelizier, J.L., Arenzana-Seisdedos, F., Beretta, A. (1998). HIV-1 resistance phenotype conferred by combination of two separate inherited mutations of the CCR5 gene. The Lancet 351, 14-18.

Rapacz, J., Hasler-Rapacz, J. (1986). Polymorphism and inheritance of swine intestinal receptors mediating adhesion of three serological variants of Escherichia coli producing K88 pilus antigen. Anim. Genet. 17, 305-321.

Rosenstein, B.J., Zeitlin, P.L. (1998). Cystic fibrosis. The Lancet 351, 277-282.

Rossman, M.G., Bella, J., Kolatkar, P.R., He, Y., Wimmer, E., Kuhn, R.J., Baker, T.S. (2000). Cell recognition and entry by rhino- and enteroviruses. Virology 269, 239-247.

Samson, M., Libert, F., Doranz, B.J., Rucker, J., Liesnard, C., Farber, C.M., Saragosti, S., Lapoummerouilie, C., Cogneaux, J., Forceille, C., Muyldermans, G., Verhofstede, C., Burtonboy, G., Georges, M., Imai, T., Rana, S., Yi, Y., Smyth, R.J., Collman, R.G., Doms, R.W., Vassart, G., Parmentier, M. (1996). Resistance to HIV-1 infection in caucasian individuals bearing mutant alleles of the CCR-5 chemokine receptor gene. Nature 382, 722-725.

Spear, P.G. (1998). A welcome mat for leprosy and Lassa fever. Science 282, 1999-2000.

Stear, M.J., Wakelin, D. (1998). Genetic resistance to parasitic infection. Rev. Sci. Tech. 17, 143-153.

Stephens, J.C., Reich, D.E., Goldstein, D.B., Shin, H.D., Smith, M.W., Carrington, M., Winkler, C., Huttley, G.A., Allikmets, R., Schriml, L., Gerrard, B., Malasky, M., Ramos, M.D., Morlot, S., Tzetis, M., Oddoux, C., di Giovine, F.S., Nasioulas, G., Chandler, D., Aseev, M., Hanson, M., Kalaydjieva, L., Glavac, D., Gasparini, P., Dean, M. et al. (1998). Dating the origin of the CCR5-Delta32 AIDS-resistance allele by the coalescence of haplotypes. Am. J. Hum. Genet. 62, 1507-1515.

Svanborg Eden, C., Hagberg, L., Hanson, L.A., Hull, S., Hull, R., Jodal, U., Leffler, H., Lomberg, H., Straube, E. (1983). Bacterial adherence-a pathogenetic mechanism in urinary tract infections caused by Escherichia coli. Prog. Allergy 33, 175-188.

Tatsuo, H., Ono, N., Tanaka, K., Yanagi, Y. (2000). SLAM (CDw150) is a cellular receptor for measles virus. Nature 406, 893-897.

Theodorou, I., Meyer, L., Magieroska, M., Katlama, C., Rouzioux, C. (1997). HIV-1 infection in an individual homozygous for CCR5 delta32. Seroco Study Group. The Lancet 349, 1219-1220.

Van den Broeck, W., Cox, E., Goddeeris, B.M. (1999). Receptor-dependent immune responses in pigs after oral immunization with F4 fimbriae. Infect. Immun. 67, 520-526.

Wu, L., Paxton, W.A., Kassam, N., Ruffing, N., Rottman, J.B., Sullivan, N., Choe, H., Sodroski, J., Newman, W., Koup, R.A., Mackay, C.R. (1997). CCR5 levels and expression pattern correlate with infectability by macrophage-tropic HIV-1 in vitro. J. Exp. Med. 185, 1681-1691.

Chapter 8.   Genes involved in innate immunity

Innate immunity refers to the first-line host defense that combats the infection in the early hours after exposure to micro-organisms. In addition to this role in the early stages of infection, the innate immunity plays a key role in stimulating the subsequent, clonal response of adaptive immunity. Key differences between the two systems are the lack of a memory function in the adaptive immune system and the large repertoire of antigen-recognition receptors in the adaptive immune system. An essential characteristic of both systems is their ability to distinguish self from infectious nonself. The innate immune system is an evolutionary very ancient, markedly conserved defense mechanism, as shown by similarities between pathogen recognition, signaling pathways, and effector mechanisms of innate immunity between mammals, lower vertebrates, and even plants (Hoffmann et al 1999). Key mechanisms of innate immunity are phagocytosis of invading micro-organisms, proteolytic cascades, leading to localized blood clotting and opsonization, and transient production of antimicrobial peptides.

Collectins, mannose-binding protein
To recognize invading micro-organisms, the mammalian host uses a wide spectrum of pattern recognition molecules or receptors, including complement, collectins, and antimicrobial peptides (see below). They may recognize patterns shared by groups of micro-organisms, such as the lipopolysaccharides (LPS) of Gram-negative bacteria, the glycolipids of mycobacteria, the lipoteichoic acids of Gram-positive bacteria, the mannans of yeast, and the double-stranded RNAs of viruses (Hoffmann et al 1999). The collectins have collagen tails and globular lectin domains and may recognize a wide spectrum of pathogens. Members of this family include the lung surfactant proteins SP-A and SP-D, locally acting in the lung to limit infection, the β-inhibitors of influenza virus infectivity, and the mannose-binding protein (MBP). SP-A and SP-D bind to the carbohydrate moiety on the surface of pathogens and stimulate their agglutination, phagocytosis and killing. They stimulate phagocytic cells by binding to specific receptors on these cells. SP-A and SP-D have originally been identified in pulmonary tissue, but they may be functional in the innate defense on a wide range of mucosal tissues. Studies have been initiated to examine whether SP-A and SP-D gene variants are associated with increased risk on lung infection (Creuwels 1999).
The β-inhibitors of influenza virus infectivity bind to carbohydrate on the surface of influenza virus and thus sterically inhibit virus interaction with host cells. They may also act as opsonins (Malhotra and Sim 1995).
Human MBP is synthesized in the liver as an acute phase reactant. During acute phase reactions the concentration of MBP in blood may increase 1.5 – 3 times. At the site of infection it interacts with the complement system and acts directly as an opsonin, using the C1q receptor on macrophages. MBP has a broad binding activity, selectively recognizing the repetitive carbohydrate patterns on the surface of bacteria, yeast, parasites, mycobacteria, and certain viruses (Anders et al 1990, Epstein et al 1996, Ezekowitz 1998, Feizi and Larkin 1990, Malhotra and Sim 1995). Despite this broad recognition pattern, MBP does not recognize self-glycoproteins. The molecular explanation for this recognition specificity has recently been elucidated (Hoffmann et al 1999). MBP activates proteolytic cascades, including complement, through activating C4 and C2, and it promotes phagocytosis. Its significance is indicated by the observa-

tion that mutations in the gene encoding for MBP results in defects of opsonisation and recurrent pyogenic infections in childhood and fatal bacterial infections in adults (Garred et al 1995, Sumiya et al 1991, Summerfield et al 1997, Super et al 1989).

Each of three different point mutations (the B, C, and D alleles; A=wild-type) in one of the coding regions of the MBP gene on chromosome 10 causes low MBP concentrations in a dominant way and deficiency in opsonization and phagocytosis. Presence of homozygous or heterozygous gene variants may lead to less than 10 - 50 % of the normal MBP concentrations in serum. The mutations are at codons 52, 54, and 57 in exon 1, coding the collagen domain, and impair assembly of the MBP homopolymer and activation of complement (Lipscombe et al 1992, Madsen et al 1994). The variants occur at frequencies of about 0.12 – 0.25 in three major ethnic groups. In sub Saharan Africa the frequency of codon 57 mutations may reach 30 % locally. In addition to polymorphisms in the coding regions, variants in the promoter region have been described that appear to further explain the large interindividual variation in MBP serum concentrations. In particular, a polymorphism in codon −221 (*X/Y* type) has a significant effect on the MBP concentration, with the *Y* promoter having high and the *X* having low MBP-expressing activity (Koch et al 2001, Madsen et al 1995).

In Western Europe, MBP deficiencies may occur in approximately 10 % of the population, thus constituting one of the most frequently occurring immune disorders. The frequency of heterozygotes for the abnormal structural alleles was not different in young and adult patients with different recurrent infections from that in the background population (approximately 36 %). In contrast, the frequency of the homozygotes for the abnormal alleles was significantly increased (8.3 % vs. 0.8 %, yielding an odds ratio of 11.1 [95 % confidence interval: 1.5 – 83.8]). The etiological fraction was 0.08, indicating that homozygosity may explain 8 % of the cases referred for recurrent and unusual infection and immunodeficiency (Garred et al 1995). Other reports, however, also indicated increased susceptibility to infections in heterozygotes (Hibberd et al 1999, Sumiya et al 1991, Summerfield et al 1997). Recently Hibberd et al (1999) found that one third of all cases of meningococcal disease could be ascribed to MBP variants (homozygous and heterozygous). In addition, a 2.08-fold (95 % CI: 1.41-3.06) increased relative risk of acute respiratory tract infection was found in MBP-insufficient children in West Greenland. This occurred especially during the vulnerable period of childhood from age 6 to 17 months, when the adaptive immune system is still immature (Koch et al 2001). Cystic fibrosis (CF) patients that also harbor MBP mutations experience more frequently infections with *Pseudomonas aeruginosa* and *Burkholderia cepacia* and have a life expectation that is 8 years shorter than patients with CF alone (Garred et al 1999). In contrast to the findings with respect to non-specific acute and recurrent infections, no evidence was found of MBP variants predisposing to malaria and tuberculosis infection (Hill 1998). However, the MBP codon 52 variant was found more frequently in association with persistent HBV infection. In The Gambia no association of MBP allelic variants with persistent HBV infection was found but the codon 52 variant is not present in that population (Thomas et al 1996, Thursz 1997).

The high frequency of variant alleles suggests, from an evolutionary point of view, that low levels of MBP may also confer some, not yet identified advantage for the host and that the apparently balanced polymorphisms may be based on heterozygous advantage (Garred et al 1994). Low MBP concentrations might for example protect against infections that use MBP or complement for cellular invasion, such as *Leishma-*

nia and *Mycobacterium* spp. Because homozygosity for MBP is a life-long risk factor, and because the deficiency may be treated by purified or recombinant MBP (Garred et al 1994), screening for MBP variants may be warranted in patients with recurrent infections.

Complement

The complement system plays a central role in the defense against micro-organisms. It includes approximately 30 proteins. The complement cascade is activated either directly or indirectly by micro-organisms and results in the synthesis of vasoactive peptides, the anaphylatoxins, the opsonization of micro-organisms for phagocytosis, or the assembly on their surface of the pore-forming membrane attack complex. There are three pathways of complement activation that differ in the initiation of the cascade leading to cleavage of the third complement component C3. The classical pathway requires antibody and the first complement components (C1q, C1r, C1s, C4, C2, and C3), the alternative pathway (C3, factor B, factor D, and properdin) is activated directly by the micro-organism, and the lectin pathway requires MBP. Binding of ligands by MBP results in the activation of the MBP-associated proteases, MASP1 and MASP2, which in turn activate the C3 convertase (Hoffmann et al 1999). The three activation routes share the terminal route (C3, C5-C9) (Fig. 6).

Genetic defects in the complement system have since long been associated with increased and often lifelong susceptibility to infection (Alper et al 1970, Guenther 1983). Isolated genetic defects of individual components of the complement system have been described in man for all the components of the classical pathway and the mem-

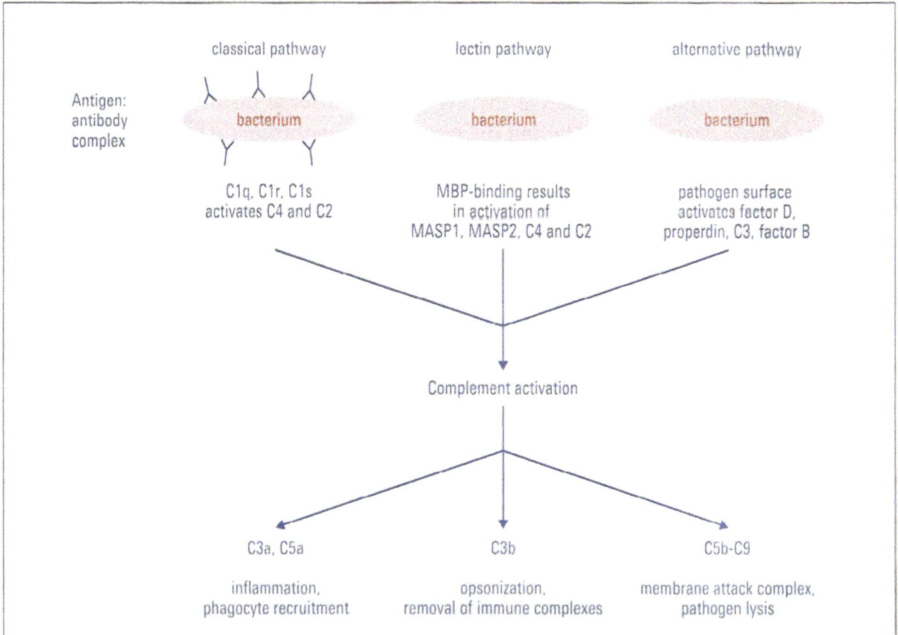

Figure 6. Simplified scheme of the three pathways of complement activation resulting in three main effector pathways.

brane attack complex, as well as for factor I, H, and properdin. Properdin deficiency is an X-linked recessive disorder. The other complement deficiencies are inherited in an autosomal co-dominant manner (Egan et al 1994, Lachmann 1984, Orren et al 1992). Although complement undoubtedly plays a role in host defense against many pathogens, it appears most important in protection against extracellular encapsulated bacteria, especially *Neisseria meningitidis,* but also *Streptococcus pneumoniae, Moraxella catarrhalis, Haemophilus influenzae,* and to a lesser extent, *Neisseria gonorrhoeae* (Figueroa and Densen 1991). The particularly frequent occurrence of neisserial infection in complement-deficient individuals points to their ability to survive in phagocytic cells so that the cytolytic activity of complement is needed to kill them. Infections with *N. meningitidis* and *N. gonorrhoeae* are therefore most typical for the late component deficiencies, with over 40 % of the homozygotes affected (Ross and Densen 1984). Mortality form meningococcal infection, however, appears lower in these patients than in immunocompetent subjects. These findings likely point to a dual role of complement in the pathogenesis of meningococcal infection (Lehner et al 1992). In a Dutch cohort, complement deficiency was present in 18% of the patients who had had experienced a meningococcal infection, mainly due to serogroups X, Y, W135, or non-groupable strains. The most commonly found deficiencies were properdin deficiency (39 %) and C8 (18 %) deficiency. Interestingly, 30 % of the complement deficient patients reported other family members having experienced meningitis (Swart et al 1993). Very recently a complement factor D deficiency was found in a young woman who had experienced a serious *Neisseria meningitidis* infection and in three of her relatives who had no history of infections. The defect was caused by a homozygous point mutation that changed the TCG codon for serine-42 into a TAG stop codon and was present in both alleles of the affected family members. There was a high degree of consanguinity in the family. The defect resulted in a complete factor-D deficiency and a very low capacity to opsonize *Escherichia coli* and *Neisseria meningitidis* for phagocytosis by normal human neutrophils, an indication (but not definite proof) that the mutation may be causative involved in the disease history. Nineteen family members with one mutant allele had decreased or low-normal factor-D levels in their serum (Biesma et al 2000).

Genes controlling macrophage function: The Nramp1 gene
The innate immune system represents the early lines of resistance once a pathogen has entered the body. Central in the innate immune system is the macrophage. Macrophages were identified as the cellular compartment responsible for genetically controlled host resistance of inbred mouse strains against infection with several mycobacterial species, such as *Mycobacterium lepraemurium, M. intracellulare, M. bovis* strain bacillus Calmette-Guérin (BCG), and other unrelated intracellular parasites, including *Salmonella typhimurium* and *Leishmania donovani* (Gros et al 1983). In mice infection with the non-pathogenic *M. bovis* strain BCG is biphasic, with an early phase, 0-3 weeks postinfection, in which there is rapid proliferation of bacteria in susceptible but not in resistant mouse strains. A single dominant gene located on mouse chromosome 1, which has been designated *Bcg, Ity, Lsh,* or *Nramp1,* regulates this differential growth. *Bcg* is present in two allelic forms in inbred mouse strains, respectively *Bcg^r* (resistant, dominant), and *Bcg^s* (susceptible, recessive) (Gros et al 1981, Skamene et al 1998). Macrophages from *Bcg^r* and *Bcg^s* mice have a differential capacity to restrict the growth of intracellular parasites. In addition, macrophages from *Bcg^r* (resistant)

mice are superior to macrophages from *Bcg^s* (susceptible) mice in functional parameters of macrophage activation. These include the production of toxic oxygen, nitrogen radicals, TNF release, and expression of surface markers associated with activation.

The *Bcg* locus was identified by an approach involving positional cloning through the generation of high-resolution linkage and physical maps of the chromosomal region carrying the gene, the cloning of the region in recombinant yeast artificial chromosomes (YAC clones), and a systematic search for candidate genes by exon trapping. One of the identified genes was expressed exclusively in macrophages and was denoted as *Nramp1* for natural resistance-associated macrophage protein 1 (Malo et al 1993, Malo et al 1993, Vidal et al 1993). The Nramp1 protein is an integral membrane protein of 548 amino acids with 12 putative transmembrane domains, a glycosylated extracellular loop, and several sites for phosphorylation by protein kinase C. Nramp1 further contains a sequence motif found in membrane transport proteins (see below), a region of identity with Src homology 3 (SH3) domains, and two putative *N*-linked glycosylation sites. Sequencing of the *Nramp1* gene in *Bcg^r* and *Bcg^s* mice revealed that the susceptible phenotype was associated with a nonconservative glycine to aspartic acid substitution at position 169 (G169D) in the predicted fourth transmembrane domain of the protein (Skamene et al 1998). The mutation results in a complete disappearance of the mature protein, suggesting that the mutation affects the normal processing of the protein. To formally proof the function of the gene, Vidal et al (1995) generated a *Nramp1* gene knockout mutant mouse on a *Bcg^r* background and this mouse displayed the susceptible phenotype.

The human homolog *NRAMP1* was isolated by low-stringency screening of a human spleen cDNA library (Cellier et al 1994). The gene is located on human chromosome 2 region q35 and shows 88 % sequence homology with its mouse counterpart. There are some functional differences between the murine *Nramp1* gene and the human NRAMP1. While the murine gene is mainly expressed in the spleen by macrophages, the human homolog is mainly expressed in blood, lung and spleen. In blood the gene is mainly expressed by polymorphonuclear leukocytes and to a lesser degree by monocytes. Maturation and differentiation of monocytes leads to enhanced expression (Cellier et al 1997).

Nramp1 belongs to a family of proteins, which is extremely conserved in evolution and which also includes the highly homologous Nramp2. The ubiquitously expressed gene Nramp2 is an Fe(2+) transporter and a mutation in this gene causes microcytic anemia in mice and rats. Homologs are found in *Drosophila*, worms, plants, yeast, and bacteria. Secondary structural analysis of these proteins indicates that they form a transmembrane pore that could be involved in transmembrane transport of divalent cations, notably iron and oxidized nitrogen radicals. Especially these oxidized nitrogen radicals are known to be crucial effectors of microbicidal activity. Immunofluorescence studies demonstrated that Nramp1 is localized in macrophages exclusively in the late endosome and lysosome, organelles that may interact with phagosomes. After phagosomes have been formed by the internalization of particulate materials, Nramp1 is recruited to the membrane of the phagosome and remains associated with this structure during its maturation to phagolysosome. Nramp1 may thus control the replication of intracellular parasites by influencing the bactericidal function of the microbe-containing phagosome (Skamene et al 1998). In addition to oxidized nitrogen radicals, proposed mechanisms of this function include the removal of iron and other divalent cations from the phagosomal space, and phagosomal acidifcation (Canonne-Hergaux 1999, Hackham et al 1998). On the basis of the SH3 homology it is also likely that Nramp1 participates in membrane signal transduction in mature macrophages. This suggestion is supported by the known pleiotropic effects of Nramp1 (McLeod et al 1995). The mechanism of resistance does not appear to be associated with the kinetics and magnitude of the cytokine responses following infection (Eckmann et al 1996). Evidently other factors, such as plasmid-virulence genes and polymorphonuclear leukocytes, interact with *Nramp1* in affecting the course of infection with intracellular bacteria (Vassiloyanakopoulos et al 1998).

The *Nramp1* gene also influences the outcome of infection with the intracellular parasite *Toxoplasma gondii* in mice. However, this effect is modulated by the inoculation route, time after infection, and genetic background of the mice with the resistant allele (Blackwell et al 1994). These findings likely further illustrate the pleiotropic effects of several macrophage products, stimulated by different primary activation routes, on the parasite rather than a direct effect of toxic oxidized nitrogen radicals alone. Interestingly, infections with *Listeria* and *Legionella*, which do not associate with Nramp1-containing phagosomes, are not affected by *Nramp-1* mutations.

A number of polymorphic variants have been identified in the human *NRAMP1* gene. These include polymorphisms in potentially regulatory sequences in the promoter and the 3' untranslated region, a polymorphism in intron 4, and a nonconservative D543N amino acid substitution in the predicted cytoplasmic C terminus of the protein (Blackwell et al 1995, Buu et al 1995, Lewis et al 1996, Liu et al 1995, Searle and Blackwell 1999, White et al 1994). Also four alleles constituting a functional Z-DNA forming repeat polymorphism have been observed in the promoter region of human *NRAMP1*. Alleles 1 and 4 are rare with gene frequencies of approximately 0.001, while alleles 2 and 3 occur at gene frequencies of 0.25 and 0.75 respectively. In the absence of exogenous stimuli, alleles 1, 2, and 4 are poor promoters of gene expression in a luciferase reporter gene system. Allele 3 drives high expression. Allele 3 shows allelic association with autoimmune disease susceptibility, and allele 2 with infectious disease susceptibility (tuberculosis, leprosy). Hence, balancing selection is likely maintaining these two alleles in human populations (Blackwell and Searle 1999). In addition, Liu et al (1995) physically linked 2 highly polymorphic microsatellite markers, D2S104 and D2S173, to *NRAMP1* on a 1.5 Mb YAC contig. These should be useful in further assessing the role of *NRAMP1* in disease susceptibility.

The contribution of *NRAMP1* polymorphisms to disease susceptibility has mainly been analyzed with regard to leprosy and tuberculosis. While studies in Polynesia, Brazil and Pakistan failed to find linkage between leprosy and *NRAMP1* polymorphisms, a study in Vietnam found evidence of multigenic control of leprosy susceptibility, including control by *NRAMP1* (Skamene et al 1998).

In a case control study in the Gambia, Bellamy et al (1998) found four *NRAMP1* polymorphisms associated with smear-positive pulmonary tuberculosis. Subjects who were heterozygous for 2 *NRAMP1* polymorphisms in intron 4 and the 3' untranslated region of the gene were particularly overrepresented among the tuberculosis patients as compared with those with the most common NRAMP1 genotype (OR = 4.1). For a further discussion on the role of NRAMP1 in tuberculosis see chapter 16.

The role of *NRAMP1* polymorphisms has also been studied in immunologic disease. For example, linkage of autoimmune disease (rheumatoid arthritis, juvenile rheumatoid arthritis, Crohn's disease) to *NRAMP1* could be demonstrated (Shaw et al 1996). As mentioned, *NRAMP-1* gene polymorphisms may have been conserved during evolution because polymorphisms that protect against infection may afford susceptibility to autoimmune disease, and vice versa.

Important for animal husbandry is the finding that in chickens, as in mice, the NRAMP1 gene is involved in natural resistance to infection with *Salmonella typhimurium* (Hu et al 1997). This infection is of major importance in the poultry industry and may give rise to zoonotic infections in humans. One G to A substitution at nucleotide 696 resulted in the nonconservative displacement of Arg223 to Gln223 and was associated with susceptibility. Both in chickens and in mice resistance to infection with *S. typhimurium* is inherited as a complex trait and explained in part by NRAMP1 and TNC (see chapter 29) (Hu et al 1997, Sebastiani et al 1998).

Other genes controlling microbial replication in macrophages
Legionella pneumophila is an intracellular pathogen that causes Legionnaires' disease in humans. Inbred mouse strains are uniformly resistant to *L. pneumophila* with the notable exception of A/J mice. This difference appears to be determined by replication of the bacterium in macrophages. A/J macrophages are permissive to *L. pneumophila* replication. A mouse locus called *Lgn1* determines differences in the replication of *L. pneumophila* intracellularly in macrophages. Candidate genes for this phenotypic difference have been located in the complex *Naip* (neuronal apoptosis inhibitory protein) gene cluster (esp. *Naip2* or *Naip5*) on chromosome 13D1-D3. This region is syntenic with the spinal muscular atrophy (SMA) locus on human chromosome 5. Interestingly, macrophages from permissive A/J mice express significantly less Naip protein than those from nonpermissive mice. Naip protein expression is increased after phagocytotic events. Furthermore, Naip protein levels increase during 6 – 48 h postinfection with either virulent or avirulent strains of *L. pneumophila*. Studies are ongoing to examine whether this locus might add in explaining the large differences in susceptibility to this bacterium in humans (Diez et al 2000, Growney and Dietrich 2000).

Phagocytic cell defects
Defects in phagocytic cell function often result in recurrent, severe infections with pyogenic bacteria, such as Gram-positive cocci, enterobacteriaceae, *Pseudomonas* spp., and fungi. The spectrum of disease in these patients may include repeated bouts of cellulitis, pharyngitis, perirectal and other abscesses, pneumonia, osteomyelitis, and bacteriemia. The infections may persist despite apparently effective cellular and humoral immune responses, and they may even persist despite antibiotic treatment. Syndromes have been described which are based on defects in the leukocyte integrins CD11a/CD18 (LFA-1), CD11b/CD18 (Mac-1/CR3), and CD11c/CD18 (CR4/gp150,95). These molecules share a common β2-subunit but have distinct α-subunits. They bind to intercellular adhesion molecules (ICAMs) as counter receptors. The integrins are responsible for cell-cell interactions during phagocytic cell adhesion to vascular endothelium, during migration across blood vessel walls, and during ingestion of opsonized bacteria. The heterogeneity of the human condition is associated with different mutations. For example, CD18 deficiency or leukocyte adhesion deficiency leads to widespread pyogenic bacterial infections (Fischer et al 1988).

A second type of leukocyte adhesion deficiency is caused by a defect in the expression of sialated Lewis-X (sLE^x or CD15-s antigen), which is the ligand for CD62E and CD62P on endothelial cells, and which is a crucial structure in the migration and extravasation of granulocytes during inflammation. The number of blood granulocytes is increased, but these cells do not attach to vascular endothelium. The patients suffer from recurrent pneumonias and other bacterial infections.

A similar disease, bovine leukocyte adhesion deficiency (BLAD), has recently been described in cattle and has even become of economic importance (Gerardi 1996). As in humans, the physiological basis for BLAD is a deficiency in the chemotactic and phagocytic properties of leukocytes and particularly neutrophils. The inhibition of diapedesis in the inflammatory response prevents normal immune reactions to invading pathogens. Chronic and recurrent necrotic and gangrenous infections of soft tissues are prevalent, in addition to secondary infections with bacteria and fungi. Dermatomycoses and impaired pus formation are also common findings. Neutrophils from calves homozygous for BLAD have a reduced phagocytotic and yeast-killing capacity, but a higher respiratory burst activity. The latter may probably be upregulated to partially compensate for the lack of typical adherence-dependent host defense functions. Homozygous calves also had extreme neutrophilia and significantly more immature neutrophils. These calves usually die when they are less than a year old. A point mutation (D128G) in the gene coding for CD18 is implicated in the etiology of BLAD (Gerardi 1996, Shuster et al 1992). The recessive genetic defect leads to a deficiency in Mac-1 (CD11b/CD18) glycoprotein expression on leukocytes (Kehrli et al 1990). Heterozygous calves do not show detectable functional differences in leukocytes (Sipes et al 1999). The disease became of major economic importance in the Holstein breed, where only a very few sires, selected for high milk yields in their offspring, fertilize many cows. Many cases could be traced back to "Osborndale Ivanhoe", a very prolific sire designated the father of the Holstein breed. In the USA there were years that about 16,000 calves with BLAD were born. Screening programs and the elimination of affected sires have been developed to reduce the incidence of BLAD.

Other defects in phagocytic cells affect their ability to kill ingested bacteria. The respiratory burst, leading to the production of toxic superoxide radical, is defective in chronic granulomatous disease. Although the severity may somewhat vary, several genetic defects in the four genes encoding the four constituent proteins of the leukocytic NADPH oxidase system cause a similar syndrome. When infections persist because phagocytic cell function is defective, CD4+ T cells may be chronically stimulated and hence granulomas may be formed. One cytosolic subunit of NADPH, needed for activating oxidase in cells ingesting micro-organisms is p47-phox. About one third of all chronic granulomatous disease patients suffer from p47-phox deficiency. Mutations in the p47 gene are hard to detect, because there are two non-active pseudogenes with very high homology to the functional p47 gene. Gene-specific PCR assays for cDNA and for genomic DNA to analyze the functional p47 gene and its peudogenes have been developed for diagnosis and for screening families for carriers of p47 mutations (Roos 1999).
Glucose-6-phosphate dehydrogenase deficiency and myeloperoxidase deficiency also lead to deficiencies in intracellular killing of bacteria resulting in similar chronic infections and granuloma formation. In these cases the oxidase does not receive

enough substrate to generate sufficient amounts of reactive oxygen species.
In Chédiak-Higashi syndrome the fusion of lysosomes with phagosomes in phagocytes is defective. The phagocytic cells in these patients have very large granules and are impaired in intracellular killing of pyogenic bacteria. The disease may result in granulomas (Janeway and Travers 1994, Rotrosen and Gallin 1987).

Natural killer cells

Resistance to herpesvirus infections is under control of natural killer (NK) cells. Experimental inoculation of mice with the murine cytomegalovirus has provided clues to the genetic restriction mediated by NK cells. While resistance to infection with murine cytomegalovirus appeared under control of both MHC and non-MHC genes, a single autosomal dominant locus, *Cmv1*, controls virus replication in the spleen through the action of NK cells. Linkage analysis in the CXB recombinant inbred mice strain panel showed that *Cmv1* is closely linked to the NK1.1 locus on chromosome 6, a region defined as the NK cell gene complex. This region encodes the CD56 antigen (also known as NKH-1 or Leu-19 for humans), which is expressed on NK cells and which is involved in adhesion to the target cells and signalling for interferon-γ production. The region further contains a high density of NK cell receptor genes. To confirm the involvement of the NK1.1 locus in herpesvirus susceptibility, a congenic mouse strain (BALB.B6-*Cmv1r*/Sc) has been produced in which the region of chromosome 6, containing *Cmv1* and the NK cell complex from the resistant C57BL/6J strain, has been introduced onto the susceptible BALB/c background. This congenic strain exhibited a level of resistance comparable to that of the C57BL/6J strain (Scalzo et al 1992, Scalzo et al 1995). By expression analysis, Brown et al (2001) established that the natural killer cell activation receptor, Ly-49H, is specifically involved in resistance to murine cytomegalovirus and accounts for the *Cmv1*^{r} phenotype. This is one of the receptors that activate cytotoxicity of NK cells.

Antimicrobial peptides

Most constitutively expressed and inducible antimicrobial peptides, including the defensins, that participate in innate immunity act by permeabilizing the cell membranes of micro-organisms. Other antimicrobial peptides act by disrupting internal signalling of micro-organisms. Such antimicrobial peptides are normally produced by the epithelial linings of the intestines, the respiratory and urogenital tracts, and by neutrophils. During inflammation and infection their production may be enhanced by inflammatory signals such as interleukin-1β and by binding of bacterial products to Toll-like receptors (see chapter 30). Apart from gene knockout experiments in mice, patients with cystic fibrosis indicate the significance of these antimicrobial peptides. In these patients the salinity of the bronchial airway fluid disrupts the function of antimicrobial peptides that are found in the respiratory epithelium, notably the human β-defensin-1. This leads to colonization and infection with organisms like *Staphylococci* and *Pseudomonas* (see chapter 13). Naturally occurring functional polymorphisms or defects have not been reported so far.

References

Alper, C.A., Abramson, N., Johnston, R.B., Jandl, J.H., Rosen, F.S. (1970). Increased susceptibility to infection associated with abnormalities of complement-mediated functions and of the third component of complement (C3). New Engl. J. Med. 282, 349-354.

Anders, E.M., Hartley, C.A., Jackson, D.C. (1990). Bovine and mouse serum beta inhibitors of influenza viruses are mannose-binding lectins. Proc. Natl. Acad. Sci. USA 87, 4485-4489

Bellamy, R., Ruwende, C., Corrah, T., McAdam, K.P.W.J., Whittle, H.C., Hill, A.V.S. (1998). Variations in the *NRAMP1* gene and susceptibility to tuberculosis in West Africans. New Engl. J. Med. 338, 640-644.

Biesma, D.H., Hannema, A.J., van Velzen-Blad, H., Mulder, L., van Zwieten, R., Kluijt, I., Roos, D. (2000). A family with complement factor-D deficiency. Nieuwsbulletin Nederlandse Vereniging voor Bloedtransfusie, 11, 37.

Blackwell, J.M., Roberts, C.W., Roach, T.I., Alexander, J. (1994). Influence of macrophage resistance gene *Lsh/Ity/Bcg* (candidate *Nramp*) on *Toxoplasma gondii* infection in mice. Clin. Exp. Immunol. 97, 107-112.

Blackwell, J.M., Barton, C.H., White, J.K., Searle, S., Baker, A.-M., Williams, H., Shaw, M.-A. (1995). Genomic organization and sequence of the human *NRAMP* gene: identification and mapping of a promoter region polymorphism. Mol. Med. 1, 194-205.

Blackwell, J., Searle, S. (1999). Genetic regulation of macrophage activation: understanding the function of *Nramp1* (=*Ity/Lsh/Bcg*). Immunol. Lett. 65, 73-80.

Brown, M., Dokun, A.O., Heusel, J.W., Smith, H.R.C., Beckman, D.L., Blattenberger, E.A., Dubbelde, C.E., Stone, L.R., Scalzo, A.A., Yokoyama, W.M. (2001). Vital involvement of a natural killer cell activation receptor in resistance to viral infection. Science 292, 934-937.

Buu, N., Cellier, M., Gros, P., Schurr, E. (1995). Identification of a tetra-nucleotide length polymorphism in the 3' untranslated region of the human *NRAMP1* gene. Immunogenetics 42, 428-429.

Canonne-Hergaux, F., Gruenheid, S., Govoni, G., Gros, P. (1999). The Nramp1 protein and its role in resistance to infection and macrophage function. Proc. Assoc. Am. Physicians 111, 283-289.

Cellier, M., Govoni, G., Vidal, S.M., Kwan, T., Groulx, N., Liu, J., Sanchez, F., Skamene, E., Schurr, E., Gros, P. (1994). Human natural resistance-associated macrophage protein: cDNA cloning, chromosomal mapping, genomic organization, and tissue specific expression. J. Exp. Med. 180, 1741-1752.

Cellier, M., Shustik, C., Dalton, W., et al (1997). The human *NRAMP1* gene as a martker of professional primary phagocytes: studies in blood cells, and in HL-60 promyelitic leukemia. J. Leukoc. Biol. 61, 96-105.

Creuwels, L.A.J.M. (1999). Personal communication.

Diez, E., Yaraghi, Z., MacKenzie, A., Gros, P. (2000). The neuronal apoptosis inhibitory protein is expressed in macrophages and is modulated after phagocytosis and during intracellular infection with *Legionella pneumophila*. J. Immunol. 164, 1470-1477.

Eckmann, L., Fierer, J., Kagnoff, M.F. (1996). Genetically resistant (Ityr) and susceptible (Itys) congenic mouse strains show similar cytokine responses following infection with *Salmonella dublin*. J. Immunol. 156, 2894-2900.

Egan, L.J., Orren, A., Doherty, J., Wurzner, R. McCarthy, C.F. (1994). Hereditary deficiency of the seventh component of complement and recurrent meningococcal infection: investigations of an Irish family using a novel haemolytic screening assay for complement activity and C7 M/N allotyping. Epidemiol. Infect. 113, 275-281.

Epstein, J., Eichbaum, Q., Sheriff, S., Ezekowitz, R.A.B. (1996). The collectins in innate immunity. Curr. Opin. Immunol. 8, 29-35.

Ezekowitz, R.A.B. (1998). Genetic heterogeneity of mannose-binding proteins: the Jekyll and Hyde of innate immunity? Am.J.Hum.Genet. 1, 6-9.

Feizi, T., Larkin, M. (1990). AIDS and glycosylation. Glycobiology 1, 17-23.

Figueroa, J.E., Densen, P. (1991). Infectious diseases associated with complement deficiencies. Clin. Microbiol. Rev. 4, 359-395.

Fischer, A., Lisowska-Grospierre, B., Anderson, D.C., Springer, T.A. (1988). Leukocyte adhesion deficiency: Molecular basis and functional consequences. Immunodef. Rev. 1, 39-54.

Garred, P., Harboe, M., Oettinger, T., Koch, C., Svejgaard, A. (1994). Dual role of mannan-binding protein in infections: another case of heterosis? Eur. J. Immunogenet. 21, 125-131.

Garred, P., Madsen, H.O., Hofmann, B., Svej-gaard, A. (1995). Increased frequency of homozygosity of abnormal mannan-binding-protein alleles in patients with suspected immunodeficiency. Lancet 346, 941-943.

Garred, P., Pressler, T., Madsen, O., Frederikson, B., Svejgaard, A. Hoiby, N., et al (1999). Association of mannose-binding lectin gene heterogeneity with severity of lung disease and survival in cystic fibrosis. J. Clin. Invest. 104, 431-437.

Gerardi, A.S. (1996). Bovine leucocyte adhesion deficiency: a review of a modern disease and its implications. Res. Vet. Sci. 61, 183-186.

Gros, P., Skamene, E., Forget, A. (1981). Genetic control of natural resistance to *Mycobacterium bovis* (BCG) in mice. J. Immunol. 127, 2417-2421.

Gros, P., Skamene, E., Forget, A. (1983). Cellular mechanisms of genetically-controlled host resistance to *Mycobacterium bovis* (BCG). J. Immunol 131, 1966-1973.

Growney, J.D., Dietrich, W.F. (2000). High-resolution and physical map of the *Lgn1* interval in C57BL/6J implicates *Naip2* or *Naip5* in *Legionella pneumophila* pathogenesis. Genome Res. 10, 1158-1171.

Guenther, L.C. (1983). Inherited disorders of complement. J. Am. Acad. Dermatol. 9, 815-839.

Hackham, D.J., Rotstein, O.D., Zhang, W., Gruenheid, S., Gros, P., Grinstein, S. (1998). Host resistance to intracellular infection: mutation of natural resistance-associated macrophage protein 1 (Nramp1) impairs phagosomal acidification. J. Exp. Med. 188, 351-364.

Hibberd, M.L., Sumiya, M., Summerfield, J.A., Booy, R., Levin, M. (1999). Association of variants of the gene for mannose-binding lectin with susceptibility to meningococcal disease. Lancet 353, 1049-1053.

Hill, A.V.S. (1998). The immunogenetics of human infectious diseases. Annu. Rev. Immunol. 16, 593-617.

Hoffmann, J.A., Kafatos, F.C., Janeway Jr., C.A., Ezekowitz, R.A.B. (1999). Phylogenetic perspectives in innate immunity. Science 284, 1313-1318.

Hu, J., Bumstead, N., Barrow, P., Sebastiani, G., Olien, L., Morgan, K., Malo, D. (1997). Resistance to Salmonellosis in the chicken is linked to NRAMP1 and TNC. Genome Res. 7, 693-704.

Janeway, C.A., Travers, P. (1994). Immunobiology, the immune system in health and disease. Current Biology, Garland Publishing, Blackwell Scientific Publications, Oxford.

Kehrli, M.E. Jr., Schmalstieg, F.C., Anderson, D.C., Van der Maaten, M.J., Hughes, B.J., Ackermann, M.R., Wilhelmsen, C.L., Brown, G.B., Stevens, M.G., Whetstone, C.A. (1990). Molecular definition of the bovine granulocytopathy syndrome: Identification of a deficiency of the Mac-1 (CD11b/CD18) glycoprotein. Am. J. Vet. Res. 51, 1826-1836.

Koch, A., Melbye, M., Sorensen, P., Homoe, P., Madsen, H.O., Molbak, K., Hansen, C.H., Andersen, L.H., Hahn, G.W., Garred, P. (2001). Acute respiratory tract infections and mannose-binding lectin insufficiency during childhood. JAMA 285, 1316-1321.

Lachmann, P.J. (1984). Inherited complement deficiencies. Philos. Trans R. Soc. Lond. B Biol. Sci. 306, 419-430.

Lehner, P.J., Davies, K.A., Walport, M.J., Cope, A.P., Wurzner, R. Orren, A., Morgan, B.P. Cohen, J. (1992). Meningococcal septicaemia in a C6-deficient patient and effects of plasma transfusion on lipopolysaccharide release. Lancet 340, 1379-1381.

Lewis, L.A., Victor, T.C., Helden, E.G.H., et al (1996). Identification of C to T mutation at position −236 bp in the human *NRAMP* gene promoter. Immunogenetics 44, 309-311.

Liu, J., Fujiwara, M., Buu, N., Sanchez. F.O., Cellier, M., Paradis, A.J., Frappier, D., Skamene, E., Gros, P., Morgan, K., Schurr, E. (1995). Identification of polymorphisms and sequence variants in the human natural resistance-associated macrophage protein (*NRAMP*) gene. Am. J. Hum. Genet. 56, 845-853.

Lipscombe, R.J., Sumiya, Hill, A.V. Lau, Y.L., Levinsky, R.J. Summerfield, J.A., Turner, M.W. (1992). High frequencies in African and non-African populations of independent mutations in the mannose-binding protein gene. Hum. Mol. Genet. 1, 709-715.

Madsen, H.O., Garred, P., Kurtzhals, J.A., Lamm, L.U. Ryder, L.P., Thiel, S., Svejgaard, A. (1994). A new frequent allele is the missing link in the structural polymorphism of the human mannan-binding protein. Immunogenetics 40, 37-44.

Madsen, H.O., Garred, P., Thiel, S., Kurtzhals, J.A., Lamm, L.U., Ryder, L.P., Svejgaard, A. (1995). Interplay between promoter-and structural gene variants control basal serum level of mannan-binding protein. J. Immunol. 155, 3013-1320.

Malhotra, R., Sim, R.B. (1995). Collectins and viral infection. Trends Microbiol. 240, 240-244.

Malo, D., Vidal, S.M., Skamene, E., Gros, P. (1993). High resolution linkage map in the vicinity of the host resistance locus *Bcg* on mouse chromosome 1. Genomics 16, 655-663.

Malo, D., Vidal, S.M., Lieman, J. et al. (1993). Physical delineation of the minimal chromosomal segment encompassing the host resistance locus *Bcg*. Genomics 17, 667-675.

McLeod, R., Buschman, E., Arbuckle, L.D., Skamene, E. (1995). Immunogenetics in the analysis of resistance to intracellular pathogens. Current Opinion Immunol. 7, 539-552.

Orren, A., Wurzner, R., Potter, P.C., Fernie, B.A., Coetzee, S., Morgan, B.P. Lachmann, P.J. (1992). Properties of a low molecular weight complement component C6 found in human subjects with subtotal C6 deficiency. Immunology 75, 10-16.

Roos, D. (1999). Scientific report Sanquin-CLB, Amsterdam.

Ross, S.C., Densen, P. (1984). Complement deficiency states and infection: epidemiology, pathogenesis and consequences of neisserial and other infections in an immune deficiency. Medicine (Baltimore) 63, 243-273.

Rotrosen, D., Gallin, J.I. (1987). Disorders of phagocyte function. Annu. Rev. Immunol. 5, 127-150.

Scalzo, A.A., Fitzgerald, N.A., Wallace, C.A., Gibbons, A.E., Smart, Y.C., Burton, R.C., Shellam, G.R. (1992). The effect of the *Cmv1* resistance gene, which is linked to the natural killer cell gene complex, is mediated by natural killer cells. J. Immunol. 149, 581-589.

Scalzo, A.A., Lyons, P.A., Fitzgerald, N.A., Forbes, C.A., Shellam, G.R. (1995). The BALB.B6-Cmv1r mouse: a strain congenic for *Cmv1* and the NK gene complex. Immunogenetics 41, 148-151.

Searle, S., Blackwell, J.M. (1999). Evidence for a functional repeat polymorphism in the promoter of the human *NRAMP1* gene that correlates with autoimmune versus infectious disease susceptibility. J. Med. Genet. 36, 295-299.

Sebastiani, G., Olien, L., Gauthier, S., Skamene, E., Morgan, K., Gros, Malo, D. (1998). Mapping of genetic modulators of natural resistance to infection with *Salmonella typhimurium* in wild-derived mice. Genomics 47, 180-186.

Shaw, M.A., Clayton, D., Atkinson, S.E. et al. (1996). Linkage of rheumatoid arthritis to the candidate gene *NRAMP1* on 2q35. J. Med. Genet. 33, 672-677.

Shuster, D.E., Kehrli, M.E. Jr., Ackerman, M.R., Gilbert, R.O. (1992). Identification and prevalence of a genetic defect that causes leukocyte adhesion deficiency in Holstein cattle. Proc. Natl. Acad. Sci USA 89, 9225-9229.

Sipes, K.M., Edens, H.A. Kehrli, M.E., Miettinen, H.M., Cutler, J.E., Jutila, M.A., Quinn, M.T. (1999). Analysis of surface antigen expression and host defense function in leukocytes from calves heterozygous or homozygous for bovine leukocyte adhesion deficiency. Am. J. Vet. Res. 60, 1255-1261.

Skamene, E., Schurr, E., Gros, P. (1998). Infection genomics: *Nramp1* as a major determinant of natural resistance to intracellular infections. Annu. Rev. Med. 49, 275-287.

Sumiya, M., Super, M., Tabona, P., et al. (1991). Molecular basis of opsonic defect in immunodeficient children. Lancet 337, 1569-1570.

Summerfield, J.A. Sumiya, M., Levin, M., Turner, M.W. (1997). Association of mutations in mannose-binding protein gene with childhood infection in consecutive hospital series. British Med. J. 314, 1229-1232.

Super, M., Thiel, S., Lu, J., Turner, M.W. (1989). Association of low levels of mannan-binding protein with a common defect of opsonisation. Lancet 2, 1236-1239.

Swart, A.G., Fijen, C.A., te Bulte, M.T., Daha, M.R., Dankert, J. Kuijper, E.J. (1993). Complement deficiencies and meningococcal disease in The Netherlands. Ned. Tijdschr. Geneesk. 137, 1147-1152.

Thomas, H.C., Foster, G.R., Smiya, M., McIntosh, D., Turner, M.W., Summerfield, J.A. (1996). Mutation of gene of mannose-binding protein associated with chronic hepatitis B viral infection. Lancet 348, 1417-1419.

Thursz, M.R. (1997). Host genetic factors influencing the outcome of hepatitis. J. Viral Hepatitis 4, 215-220.

Vassiloyanakopoulos, A.P., Okamoto, S., Fierer, J. (2000). The crucial role of polymorphonuclear leukocytes in resistance to *Salmonella dublin* infections in genetically susceptible and resistant mice. Proc. Natl. Acad. Sci USA 95, 7676-7681.

Vidal, S.M., Malo, D., Vogan, K., Skamene, E., Gros, P. (1993). Natural resistance to infection with intracellular parasites: isolation of a candidate for *Bcg*. Cell 73, 469-485.

Vidal, S., Tremblay, M.L., Govoni, G. et al. (1995). The *Ity/Lsh/Bcg* locus: natural resistance to infection with intracellular parasites is abrogated by disruption of the Nramp1 gene. J. Exp. Med. 182, 655-666.

White, J.K., Shaw, M.-A., Barton, C.H., Cerretti, D.P., Williams, H., Mock, B.A., Carter, N.P., Peacock, C.S., Blackwell, J.M. (1994). Genetic and physical mapping of 2q35 in the region of the *NRAMP* and *IL8R* genes: identification of a polymorphic repeat in exon 2 of *NRAMP*. Genomics 24, 295-302.

Chapter 9.   Genes involved in antigen processing and presentation

Clearance of infections is effected by the cellular and humoral immune systems. For that purpose antigen-presenting cells display fragments of antigens to CD4+ T-helper cells in the cleft of the major histocompatibility (MHC) complex class II molecules. These are cell surface α/β heterodimeric proteins whose function is to present processed antigens to T cells. CD4+ T-helper cells may respond by proliferating and by secreting cytokines, which support the effector arm of the immune system, especially antibody producing B cells, and CD8+ cytotoxic T cells. *De novo* synthesized intracellular antigens are presented to CD8+ T lymphocytes by MHC class I molecules. Proteins in cells are therefore degraded continuously by a large multicatalytic protease complex consisting of 28 subunits (each between 20 to 30 kDa), which is called the proteasome. Genes of the proteasome complex are linked to the MHC. Variations in MHC amino acid sequences determine which antigenic fragments are presented to T cells, as well as the efficiency of this process, and they may thus control the quality and quantity of the T cellular response. Polymorphisms of the MHC class I and II molecules (including the proteasome complex) may therefore be associated with variable outcomes of many infectious and autoimmune diseases.

The human MHC or Human Leukocyte Antigen (HLA) System comprises a series of closely linked genetic loci on the short arm of chromosome 6, which constitute the most polymorphic genetic system known. For an overview of the HLA nomenclature see Bodmer et al 1999. Most centromeric on chromosome 6p is the class II region, which contains the 17 known HLA class II genes and pseudogenes. Contagious to that is the class III region, which encodes several of the components of the complement system. Telomeric to the class III region is the class I region, which encodes more than 18 HLA class I-related genes and pseudogenes. Recently a number of genes putatively involved in inflammation and infection have been identified in the central MHC, at the telomeric end of the class III region. This region has tentatively been designated as MHC class IV region and includes, among others, the genes encoding tumor necrosis factor-α, lymphotoxin α, and lymphotoxin β (Gruen and Weissman 1997). Several of these genes are still of unknown function and of an unknown degree of polymorphism. So their potential contribution to infectious disease remains to be established. Because there are many alleles at each MHC locus and many loci in each individual, a large repertoire of antigenic recognition is available to a species. Indeed hundreds of alleles (> 700) have been described so far. Many thousands of different MHC phenotypes are possible. In addition, heterozygosity at MHC class I and II loci may be as high as 80 – 90 %. This large repertoire helps to circumvent molecular mimicry by pathogens, ensures a large diversity of antigenic recognition, and maintains heterozygosity.

The MHC class I region has been conserved in evolution and there is evidence that infectious disease is a major selective force in maintaining the polymorphism (see chapter 2). The polymorphism is located mainly around the groove that binds foreign peptides and that presents them to the T cell receptor (Bjorkman et al 1987). Some MHC alleles are found in different species and these must therefore have been arisen millions of years ago before speciation, suggesting that selection has been directed at maintaining the polymorphisms by overdominant selection (Figueroa et al 1988, Lawlor et al 1988). This "trans-species" and overdominant selection hypothesis of MHC evolution is further corroborated by the increased rate of non-synonymous nucleotide substitutions in the peptide-binding groove and its surroundings (Hughes and Nei 1988).

Also in class II loci most of the β-chain polymorphism is located in the "hypervariable" regions (HVRs). In these HVR regions amino acid sequence similarity between distantly related species is observed. This represents direct descent of ancestral sequences rather than convergent evolution. Furthermore, half the sequence polymorphism in class II β-chain genes of mice persists in evolution and is encoded by the same DNA sequence in humans. No evidence for increased mutation rate within this HVR was found. The HVR is therefore probably a genetic unit of recombination, and selection for HVR sequences and combinations is likely restrained by functional considerations (Brown et al 1988, Lundberg and McDevitt 1992).

Another characteristic is the strong linkage disequilibrium between particular alleles of genes across the MHC. Because recombination over the whole of the MHC is not significantly different from that of other regions of the human genome, the strong linkage disequilibrium might be due to selection by infectious agents (Hill et al 1991). Evidently the evidence that infectious diseases have been a necessarily major selective force in MHC selection is necessarily fragmented, anecdotal and more or less archeological. De Vries et al (1979) provide an example. In the middle of the last century a group of Dutch farmers emigrated to Surinam in South America and suffered form typhoid fever, killing 50 % of them. A few years later they suffered from yellow fever, killing another 20 %. A few of 26 examined polymorphisms showed differences in frequencies between descendants from the emigrants and from the general Dutch population. These were unlikely to be due to drift. Descendants of the emigrants completely lacked HLA-DR2, which has a gene frequency of 0.17 among the Dutch population, and they had significantly increased frequencies of HLA-DR4 and HLA-DRw13. DR2 thus appears associated with mortality, and DR4 and DRw13 appear to have conferred protection against typhoid fever-associated mortality. This suggestion is corroborated by the finding that DR4 and DRw13 present a presumed protective T cell epitope. The example illustrates that gene frequencies in a population can change within a relatively short period of time. Evidently episodes of selection by fatal infections may be intervened by long periods of selective neutrality.

Further indications for selection working at the MHC are the occurrence of heterozygous advantage (overdominance), which may be advantageous both at the level of the individual and the population in conferring resistance against a wide variety of infectious diseases. Thursz et al (1997), for example, provided evidence supporting models of overdominant selection in which MHC homozygotes are less likely to clear an HBV infection and thus more likely to become persistently infected. In their study in the Gambia, significantly fewer subjects with persistent infection were heterozygotes for haplotypes of the HLA class II region genes, HLA-DR and HLA-DQ. Heterozygote advantage is also documented for HIV infection. Maximum HLA heterozygosity of class I loci (A, B, and C) delayed AIDS onset among American patients infected with HIV-1, whereas individuals who were homozygous for one or more loci progressed rapidly to AIDS and death (Carrington et al 1999). In the same study the HLA class I alleles B*35 and Cw*04 were also consistently associated with rapid development of AIDS. (Note: the numbers following the asterisk are allelic designations.) Evidently, these examples suggest that in the presence of numerous infectious pathogens, maximum heterozygosity of different alleles could afford heterozygote advantage overall. Initially many efforts have been devoted to finding MHC disease associations. However, despite the important physiological role of the MHC there have been very few direct relationships between particular MHC alleles and resistance to infectious disease. Moreover, as outlined above (chapter 4), a disease association is in essence a statistical phenomenon, which does not indicate a causal relationship between the allelic MHC gene product and the disease. A significant association between a disease and a given allele does therefore not imply that the causative gene has been identified. Another gene in linkage disequilibrium with the first gene may yield an even better correlation.

If the MHC polymorphism has been generated during many generations due to small heterozygous advantages, then the differences between allelic effects may be small (Klein and Figueroa 1986). In addition, populations may have been selected for disease resistance for many generations, making resistance genes with major effects at the present time very rare. Nonetheless modest effects may be evolutionary important and sufficient enough to select and maintain polymorphism. When the disease is influenced by several loci possibly showing epistatic interaction, or when MHC loci are in linkage disequilibrium with a major disease locus, detailed studies of inheritance are needed to elucidate the contribution of the different loci. However, despite these drawbacks, finding MHC alleles strongly associated with disease may give basic insight in disease mechanisms and etiology.

In addition, the difficulties in finding clear MHC disease associations may be no surprise as pathogens likely present very many T cell epitopes, which will have varying MHC restriction patterns. So the existence of only few particular MHC disease associations would indicate that most epitope-specific responses are of limited protective value and that only a small number of these responses may be protective. Knowledge of these MHC disease associations could nonetheless lead to the identification of protective epitopes and responses. Consequently, the lack of a MHC association may point to a type of immunity that is based on the recognition of a large number of equally protective epitopes. The same is true when disease susceptibility or (vaccine-induced) immune responses associate with MHC heterozygosity (see chapter 4).

Another reason that significant and consistent MHC associations have been difficult to prove may be that in different populations different alleles may be associated with particular infectious diseases. For example, associations of MHC alleles and resistance to falciparum malaria are different in populations in The Gambia and East Africa (Wills 1996). Ethnic studies may therefore make available recombination events that alter haplotypes and thus allow the identification of susceptibility determinants in a given population. HLA associations may also occur with molecularly defined strains of a pathogen only. In any case, the rarity of clear MHC disease associations appears to indicate that major selective forces have favored presentation of a T cell repertoire that provides protection against a wide range of pathogens. Other explanations for difficulties in finding MHC-disease associations may be the genetic variation of the pathogen, small effects, ethnic differences, and the low frequency of protective MHC alleles, which requires large studies for assessing their role in disease. Finally, selective forces on highly polymorphic T-cell epitopes of the pathogen may result in MHC-disease associations that likely fluctuate both spatially and temporally. Thus the interaction between MHC and infectious disease appears very complex, making it often difficult to find clear-cut relations between a particular MHC allele and susceptibility to a specific disease.

Despite the difficulties, a number of studies is accumulating that report an association between a MHC specificity and a given disease, which is nearly always bound up with immunological processes (Table 4). As may be expected from its function, MHC associations are found more often with the progression of infectious disease, such as in leprosy and hepatitis C virus infection, rather than with its initiation.

Many of strongest MHC disease associations occur in autoimmune disease. For example the strong association of HLA-B27 with ankylosing spondylitis is remarkable. This occurs in very different ethnic groups with very different HLA-B27 frequencies. In Caucasians up to 95 % of the patients are of B27 phenotype in contrast to approximately 5 % in controls. However, only 4 % of HLA-B27-positive individuals develop the disease, indicating that other genes and environmental triggers (for example infections) are important in the pathogenesis. Remarkably, some of the arthritis syndromes seen after bacterial infections are also associated with HLA-B27 (see Table 4). So far the mechanisms of B27 involvement in reactive arthritis are unexplained.

Immune evasion

Indirect evidence for the significance of the MHC complex in host defense comes from the several strategies employed by micro-organisms to circumvent its action, often leading to specialization of the pathogen to a certain ecological niche and a particular host, and reduced opportunities for transmission. For example *Plasmodium* spp. infect red blood cells which do not express MHC alleles. The spirochete *Treponema pallidum*, the cause of syphilis, evades immune recognition by coating itself with an antigenically unreactive cell surface rich in lipids and host proteins. In this man-

Table 4. MHC allele associations with infectious diseases

Disease phenotype	Association with MHC phenotype	Odds ratio	Ref.
Class I associations			
Reduced HIV-1 susceptibility	HLA-Aw28	0.21	
(alloantigenic responses)	HLA-Bw70	0.3	Fowke et al 1996
Reduced HTLV-1-associated myelopathy	HLA-A*02	0.5	Jeffery et al 1999
Reduced severe malaria	HLA-B53	0.59	Hill et al 1991
Infectious mononucleosis	HLA-B35	1.69	Tiwari and Terasaki 1985
Post-*Salmonella* arthritis	HLA-B27	29.7	
Post-*Shigella* arthritis	HLA-B27	20.7	
Post-*Yersinia* arthritis	HLA-B27	17.6	
Post-gonococcal arthritis	HLA-B27	14.0	Roitt 1997
Class II associations			
Clinical tuberculosis	HLA-DQB1*0503	7.3	Goldfeld et al 1998
Tuberculoid leprosy	HLA-DR2	8.2	Roitt 1997
Persistent hepatitis C virus infection	HLA-DRB1*0701	2.04	
	HLA-DRB4*0101	2.38	
Self-limiting hepatitis C virus infection	HLA-DRB1*1101	2.14	
	HLA-DQB1*0301	2.22	Thursz et al 1999
Herpes virus infections	HLA-DR3	13.35	Tiwari and Terasaki 1985

ner only antigens from dead spirochetes are presented to the host's immune system, and in most instances the immune response is not sufficiently effective to eradicate the disease (Mimms et al 1993). Finally, the herpesviruses have developed several mechanisms interfering with class I-restricted antigen presentation, probably enabling the virus from elimination during acute and latent infection. Specific herpes viral proteins may destroy MHC class I molecules, block their transport from the endoplasmic reticulum to the cell surface, or block peptide transport by the transporters associated with antigen presentation (TAP) (Ploegh 1998).

Defects and polymorphisms in antigen presentation
Classical class I molecules (or "class Ia genes") are *HLA-A, -B, -C* in humans, and *H2-K, -D, -L* in mice. They are expressed on almost all somatic cells where they present antigenic peptides to the T cell receptors on cytotoxic T lymphocytes. MHC class I molecules are assembled in the endoplasmic reticulum where they bind peptides from intracellularly cleaved and processed protein antigens (by the proteasome) (Fig. 7). The antigenic peptides are transported from the cytosol into the lumen of the endoplasmic reticulum by the transporters of antigenic peptides (TAP). In the endoplasmic reticulum the processed peptides associate with the class I MHC and β_2-microglobulin proteins to form a trimeric complex. The peptide transporter complex constitutes of two subunits, TAP1 and TAP2. TAP 1 and TAP2 form a pore in the endoplasmic reticulum, which allows endogenous antigenic peptides to pass into the lumen of the endo-

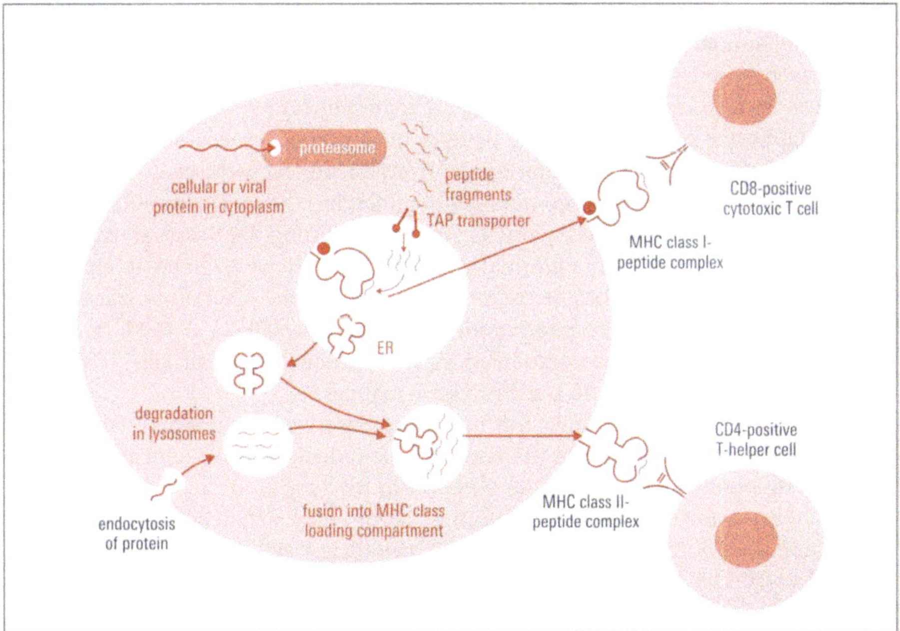

Figure 7. Schematic representation of antigen processing of de novo synthesized proteins (above), resulting in antigen presentation by MHC class I molecules, and of antigen processing of proteins taken up by endocytosis or pinocytosis from the surroundings of the cell, resulting in antigen presentation by MHC class II molecules (below).
TAP: transporter of antigenic peptides
ER: endoplasmic reticulum

plasmic reticulum where they are loaded onto MHC class I molecules. TAP1 and TAP2 are thus involved in epitope selection. Absence of either TAP1 or TAP2 abolishes the transport of peptides into the endoplasmic reticulum, and subsequently their presentation by MHC class I molecules on the cell surface. During the assembly of MHC class I molecules TAP1, but not TAP2, interacts with class I heavy chains. This observation is consistent with the finding that TAP1 alone, but not TAP2, may facilitate the transport or binding of some peptides (Androlewicz et al 1994). The TAP genes are located in the MHC, between the HLA-DQ and-DP genes. This location may point to an evolutionary strategy for the co-selection of functionally related genes or to a requirement for co-regulation (probably by interferon-γ).

Proteins from outside the cell may enter the cell by endocytosis from the cell surface or by pinocytosis. The proteins are degraded by proteolytic enzymes to short peptides, which may enter a specialized MHC-class II-loading compartment. In this compartment the short peptides bind to the MHC class II molecule. The complex is then transported to the cell surface, where the antigenic peptides are presented to CD4+ T helper cells.

Only very few cases of human HLA class I deficiency have been reported (de la Salle et al 1998). This deficiency can be based on a transcription defect, or on the absence of a functional TAP1 or TAP2 subunit. HLA class I deficiency based on a defect in transcription is characterized by a tenfold reduction in the number of HLA class I molecules. This defect does not lead to severe immunodeficiency, and can partially be overcome by inflammatory cytokines (de la Salle et al 1998, Sullivan et al 1985).
A homozygous mutation leading to functional TAP2 deficiency results in a reduction in the cell surface expression of HLA class I molecules to 1 – 3 % of normal levels ("bare lymphocyte syndrome"). These patients have normal levels of mRNA encoding MHC class I molecules and normal levels of production of MHC class I molecules. However, the class I molecules do not load peptides, are therefore unstable, and cannot reach the cell surface. These patients are healthy during the first years of life and do not appear to suffer from abnormal viral infections. However, later in childhood they begin to suffer from chronic bacterial infections of the respiratory tract, which leads to bronchiectasis and respiratory insufficiency (de la Salle et al 1994, Donato et al 1995). The syndrome is accompanied by dysfunction of natural killer cells. The defect is transmitted in an autosomal recessive manner.
TAP1-deficiency (based on frame shift mutations) also leads to unstable HLA class I molecules and their retention in the endoplasmic reticulum. TAP1-deficient patients display skin lesions and vasculitis, in addition to the typical lung syndrome of HLA class I deficiency (de la Salle et al 1999). These lesions may be based on an autoimmune reaction. One of the patients displayed no cytotoxic T cell response to autologous herpesvirus-infected cells, although antibodies were found in the serum. This finding is in accordance with the expected inefficient HLA class I-restricted cytotoxic response of CD 8+ αβ T cells, because in these TAP-deficient patients the HLA class I-restricted T cell response can only be directed against TAP-independent epitopes. Such epitopes are probably less numerous than the TAP-dependent epitopes. Thus, remarkably, the few described TAP1- and TAP2-deficient patients did not seem to have a particular high susceptibility to viral infections (de la Salle et al 1999). This finding probably stresses the *in vivo* significance of TAP-independent antigen presentation in viral

infections. The recurrent bacterial infections in these patients point to a physiological significance of the presentation of bacterial antigens by class I molecules, probably on macrophages. The finding was further substantiated in TAP1-deficient-, as well as in β_2-microglobulin-deficient mice. These mice lacked class I MHC-restricted CD8+ T cells and died rapidly after infection with *Mycobacterium tuberculosis* (but not after infection by *Leishmania* and *Toxoplasma*) (Behar et al 1999).

In addition to the severe, but seldom TAP deficiencies, several TAP polymorphisms have been described and examined for involvement in infectious disease. TAP1 and TAP2 have two and three polymorphic sites respectively. So far, conflicting results have been reported regarding the association between chronic hepatitis C disease with TAP2 alleles (Hohler et al 1996, Kuzushita et al 1999). Interestingly, Kaslow et al (1996) have presented evidence that particular combinations of HLA class I and II alleles together with specific TAP alleles are associated with the course of HIV-1 infection. Similarly, a study from Rajalingam et al (1997) found evidence that alleles in the TAP2 region in conjunction with HLA-DR15 influence susceptibility to tuberculoid leprosy and pulmonary tuberculosis.

Lack of expression of MHC class II gene products leads to a particular form of severe combined immune deficiency (SCID) or bare lymphocyte syndrome. At least four complementing defects in genes encoding trans-acting factors regulating MHC gene expression have been identified in these patients, indicating that at least four different genes are required for normal MHC class II expression (Janeway and Travers 1994). The mutations appear to affect the accessibility of the class II promoters within the environment of the MHC (Kara and Glimcher 1993). Because the thymus during development also lacks MHC class II molecules, there is no positive selection of CD4+ T cells and only few develop. However, these few cells cannot be stimulated due to the lack of MHC class molecules on their antigen presenting T cells. In these patients MHC class I expression and CD8+ T cell development is normal.

References

Androlewicz, M.J., Ortmann, B., van Endert, P.M., Spies, T., Cresswell, P. (1994). Characteristics of peptide and major histocompatibility complex class I/β_2-microglobulin binding to the transporters associated with antigen processing (TAP1 and TAP2). Proc. Natl. Acad. Sci. USA 91, 12716-12720.

Behar, S.M., Dascher, C.C., Grusby, M.J., Wang, C.-R., Brenner, M.B. (1999). Susceptibility of mice deficient in CD1D or TAP1 to infection with *Mycobacterium tuberculosis*. J. Exp. Med. 189, 1973-1980.

Bjorkman, P.J., Saper, M.A., Samraoui, B., Bennett, W.S., Strominger, J.L., Wiley, D.C. (1987). The foreign antigen binding site and T cell recognition regions of class I histocompatibility antigens. Nature 329, 512-518.

Bodmer, J.G., Marsh, S.G., Albert, E.D., Bodmer, W.F., Bontrop, R.E., Dupont B., Erlich, H.A., Hansen, J.A., Mach, B., Mayr, W.R., Parham, P., Petersdorf, E.W., Sasazuki, T., Schreuder, G.M., Strominger, J.L., Svejgaard, A., Terasaki, P.I. (1999). Nomenclature for factors of the HLA System, 1998. Tissue Antigens 53, 407-446.

Brown, J.H., Jardetzky, T., Saper, M.A., Samraoui, B., Bjorkman, P.J. Wiley, D.C. (1988). A hypothetical model of the foreign antigen binding site of class II histocompatibility molecules. Nature, 322, 845-850.

Carrington, M., Nelson, G.W., Martin, M.P., Kissner, T., Vlahov, D., Goedert, J.J., Kaslow, R., Buchbinder, S., Hoots, K., O'Brien, S.J. (1999). HLA and HIV-1: heterozygote advantage and B*35-Cw*04 disadvantage. Science 283, 1748-1752.

de la Salle, H. et al. (1994). Homozygous human TAP peptide transporter mutation in HLA class I deficiency. Science 265, 237-241.

de la Salle, H., et al. (1998). HLA class I deficiencies. In: Primary immunodeficiency diseases: a molecular approach. Ed. by H.D. Ochs, C.E.I. Smith, J.M. Puck. Oxford University Press, New York, pp. 181-188.

de la Salle, H., Zimmer, J., Fricker, D., Angenieux, C., Caenave, J.-P., Okubo, M., Maeda, H., Plebani, A., Tongio, M.-M., Dormoy, A., Hanau, D., (1999). HLA class I deficiencies due to mutations in subunit 1 of the peptide transporter TAP1. J. Clin. Investig. 103, R9-R13.

de Vries, R.R.P., Meera Khan, P., Bernini, L.F., Van Loghem, E., Van Rood, J.J. (1979). Genetic control of survival to epidemics? J. Immunogenet. 6, 271-287.

Donato, L., de la Salle, H., Hanau, D., Tongio, M.-M., Oswald, M., Vandevenne, A., Geistert, J.. (1995). Association of HLA class I antigen deficiency to a TAP2 gene mutation with familial bronchiectasis. J. Pediatr. 127, 895-900.

Figueroa, F., Günther, E., Klein, L. (1988). MHC polymorphism pre-dating speciation. Nature 335, 265-267.

Fowke, K.R., Nagelkerke, N.J.D., Kimani, J., et al. (1996). Resistance to HIV-1 infection among persistently seronegative prostitutes in Nairobi, Kenya. Lancet 348, 1347-1351.

Goldfeld, A.E., Delgado, J.C., Thim, S., Bozon, Ugilaloro, A.M., Turbay, D., Cohen, C., Yunis, E.J. (1998). Association of an HLA-DQ allele with clinical tuberculosis. JAMA 279, 226-228.

Gruen, J.R., Weissman, S.M. (1997). Evolving views of the major histocompatibility complex. Blood 11, 4252-4265.

Hill, A.V.S., Allsopp, C.E.M., Kwiatkowski, D., Anstey, N.M., Twumasi, P., Rowe, P.A., Bennet, S., Brewster, D., McMichael, A.J., Greenwood, B.M. (1991). Common West African HLA antigens are associated with protection from severe malaria. Nature 352, 595-600.

Hohler, T., Gerken, G., Schneider, P.M., Meyer zum Buschenfelde, K.H., Rittner, C. (1996). Antigen-processing polymorphism in chronic hepatitis C infection. Exp. Clin. Immunogenet, 13, 7-11.

Hughes, A.L., Nei, M. (1988). Pattern of nucleotide substitution at major histocompatibility complex I loci reveals overdominant selection. Nature, 335, 167-170.

Janeway, C.A., Travers, P. (1994). Immunobiology, the immune system in health and disease. Current Biology, Garland Publishing, Blackwell Scientific Publications, Oxford.

Jeffery, K.J.M., Usuku, K., Hall, S.E., Matsumoto, W., Taylor, G.P., Procter, J., et al (1999). HLA alleles determine human T-lymphotropic virus-I (HTLV-I) proviral load and the risk of HTLV-I associated myelopathy. Proc. Natl. Acad. Sci. 96, 3848-3853.

Kara, C.J., Glimcher, L.H. (1993). Promoter accessibility within the environment of the MHC is affected in class II-deficient combined immunodefiency. EMBO J. 12, 187-193.

Kaslow, R.A., Carrington, M., Apple, R., Park, L., Munoz, A., Saah, A.J., Goedert, J.J., Winkler, C., O'Brien, S.J., Rinaldo, C., Detels, R., Blattner, W., Phair, J., Erlich, H., 7Mann, D.L. (1996). Influence of combinations of human major histocompatibility complex genes on the course of HIV-1 infection. Nat. Med. 2, 405-411.

Klein, J., Figueroa, F. (1986). Evolution of the major histocompatibility complex. CRC Crit. Rev. Immun. 6, 295-386.

Kuzushita, N., Hayashi, N., Kanto, T., Takehara, T., Tatsumi, T., Katayama, K., Ohkawa, K., Ito, A., Kasahara, A., Moribe, T., Sasaki, Y., Hori, M. (1999). Involvement of transporter associated with antigen processing 2 (TAP2) gene polymorphism in hepatitis C virus infection. Gastroenterology 116, 1149-1154.

Lawlor, D.A., Ward, F.E., Ennis, P.D., Jackson, A.P., Parham, P. (1988). HLA-A and B polymorphism predate the divergence of humans and chimpanzees. Nature 335, 268-271.

Lundberg, A.S., McDevitt, H.O. (1992). Evolution of major histocompatibility complex class II diversity: direct descent in mice and humans. Proc. Nat. Acad. Sci. USA. 89, 6545-6549.

Mims, C.A, Playfair, J.H.L., Roitt, I.M., Wakelin, D., Williams, R. (1993). Medical microbiology, Mosby-Year Book Europe Ltd.

Ploegh, H. (1998). Viral strategies of immune evasion. Science 280, 248-253.

Rajalingam, R., Singal, D.P., Mehra, N.K. (1997). Transporter associated with antigen-processing (TAP) genes and susceptibility to tuberculoid leprosy and pulmonary tuberculosis. Tissue Antigens 49, 168-172.

Roitt, I.M. (1997). Essential immunology, 9th ed. Blackwell scientific publications.

Sullivan, K.E., Stobo, J.D., Peterlin, B.M. (1985). Molecular analysis of the bare lymphocyte syndrome. J. Clin. Invest. 76, 75-79.

Thursz, M.R., Thomas, H.C., Greenwood, B.M., Hill, A.V.S. (1997). Heterozygote advantage for HLA class-II type in hepatitis B virus infection. (Letter) Nature Genet. 17, 11-12.

Thursz, M., Yallop, R., Goldin, R., Trepo, C., Thomas, H.C. (1999). Influence of MHC class II genotype on outcome of infection with hepatitis C virus. Lancet 354, 2119-2124.

Tiwari, J.L., Teraskai, P.I. 1985. HLA and disease associations. Springer-Verlag, New York.

Wills, C. 1996. Plagues. Harper Collins Publishers, London.

Chapter 10. Genes regulating immune and inflammatory responses

Pro- and anti-inflammatory cytokines
The genes that control inflammatory responses are cytokine genes. Cytokines function within a complex network, in which they may induce or inhibit their own production, as well as the production of other cytokines or cytokine receptors. If an individual produces high levels of pro-inflammatory cytokines or low levels of their inhibitors, this would likely result in a strong and ongoing inflammatory response to any stimulus. Proinflammatory cytokines are thus part of the innate immune response that combats infection, but if their actions are not controlled properly they can themselves cause extensive tissue damage even leading to death.

IL-1 and TNF-α are primary cytokines, which can stimulate their own production, as well as the production of a large number of secondary cytokines (di Giovine and Duff 1990). Two groups of inhibitors appear to be central to the control of IL-1 and other cytokines. Type I inhibitor proteins are structural homologues of a cytokine, for example the IL1 receptor antagonist (IL-1RA). They compete with the cytokine for binding to its receptor without producing detectable cell activation. Type II inhibitors are soluble receptors, shed from the cell surface, which can bind cytokines and so prevent their binding to target cells (Symons et al 1995).

The cytokines IL-1, IL-2, IL-6, IL-8 and tumor necrosis factor-α (TNF-α) are pro-inflammatory. IL-4, IL-10, and the soluble TNF receptors p55 and p75 are the key anti-inflammatory cytokines. Together with other cytokine receptors and inhibitors, such as the IL-1 receptor (IL-1R), the IL1-RA, and the IL-6 receptor (IL-6R), they function as control elements. Complex interactions between polymorphic molecules may thus occur.

Cytokines as modulators of the Th1/Th2 balance
T helper cells produce, upon activation, restricted and stereotyped patterns of lymphokines, which discriminate them in Th1 and Th2 cells. The balance between Th1 and Th2 cells regulates the type of inflammatory and immune responses following infectious and non-infectious stimuli. This balance appears to be influenced by many identified and non-identified factors, including, among others, the type and dose of immunogen, age, the genetic make-up of the host, preceding infections, the type of antigen-presenting cells, promoter polymorphisms in MHC class II genes, and cytokines. Precursor T cells bearing the CD4 antigen and activated by antigen in the presence of IL-12 and interferons develop predominantly into Th1 cells, whereas those activated in the presence of IL-4 develop predominantly into Th2 cells. IL-4 is one of the major cytokines involved. IL-4 is a pleiotropic cytokine and an important regulator of the immune system. It is considered crucial for the development of Th2 cell responses.

Th1 and Th2 cells both produce cytokines that serve as their own growth factor and that promote the differentiation of naive T cells to that subset. Cytokines released by one T helper subset cross-regulate the development of the other subset. Th2 cells suppress Th1 cells by secreting IL-4, whereas IFN-γ inhibits Th2 cell development. Th1 cells secrete IL-2, IFN-γ, and TNF-β, but not IL-4, IL-5, and IL-13. In contrast, Th2 cells secrete IL-4, IL-5, IL-6, IL-13, but not IL-2 and IFN-γ.

Despite this, the basic mechanisms by which naive T cells commit to a Th1 or Th2 type of cytokine expression during differentiation into mature effector cells are incompletely defined. The binding of cytokines to their receptors typically results in receptor homo- or heterodimerization, which triggers intracellular signals leading to specific transcriptional activation of target genes. These include the Jak-signal transducer and the activator of the transcription (STAT) signalling pathway. STAT factors are rapidly tyrosine-phosphorylated after stimulation with cytokines and subsequently dimerize and translocate to the nucleus where they can activate transcription. IL-12 activates three putative STATs, Stat1, Stat3, and Stat4. Of these, Stat4 appears selectively activated by IL-12. If IL-12 and IFN-γ predominate, the Stat4 pathway is activated, promoting the development of a Th1 response. In most infections IL-12 regulates the magnitude of the IFN-γ response. In contrast, signalling by IL-4 occurs through activation of Stat6 and results in the development of Th2 cells (Romagnani 1996) (Fig. 8).

Proof for a genetic influence in the development of the Th1/Th2 balance comes for example from the observation that activated CD4+ T cells from BALB/c mice, in contrast to most strains of mice, are committed towards IL-4 production and Th2 commitment (Bix et al 1998). A locus on murine chromosome 11, called *T-cell phenotype switch-1* (*Tps-1*), has been suggested to be involved in this Th-cell differentiation, because it controls the maintenance of IL-12 responsiveness and therefore the subsequent Th1/Th2 response (Romagnani 1996). In addition, a so-called "Th2 cytokine

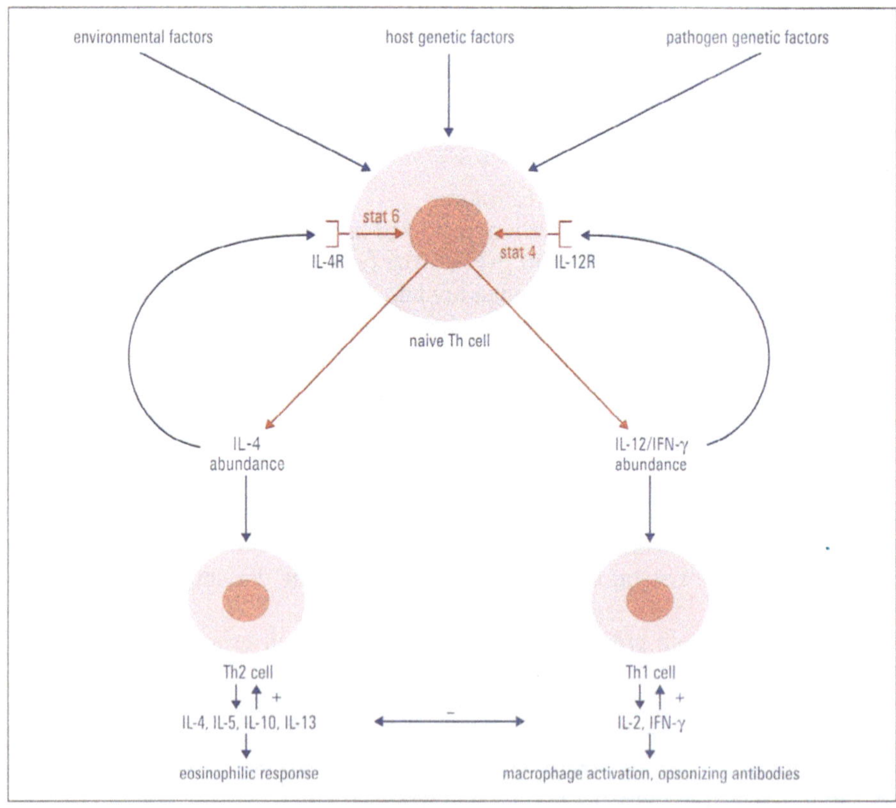

Fig. 8. Simplified schematic view of the factors and pathways leading to Th1 and Th2 cell development.

gene cluster" has been identified on the proximal arm of murine chromosome 11. The region shows conserved synteny (i.e. homology across species) with a region on human chromosome 5q and contains the genes encoding IL-3, IL-4, IL-5, IL-9, IRF1, GM-CSF, and CD14. These cytokines are important for the initiation and maintenance of the allergic inflammatory response through isotype switching of B cells to IgE synthesis (IL-4), the selective maturation of Th2 cells expressing this gene cluster (IL-4), and through growth, maturation, and activation of eosinophils (IL-3, IL-5, GM-CSF) (Kay et al 1991). In mice the region controls *Leishmania major* infection (see below). Although the region in humans appeared not linked to susceptibility for leishmanial and mycobacterial infection, the region shows linkage to and highly significant allelic association with the ability to mount an immune response to mycobacterial antigens (Blackwell et al 1997).

The Th1/Th2 balance and the outcome of Leishmania major infection
The Th1 subset efficiently induces cell-mediated responses and protection against intracellular pathogens through its ability to activate cytotoxic and phagocytic cells, while the Th2 subset provides efficient support for antibody production by B cells and stimulates the maturation and activation of eosinophils. The cytokines produced by the Th2 subset have further been shown to be important in the initiation and perpetuation of allergic responses. The type of Th cell response therefore may determine susceptibility or resistance to pathogens and the course of the infection, for example in experimental murine leishmaniasis. Mice of the BALB/c strain respond with IL-4-producing Th2 cells to *Leishmania major* that fail to promote resistance through macrophage activation, while other strains, including B10.D2, respond with IFN-γ-producing Th1 cells and are resistant. BALB/c mice develop progressive cutaneous *leishmaniasis* and fail to control the replication of the intracellular parasites. Manipulation of the immune response in BALB/c mice toward a Th1 response promotes cure (Güler et al 1996). Natural killer cells and IL-12 are critical early determinants of this effective Th1 response. However, recent results indicate that the effect of Th1 cells on resolution of lesions in genetically resistant mice also requires a functional Fas-FasL pathway of cytotoxicity. CD40, a co-stimulatory molecule on antigen-presenting cells, appears critical for the development of a Th2 response. A particular subset of TCR Vβ4-Vα8 CD4+ T cells, recognizing a major protective antigen on *L. major*, is responsible for the production of IL-4 that is required for Th2 development. This IL-4 is produced during the first two days after parasite inoculation in susceptible BALB/c strains and renders parasite-specific CD4+ T helper cells unresponsive to IL-12. The genetic control of protective and exacerbating responses to *Leishmania* appears to be multigenic, with perhaps six or more murine genetic loci controlling the outcome of infection (Louis et al 1998).

Cytokine gene polymorphism
Single nucleotide polymorphisms (SNPs) in regulatory and coding regions of cytokines have been associated with susceptibility to a number of complex disorders. A number of candidate cytokine genes that may impact the type of immune response following infectious and non-infectious stimuli have thus been identified (Daser et al 1996, Rosenwasser and Borish 1998). Most of the polymorphisms so far detected are in the regulatory sequences and introns of these genes, rather than in the exons. These polymorphisms thus likely result in variable time courses and expression profiles of cyto-

kine responses. This variability may likely have evolved in response to infection (Daser et al 1996). Nonetheless most studies so far focussed on the association between cytokine gene polymorphisms and various inflammatory, autoimmune, and allergic disorders, rather than on infectious diseases as such.

Among others, polymorphisms have been identified in the genes encoding IFN-γ, IL-1, IL-4, IL-4Rα, IL-8, IL-9, IL-10, IL-12, IL-13, TNF-α, TNF-β, and RANTES (a C-C class chemokine) (Messer et al 1991, Pociot et al 1992, Rosenwasser and Borish 1998). However, while this field is rapidly emerging, newly recognized polymorphisms and their possible association with disease are reported almost continuously (Bidwell et al 1999, http://www.pam.bris.ac.uk/services/GAI/TNFA_LTA.html). A summary of several reported associations of cytokine gene polymorphisms with infectious disease susceptibility is given in Table 5.

As mentioned above, considerable attention has been focussed on the genetic regulation of CD4+ Th1- and Th2-type helper T lymphocytes. Several candidate genes appear to be involved in the recognition and activation of CD4+ Th2-type T lymphocytes, with subsequent development of cytokines that generate mast cell, eosinophil and basophil activity. A number of these genes, associated with asthma and atopy and involved in the generation of Th2 responses, has been localized to chromosome 5q31-33 in humans. For example, cytokine genes, genes encoding growth factors and their receptors are within this region. In this region are also the genes encoding IL-4 and IL-9 that in particular promote T-cell growth, mast cell growth, and IgE synthesis. Although most attention has been directed on their association with asthma and allergic conditions, it is very likely that these genes also affect the quality of the immune response and thereby the clinical outcome following infection.

The interleukin-1 gene cluster
There are two classes of IL-1, IL-1α and IL-1β, each encoded by a separate gene, but they act on common receptors. The genes for IL-1α, IL-1β, and IL-1RA are located on a 430 Kbp stretch of DNA on the long arm of chromosome 2 (Nicklin et al 1994). Several polymorphisms have been characterized in this region: single nucleotide polymorphisms (SNP) in IL-1α (at position -889), IL-1β (at positions −511 and +3953), and a variable number of a tandem repeat (VNTR) in intron 2 of IL-1RA (Cork et al 1996, de Giovine et al 1992, McDowell et al 1995, Tarlow et al 1993).

The pro-inflammatory cytokine IL-1β and its receptor antagonist, the IL-1RA, are strongly induced by *M. tuberculosis*. The haplotype defined by the polymorphism at position +3953 of the gene encoding IL-1β appears to define the delayed-type hypersensitivity or Mantoux response and disease expression in human tuberculosis (Wilkinson et al 1999).

Among persons who remained seronegative for Epstein-Barr virus a significantly higher number of non-carriers of a composite genotype including the IL-1RA allele 2 and the IL-1β -511 allele 2 were found. Several other associations between alleles in this gene cluster and inflammatory and autoimmune diseases have been reported (Hurme et al 1998).

Transforming growth factor-β
Transforming growth factor-β (TGF-β) is a cytokine that is constitutively produced in healthy airways and presumably able to regulate the generation of pro-inflammatory and immunoregulatory cytokines. A C to T base substitution at position −509 within

the promoter region was detected as single-stranded conformational polymorphism and was confirmed by DNA sequencing (Rosenwasser and Borish 1998). Although TGF-β is involved in mediating hepatic fibrogenesis in chronic hepatitis C infection (Nelson et al 1997), the function of this polymorphism has so far not been examined in HCV and in any other infection.

Tumor necrosis factor-α

TNF-α is one of the major proinflammatory cytokines. These proinflammatory properties of TNF-α are also responsible for much of the clinical symptomatology of infections. TNF-α plays for example an important role in the pathogenesis of malaria, wherein it is acting to suppress parasitic growth and to cause clinical symptoms. Fatal cerebral malaria is associated with high levels of circulating of TNF-α. High TNF-α levels are also associated with a poor prognosis in endotoxemia. TNF-α is mainly produced by activated macrophages, but also by T-cells, B-cells, and natural killer cells. TNF-α, especially in combination with interferon-γ, is further associated with the *in vivo* and *in vitro* killing of tumor cells. TNF-α provides a rapid form of host defense against infection, but as indicated above may be fatal in excess. Because TNF-α is involved in the response to a wide range of pathogens, each with its own pathogenic mechanisms, it is likely that this would favor diversity in the genetic elements that control its production. Family studies show that up to 60 % of the variability in TNF production between individuals may be genetically determined (Knight and Kwiatkowski 1999). Genetic control not only works at the level of TNF-α and its receptors, but also at the level of other cytokines and cytokine receptors that control the level of TNF production, such as interleukin-10 and interferon-γ.

The gene for TNF-α lies on chromosome 6 within the polymorphic major histocompatibility complex (MHC) region. Its location between the MHC class I and III loci further indicates a role of TNF-α in the etiology of MHC-linked diseases, particularly those diseases which have an autoimmune or inflammatory component.

Variation in the TNF-α promoter region was, among others, found to be associated with susceptibility to malaria, leishmaniasis, scarring trachoma, and lepromatous leprosy (Knight and Kwiatkowski 1999, McGuire et al 1994). The first polymorphism studied involves the substitution of guanine by adenosine at position −308, which lies in a consensus sequence for binding of the transcription factor AP-2 (Wilson et al 1992). Functional significance of the polymorphism is further corroborated by studies showing that the G to A nucleotide change at −308 indeed enhances transcription factor binding (Abraham and Kroeger 1999). This less common TNF2 allele appears associated with an increased constitutive and inducible transcription rate for TNF-α and is thus associated with higher expression of TNF-α upon stimulation (Wilson and Duff 1995). However, other studies failed to confirm functional significance of this polymorphism (Bidwell et all 1999). In addition, because the TNF response correlates with a number of genetic markers that can all be mapped into a region with strong linkage disequilibrium, it may well be that other loci in the region further regulate the TNF response.

In the Gambia, children homozygous for the TNF2 allele had a relative risk of 7 for fatal or cerebral malaria. Although, as mentioned, the TNF2 allele is in linkage disequilibrium with several neighboring HLA alleles, this association appeared independent of HLA class I and class II variation. Heterozygosity for the TNF2 allele was not a risk factor. Interestingly, the TNF2 allele is maintained at a similar gene frequency (of approximately 0.16) in both West African and Northern European populations. The increased transcription of this gene associated with TNF2 may therefore have some biologic advantage in other inflammatory conditions. Perhaps it has beneficial effects in other important infectious diseases such as measles, leprosy, or tuberculosis, or autoimmune diseases such as systemic lupus erythmatosus. There may also be heterozygous advantage.

Possession of the TNF2 allele was also associated with a poor outcome of meningococcal disease (Nadel et al 1996), which may corroborate the finding that TNF-α plays a central role in the pathophysiology of sepsis. Indeed levels of TNF-α are directly correlated with the severity of meningococcal disease.

Recently a role of TNF-α (and its down-stream inflammatory mediators) has also been suggested to play an important role in a viral infection, i.e. enterovirus 71 infections. Enterovirus 71 is a common cause of hand, foot and mouth disease and of central nervous system infections in children. Large epidemics with a high fatality rate have occurred recently in several Asian countries. The TNF2 allele (-308A), as indicated above probably associated with higher expression of TNF-α upon stimulation, was found in a higher frequency in children experiencing a fatal outcome, and in their parents, than in the general population (Ho 2000).

Other SNPs were identified in the upstream region of the TNF-α gene at positions -857, -851, -238, -376 from the first transcribed nucleotide, and one was found in a non-translated region at position +691 (D'Alfonso and Richardi 1994, Herrmann et al 1998, Knight et al 1999). The A for G substitution at position −376 was extensively characterized by Knight al (1999). Binding experiments indicated that this substitution causes the transcription factor OCT1 to bind to a region of complex protein-DNA interactions and alters gene expression in human monocytes. This OCT-1 binding genotype is found in approximately 5 % of Africans, and is associated with a fourfold-increased susceptibility to cerebral malaria in large case-control studies in West and East African populations. The −238 A allele (or separate genetic factors near the TNF-α gene) appeared to affect different disease phenotypes of malaria in opposite ways (i.e. cerebral malaria and reinfection interval on the one hand and anemia on the other hand) (Knight et al 1999, McGuire et al 1999). These findings may relate to the different pathogenesis of these disease phenotypes. Severely anemic children with *P. falciparum* infection have low TNF-α levels, in contrast to the high levels found in cerebral malaria.

In conclusion, the significance of TNF-α in various infectious and inflammatory disease is illustrated by the finding that variation in the TNF-α promoter region was found to be associated with susceptibility to many infectious and inflammatory conditions, among others, malaria, mucocutaneous leishmaniasis, lepromatous leprosy (but not tuberculoid leprosy), scarring trachoma, meningococcal disease, enterovirus infections, and asthma (Cabrera et al 1995, Moffatt and Cookson 1997, Conway et al 1997, Roy et al 1997). In many infections TNF-α therefore apparently plays a major role in causing tissue damage. Because TNF promoter variants are in linkage disequilibrium with particular HLA class I and II alleles, it is important to consider the possible influence of TNF variation on HLA associations with infectious disease, such as for example reported for HIV (Hill 1992, Steel 1988). Regulation of TNF activity appears also under control of the TNF receptors. Future studies should therefore include the influence of polymorphisms in the receptors.

Interleukin-4 and interleukin-4 receptor-α

IL-4 is a pleiotropic cytokine that acts on the differentiation and expansion of Th2 lymphocytes, the production of IgE by B cells, the expansion and differentiation of mast cells, and the expression of endothelial vascular cell adhesion molecule 1 (VCAM-1), which in turn leads to the recruitment of eosinophilic leukocytes. IL-4 is mitogenic for T cells and enhances the expression of the Fcε receptor on B cells (Dizier et al 1999, Pan and Rothman 1999). The gene is located on chromosome 5 in the gene cluster at 5q31-33 that is linked with the regulation of serum IgE levels and the development of asthma and atopy (Marsh et al 1994). Other genes in this cluster encode for IL-3, IL-5, IL-5, IL-9, IL-13, and granulocyte macrophage colony stimulating factor (GM-CSF). A SNP has been described in the promoter region of the IL-4 gene at position −590 (C-590T). The T allele is associated with increased transcription in an *in vitro* assay and with elevated IgE levels in asthmatic families (Rosenwasser et al 1995). The frequency of the T allele is much higher in the Japanese (0.7) than in the Caucasian population (0.27) (Noguchi et al 1997).

IL- 4 acts on two receptors, the 64 kDa "common" sub-unit (γc) receptor, which is used by IL-2, IL-7, IL-9, and IL-15, and the 140 kDa IL-4 receptor-α (IL-4Rα). IL-4Rα is expressed on the surface of T cells, B cells, mast cells, basophilic leukocytes, and macrophages. The 140 kDa α subunit binds IL-4 and transduces its growth promoting and transcription activating functions. IL-4Rα is also a component of the IL-13 receptor. The IL-4Rα gene is located on the short arm of chromosome 16. Two SNPs have been described, respectively at positions 50 and 1902, which both appear functional in signal transduction and upregulated IgE synthesis (Hershey et al 1997, Mitsuyasu et al 1998).

Interleukin-8

Interleukin-8 (IL-8) is produced by various cell types, such as activated monocytes, macrophages, fibroblasts, and endothelial cells. It is one of the most potent attractants and activators of neutrophilic granulocytes. The role of IL-8 polymorphism has been investigated in respiratory syncytial virus (RSV) infection. RSV-infected epithelial cells produce IL-8, and IL-8 is found in plasma and nasal secretions of children with RSV-induced bronchiolitis. RSV infects nearly all children by the end of their second winter. However only 1 % of the infected children develop disease severe enough to require hospital admission and sometimes forced ventilation. The disease mechanisms leading to this severe form of disease are not completely understood, but likely involve immune-mediated mechanisms. An A to T polymorphism at position −251 bp upstream of the transcription start site of the IL-8 gene, with each allele occurring at approximately similar frequencies, was examined for association with the severity of RSV bronchiolitis. The A allele at this position was shown to be associated with higher IL-8 production in LPS-stimulated whole blood assays. When the frequency transmission of the −251A allele was analyzed using the transmission disequilibrium test (see chapter 4), an increased likelihood of transmission to infants with bronchiolitis was observed (OR = 1.6, *p* = 0.014). In infants with the most severe disease, the likelihood of transmission was increased further (OR = 2.46, *p* = 0.005). This effect was also greater in infants without other known risk factors for severe disease, such as prematurity and congenital heart disease (OR = 3.5, *p* = 0.004). These results suggest that a severe course of RSV bronchiolitis is associated with a genetic variant that increases IL-8 production and that neutrophil influx may contribute to airway obstruction in RSV bronchiolitis (Hull et al 2000). Evidence for a genetic influence on the course of RSV disease also comes from twin studies (see chapter 1), and racial differences in hospitalization rates. American Indian and Alaska Native children have a bronchiolitis-associated hospitalization rate that is approximately twice as high as that from all US infants (62 vs. 34 per 1,000). Evidently a decreased threshold for hospitalization among American Indian and Alaska Native children (Lowther et al 2000) might also explain the latter finding.

Interleukin-10

IL-10 is produced by various cell types, including monocytes, B cells and T cells. IL-10 is associated with differentiation of Th2 cells. IL-10 is an anti-inflammatory cytokine that inhibits T and B cells, including the production of IgE by B cells, and mast cell proliferation. It further induces apoptosis in eosinophils. Promoter polymorphisms have been described at positions −571 (C-571A), -651 (G-651A), and −1082 (G-1082A). Because the -571 position is located between the binding boxes of two groups of transcription factors, the polymorphism is probably functional. Homozygotes and heterozygotes with the A allele at position −571 have high IgE levels in their serum, suggesting a dominant effect of the A allele (Hobbs et al 1998).

Peripheral blood mononuclear cells from homozygotes for the allele characterized by G at position −1082 produce higher levels of IL-10 *in vitro* than those from carriers of the A allele. Because there were indications that IL-10 has a regulatory role in Epstein-Barr virus (EBV)-induced infections, Helminen et al (1999) determined the frequencies of the G and A alleles at position −1082 in healthy EBV-seronegative and seropositive adults, and in patients hospitalized because of a severe EBV infection. The G allele, associated with a high IL-10-producing capacity, was found more frequently in the healthy subjects. Thus the data suggest that high IL-10 levels protect individuals against EBV infection, and that low IL-10 producing capability makes individuals more susceptible to a severe EBV infection. How the increased production of IL-10 protects against EBV infection is unclear. In contrast to this model, Westendorp et al (1997) have demonstrated that a genetically determined high capacity to produce IL-10 increases mortality in meningococcal disease, illustrating varying defense mechanisms against different infections. Finally, the A allele at position −651 appeared associated with longer intervals to malaria reinfection. This result may illustrate the protective effect of IL-10 against malaria as demonstrated in mice (Luty 2001).

Interleukin-12 and interleukin-12 receptor

An important and potent cytokine which promotes the development of Th1 cells and IFN-γ production (see above) is IL-12. IL-12 is produced by dendritic cells, monocytes, macrophages, and epithelial cells, particularly in response to intracellular bacteria and parasites, or through ligation of CD40, one of the co-stimulatory molecules on antigen-presenting cells (Hsieh et al 1993, Magram et al 1996, Manetti et al 1993, Trinchieri 1995). IL-12 is a heterodimeric 70 kDa cytokine, composed of p35 and p40 subunits. It binds to the high-affinity β1β2 heterodimeric IL-12 receptor (IL-12R). IL-12 activates natural killer cells, stimulates naive Th0 cells to develop into Th1 cells, stimulates the production of IFN-γ by Th1 cells, and inhibits the production of IL-3, IL-5, and Il-10 by Th2 cells. IL-12 together with IFN-γ appears to be essential for the development of protective cell-mediated immunity to tuberculous mycobacteria and other intracellular bacteria. Individuals that lack IL-12Rβ1 chain expression have increased susceptibility to severe atypical mycobacterial disease and *Salmonella* infections. Their cells are deficient in IL-12R signaling and hence IFN-γ production. The gene encoding IL-12Rβ1 is located on chromosome 19p13.1. It encodes a 662 amino acid protein that is expressed by natural killer cells and activated T lymphocytes. Sequence analysis revealed several homozygous mutations that resulted in premature stop codons in the extracellular domain of IL-12Rβ1. Also a child suffering from Bacillus Calmette-Guérin (BCG) infection and with a recessive mutation in the gene encoding the p40 subunit of IL-12 has been reported (Altare et al 1998). The frequency of mutant alleles in different populations is unknown. Thus these patients emphasize the significance of the interferon-γ/IL-12 pathway in resistance to intracellular bacteria, while resistance to other viral, bacterial or parasitic pathogens appeared largely unaffected (de Jong et al 1998). It is at the same time however also a weak point in the defense to intracellular bacteria because there is no second line of defense. Experimental therapies consist of IFN-γ suppletion and stem cell transfer. In addition to these seldom and severe mutations, several polymorphisms in the IL-12 gene have been described, but their functional influence is so far unknown.

Interferon-γ and interferon-γ receptor 1

Inactivating mutations in the interferon-γ receptor 1 (IFN-γR1) gene (located on 6q24.1) have been identified that result in partial or complete deficiencies in IFN-γR expression or function. One of them is a dominant negative mutation caused by a tetranucleotide deletion at position 818 (818Δ4), resulting in defective activation of the Stat4 signalling pathway in macrophages. The mutation is dominant negative, because recycling of the defective receptor is disturbed, which leads to defective clearance and overexpression of the defective receptor on the cell membrane. The phenotype of IFN-γR1-deficient patients is usually more severe than those of IL-12Rβ1-deficient patients, which is likely explained by residual type 1 immunity in these IL-12Rβ1-deficient patients (Verhagen et al 2000). Many IFN-γR1-deficient patients die before the age of 10 years. These [T1]mutations further confirm the significance of the interferon-γ/IL-12 pathway in resistance to intracellular pathogens such as mycobacteria. Patients with such mutations are highly susceptible to *Salmonella* infections and idiopathic disseminating mycobacterial infections due to non-tuberculous species, such *as Mycobacterium avium, M. fortuitum, M. chelonei, M. smegmatis,* and BCG. Such infections may run a fatal course in these patients. However, it is unknown whether these individuals also have an increased susceptibility to tuberculosis and leprosy. Interestingly, parents of these patients are often (up to 30 %) related to each other, and often more family members are affected (Altare et al 1998, Jouanguy et al 1997, Levin et al 1995, Newport et al 1996, Van Dissel 2000). The occurrence of severe infections associated with respiratory syncytial virus (RSV) and parainfluenza virus type 3 (PI3) in a few patients with IFN-γR1-deficiency further indicates an important role of the IFN-γ pathway in limiting the pathology associated with these viral infections as well (Lammas et al 2000). This might be due to enhanced viral replication, enhanced Th2-associated pathology, or both.

Table 5. Cytokine gene polymorphisms associated with infectious disease susceptibility

Gene or locus	Location	Polymorphism	Frequency	Population	Study	Phenotype	Ref.
TNF-α	-308	G to A	0.16	Gambia	retrospective case control	increased risk on fatal or cerebral malaria (OR = 7)	McGuire et al 1994
TNF-α	-308	G to A			case control	increased risk on fatal meningococcal disease (OR = 2.5)	Nadel et al 1996
TNF-α	308	G to A	0.16	Gambia	case control	increased risk on ocular chlamydial infection	Conway et al 1997
TNF-α	-308	G to A			case control	increased risk on lepromatous leprosy (OR = 3)	Roy et al 1997
TNF-α	-308	G to A		Venezuela	case control	increased risk on mucocutaneous leishmaniasis (OR = 3.5)	Cabrera et al 1995
TNF-α	-308	G to A		Taiwan	retrospective case control	increased risk on fatal outcome enterovirus 71 infection	Ho 2000
TNF-α	376	A to G	0.05	Gambia, Kenya	retrospective case-control	increased risk on cerebral malaria (OR = 4.3); altered gene expression	Knight et al 1999

continue table 5. Cytokine gene polymorphisms associated with infectious disease susceptibility

Gene or locus	Location	Polymorphism	Frequency	Population	Study	Phenotype	Ref.
IL-8	-251	T to A	0.5	Caucasian	family-based	A allele associated with increased severity of RSV bronchiolitis (OR = 3.5)	Hull et al 2000
IL-10	-1082	A to G			case-control	G allele is associated with high IL-10 and protection against EBV infection	Helminen et al 1999
IL-1β/IL-1RA	+3953				case-control	Mantoux response and disease expression in tuberculosis	Wilkinson et al 1999

References

Abraham, L.J., Kroeger, K.M. (1999). Impact of the −308 TNF promoter polymorphism on the transcriptional regulation of the TNF gene: relevance to disease. J. Leukocyte Biol. 6, 562-566.

Altare, F., Jouanguy, E., Lamhamedi-Cherradi, S., Fondaneche, M.C., Fizame, C., Ribierre, F., Merlin, G., Dembic, Z., Schreiber, R., Lisowska-Grospierre, B., Fischer, A., Seboun, E., Casanova, J.L. (1998). A causative relationship between mutant IFNγR1 alleles and impaired cellular response to IFNγ in a compound heterozygous child. Am. J. Hum. Genet. 62, 723-726.

Bidwell, J., Keen, L., Gallagher, et al. (1999). Cytokine gene polymorphism in human disease: on-line databases. Genes and Immunity 1, 3-19.

Bix, M., Wang, Z.E., Thiel, B., Schork, N.J., Locksley, R.M. (1998). Genetic regulation of commitment to interleukin 4 production by a CD4(+) T cell-intrinsic mechanism. J. Exp. Med. 188, 2289-2299.

Blackwell, J.M., Black, G.F., Peacock, C.S., Miller, E.N., Sibthorpe, D., Gnananandha, D., Shaw, J.J., Silveira, F., Lins-Lainson, Z., Ramos, F., Collins, A., Shaw, M.A. (1997). Immunogenetics of leishmanial and mycobacterial infections: the Belem Familiy Study. Philos. Trans R. Soc. Lond. B. Biol. Sci. 352, 1331-1345.

Cabrera, M., Shaw, M.A., Sharples, C., Williams, H., Castes, M., Convit, J., Blackwell, J.M. (1995). Polymorphism in tumor necrosis factor genes associated with mucocutaneous leishmaniasis. J. Exp. Med. 182, 1259-1264.

Conway, D.J., Holland, M.J., Bailey, R.L., Campbell, A.E., Mahdi, O.S., Jennings, R., Mbena, E., Mabey, D.C. (1997). Scarring trachoma is associated with polymorphism in the tumor necrosis factor α (TNF-α) gene promoter and with elevated TNF-α levels in tear fluid. Infect. Immun. 65, 1003-1006.

Cork, M.J., Crane, A.M., Duff, G.W. (1996). Genetic control of cytokines. Cytokine gene polymorphisms in alopecia areata. Dermatol. Clin. 14, 671-678.

D'Alfonso, S., Richiardi, P.M. (1994). A polymorphic variation in a putative regulation box of the TNFA promoter region. Immunogenetics 39, 150-154.

Daser, A., Mitchison, H., Mitchison, A., Müller, B. 1996). Non-classical-MHC genetics of immunological disease in man and mouse. The key role of pro-inflammatory cytokine genes. Cytokine 8, 593-597.

De Jong, R., Altare, F., Haagen, I.-A., Elferink, D.G., de Boer, T., van Breda Vriesman, P.J.C., Kabel, P.J., Draaisma, J.M.T., van Dissel, J.T., Kroon, F.P., Casanova, J.-L., Ottenhoff, T.H.M. (1998). Severe mycobacterial and *Salmonella* infections in interleukin-12 receptor-deficient patients. Science 280, 1435-1438.

di Giovine, F.S., Duff, G.W. (1990). Interleukin-1: The first interleukin. Immunol. Today 11, 13-20.

di Giovine, F.S., Takhsh, E., Blakemore, A.I., Duff, G.W. (1992). Single base polymorphism at −511 in the human interleukin-1 β gene (IL1 β). Hum. Mol. Genet. 1, 450.

Dizier, M., Sandford, A., Walley, A., Philippi, A., Cookson, W., Demenais, F. (1999). Indication of linkage of serum IgE levels to the interleukin-4 gene and exclusion of the contribution of the (-590 C to T) interleukin-4 promoter polymorphism to IgE variation. Genet. Epidemiol. 16, 84-94.

Güler, M.L., Gorham, J.D., Hsieh, C.-S., Mackey, A.J., Steen, R.G., Dietrich, W.F., Murphy, K.M. (1996). Genetic susceptibility to *Leishmania*: IL-12 responsiveness in Th1 cell development. Science 271, 984-987.

Helminen, M., Lahdenpohja, N., Hurme, M. (1999). Polymorphism in the interleukin-10 gene is associated with susceptibility to Epstein-Barr virus infection. J. Infect. Dis. 180, 496-499.

Hermann, S.-M., Ricard, S., Nicaud, V., Mallet, C., Arveiler, D., Evans, A., Ruidavets, J.-B., Luc, G., Bara, L., Parra, H.-J. Poirier, O., Cambien, F. (1998). Polymorphisms of the tumour necrosis factor-α gene, coronary heart disease and obesity. Eur. J. Clin. Invest. 28, 59-66.

Hershey, G.K.K., Friedrich, M.F., Esswin, L.A., Thomas, M.L., Chatila, T.A. (1997). The association of atopy with gain-of-function mutation in the α subunit of the interleukin-4 receptor. N. Engl. J. Med. 11, 1720-1725.

Hill, A.V.S. (1996). Genetic susceptibility to malaria and other infectious diseases: from the MHC to the whole genome. Parasitology 112 (Suppl.), S75-S84.

Ho, M.-S. (2000). Enterovirus 71 infection: a great mimicry of poliomyelitis. Abstracts of the meeting on Progress in Polio Eradication: Vaccine strategies for the end game. 28 – 30 June 2000, Institut Pasteur, Paris.

Hobbs, K., Negri, J., Klinnerst, M., Rosenwasser, L.J., Borish, L. (1998). Interleukin-10 and transforming growth factor β promoter polymorphisms in allergies and asthma. Am. J. Respir. Care Med. 158, 1958-1962.

Hsieh, C.S., Macatonia, S.E., Tripp, C.S., Wolf, S.F., O'Garra, A., Murphy, K.M. (1993). Development of Th1 CD4+ T-cells through IL-12 produced by *Listeria*-induced macrophages. Science 260, 547-549.

Hull, J., Thomson, A., Kwiatkowski, D. (2000). Genetic variants of interleukin-8 influence disease severity in respiratory syncytial virus bronchiolitis. Abstract presented at the Meeting of the American Thoracic Society.

Hurme, M., Lahdenpohja, N., Santtila. (1998). Gene polymorphisms of interleukins 1 and 10 in infectious and autoimmune diseases. Ann. Med. 30, 469-473.

Jouanguy, E., Lamhamedi-Cherradi, S., Altare, F., Fondaneche, M.C., Tuerlinckx, D., Blanche, S., Emile, J.F., Gaillard, J.L., Schreiber, R., Levin, M., Fischer, A., Hivroz, C., Casanova, J.L. (1997). Partial interferon-gamma receptor 1 deficiency in a child with tuberculoid bacillus Calmette-Guérin infection and a sibling with clinical tuberculosis. J. Clin. Investig. 100, 2658-2664.

Kay, A.B., Ying, S., Varney, M., Gaga, S.R., Durham, R., Moqbel, R., Wardlaw, A.J., Hamid, Q. (1991). Messenger RNA expression of the cytokine gene cluster, interleukin-3 (IL-3), IL-4, IL-5, and granulocyte macrophage colony-stimulating factor in allergen-induced late-phase cutaneous reactions in atopic subjects. J. Exp. Med. 173, 775-778.

Knight, J.C., Kwiatkowski, D. (1999). Inherited variability of tumor necrosis factor production and susceptibility to infectious disease. Proc. Ass. Am. Physicians 111, 290-298.

Knight, J.C., Udalova, I., Hill, A.V.S., Greenwood, B.M., Peshu, N., Marsh, K., Kwiatkowski, D. (1999). A polymorphism that affects OCT-1 binding to the TNF promoter region is associated with severe malaria. Nature Genet. 22, 145-150.

Lammas, D.A, Casanova, J.L., Kumararatne, D.S. (2000). Clinical consequences of defects in the IL-12-dependent interferon-γ (IFN-γ) pathway. Clin. Exp. Immunol. 121, 417-425.

Levin, M., Newport, M.J., D'Souza, S., Kalabalikis, P., Brown, I.N., Lenicker, H.M., Agius, P.V., Davies, E.G., Thrasher, A., Klein, N., et al. (1995). Familial disseminated atypical mycobacterial infection in childhood: a human mycobacterial susceptibility gene? Lancet 345, 79-83.

Louis, J.A., Conceicao-Silva, F., Himmelrich, H., Tacchini-Cottier, F., Launois, P. (1998). Anti-Leishmania effector functions of CD4+ Th1 cells and early events instructing Th2 cell development and susceptibility to *Leishmania major* in BALB/c mice. Adv. Exp. Med. Biol. 452, 53-60.

Lowther, S.A., Shay, D.K., Holman, R.C., Clarke, M.J., Kaufman, S.F., Anderson, L.J. (2000). Bronchiolitis-associated hospitalizations among American Indian and Alaska Native children. Pediatr. Infect. Dis. J. 19, 11-17.

Luty, A.J.F. (2001). Personal communication.

Magram, J., Connaughton, S.E., Warrier, D.M., Carvajal, D.M., Wu, C.Y., Ferrante, J., Steart, C., Sarmiento, U., Faherty, D.A., Gately, M.K. (1996). IL-12-deficient mice are defective in IFNγ production and type 1 cytokine responses. Immunity 4, 471.

Manetti, R., Parronchi, P., Piccinni, M.P., Maggi, E., Trinchieri, G., Romagnani, S. (1993). Natural killer cell stimulatory factor (interleukin-12 [IL-12]) induces T-helper type-1 (Th1)-specific immune responses and inhibits the development of IL-4-producing Th cells. J. Exp. Med. 177, 1199-1204.

Marsh, D.G., Neely, J.D., Braezeale, D.R., Gosh, B., Freidhoff, L.R., Erlich-Kautzkye, E., Schou, C., Krishnaswamy, G., Beaty, T.H. (1994). Linkage analysis of IL-4 and other chromosome 5q31.1 markers and total serum immunoglobulin E concentrations. Science 264, 1152-1156.

McDowell, T.L., Symons, J.A., Ploski, R., Forre, O., Duff, G.W. (1995). A genetic association between juvenile rheumatoid arthritis and a novel interleukin-1 alpha polymorphism. Arthr. Rheum. 38, 221-228.

McGuire, W., Hill, A.V., Allsopp, C.E., Greenwood, B.M., Kwiatkowski, D. (1994). Variation in the TNF-alpha promoter region associated with susceptibility to cerebral malaria. Nature 371, 508-510.

McGuire, W., Knight, J.C., Hill, A.V., Allsopp, C.E., Greenwood, B.M., Kwiatkowski, D. (1999). Severe malaria anemia and cerebral malaria are associated with different tumor necrosis factor promoter alleles. J. Infect. Dis. 179, 287-290.

Messer, G., Spengler, U., Jung, M.C., Honold, G., Blomer, K., Pape, G.R., Riethmuller, G., Weiss, E.H. (1991). Polymorphic structure of the tumor necrosis factor (TNF) locus: an *NcoI* polymorphism in the first intron of the human TNF-β gene correlates with a variant amino acid position in position 26 and a reduced level of TNF-β production. J. Exp. Med. 173, 209-219.

Mitsuyasy, H., Izuhara, K., Mao, X.-G., Arinobu, Y., Enomoto, T., Kawai, M., Sasaki, S., Dake, Y., Hamaskai, N., Shirakawa, T., Hopkin, J.M. (1998). Ile50Val variant of IL-4Rα upregulates IgE synthesis and associates with atopic asthma. Nature Genetics 19, 119-120.

Moffat, M.F., Cookson, W.O.C.M. (1997). Tumour necrosis factor haplotypes and asthma. Hum. Molec. Genet. 6, 551-554.

Nadel, S., Newport, M.J., Levin, M. (1996). Variation in the tumor necrosis factor-alpha gene promoter region may be associated with death from meningococcal disease. J. Infect. Dis. 174, 878-880.

Nelson, D.R., Gonzalez-Peralta, R.P., Qian, K., Xu, Y., Marousis, C.G., Davis, G.L., Lau, J.Y. (1997). Transforming growth factor-β 1 in chronic hepatitis C. J. Viral Hepat. 4, 29-35.

Newport, M.J., Huxley, C.M., Huston, S., Hawrylowicz, C.M., Oostra, B.A., Williamson, R., Levin, M. (1996). A mutation in the interferon-γ-receptor gene and susceptibiity to mycobacterial infection. N. Engl. J. Med. 335, 1941-1949.

Nicklin, M.J.H., Weith, A., Duff, G.W. (1994). A physical map of the region encompassing the human interleukin-1α, interleukin-1β, and interleukin-1 receptor antagonist genes. Genomics 19, 382-384.

Noguchi, E., Shibasaki, M., Arinami, T., et al (1997). Association of asthma and the interleukin-4 promoter gene in Japanese. Clin. Exp. Allergy 28, 449-453.

Pan, P.-Y., Rothman, P. (1999). IL-4 receptor mutations. Curr. Opin. Immunol. 11, 615-620.

Pociot, F., Molvig, J., Wogensen, L., Worsaae, H., Nerup, J., (1992). A TaqI polymorphism in the human interleukin-1 β (IL-1 β) gene correlates with IL-1 beta secretion in vitro. Eur. J. Clin. Invest. 22, 396-402.

Romagnani, S. (1996). Understanding the role of Th1/Th2 cells in infection. Trends Microbiol. 4, 470-473.

Rosenwasser, L.J., Klemm, D.J., Dreback, J.K., Inamura, H., Mascall, J.J., Klinnert, M., Borish, L. (1995). Promoter polymorphisms in the chromosome 5 gene cluster in asthma and atopy. Clin. Exp. Allergy 25, Suppl.2, 74-78.

Rosenwasser, L.J., Borish, L. (1998). Promoter polymorphisms predisposing to the development of asthma and atopy. Clin. Exp. Allergy 28, Suppl. 5, 13-15.

Roy, S., McGuire, W., Mascie-Taylor, C.G., Saha, B., Hazra, S.K., Hill, A.V., Kwiatkoski, D. (1997). Tumor necrosis factor promoter polymorphism and susceptibility to lepromatous leprosy. J. Infect. Dis. 176, 530-532.

Steel, C.M., Ludlam, C.A., Beatson, D., Peutherer, J.F., Cuthbert, R.J., Simmonds, P., Morrison, H., Jones, M. (1988). HLA haplotype A1 B8 DR3 as a risk factor for HIV-related disease. Lancet 1, 1185-1188.

Symons, J.A., Young, P.R., Duff, G.W. (1995). Soluble type II interleukin-1 (IL-1) receptor binds and blocks processing of IL-1β precursor and loses affinity for IL-1 receptor antagonist. Proc. Natl. Acad. Sci. USA 92, 1714-1718.

Tarlow, J.K., Blakemore, A.I., Lennard, A., Solari, R., Hughes, H.N., Steinkasserer, A., Duff, G.W. (1993). Polymorphism in human IL-1 receptor antagonist gene intron 2 is caused by variable numbers of an 86-bp tandem repeat. Hum. Genet. 91, 403-404.

Trinchieri, G. (1995). Interleukin-12: a proinflammatory cytokine with immuno-regulatory functions that bridge innate resistance and antigen-specific adaptive immunity. Annu. Rev. Immunol. 13, 251.

Van Dissel, J. (2000). Personal communication.

Verhagen, C.E., de Boer, T., Smits, H.H., Verreck, F.A., Wierenga, E.A., Kurimot, M., Lammas, D.A., Kumararatne, D.S., Sanal, O., Kroon, F.P., van Dissel, J.T., Sinigaglia, F., Ottenhoff, T.H. (2000). Residual type 1 immunity in patients genetically deficient for interleukin 12 receptor β1 (IL-12Rβ1): evidence for an IL-12Rβ1-independent pathway of IL-12 responsiveness in human T cells. J. Exp. Med. 192, 517-528.

Westendorp, R.G.J., Langermans, J.A.M., Huizinga, T.W.J., et al (1997). Genetic influence on cytokine production and fatal meningococcal disease. Lancet 349, 170-173.

Wilkinson, R.J., Patel, P., Llewelyn, M., Hirsch, C.S., Pasvol, G., Snounou, G., Davidson, R.N., Toossi, Z. (1999). Influence of polymorphism in the genes for the interleukin (IL)-1 receptor antagonist and IL-1β on tuberculosis. J. Exp. Med. 189, 1863-1874.

Wilson, A.G., de Giovine, F.S., Blakemore, A.I., Duff, G.W. (1992). Single base polymorphism in the human tumour necrosis factor α (TNF α) gene detectable by NcoI restriction of PCR product. Hum. Mol. Genet. 1, 353.

Wilson, A.G., Duff, G.W. (1995). Genetic traits in common diseases. Br. Med. J. 310, 1482-1483.

Chapter 11. Genes influencing lymphocyte function

Inherited immunodeficiency diseases due to recessive gene defects
Several seldom-occurring inherited immunodeficiency disorders manifest themselves by recurrent or very severe infections during childhood. Because these infections were often life threatening, most of these disorders have been described only since the advent of the antibiotics in the 1950's. Almost 100 genetic defects of the immune system have now been identified. Many of these inherited immunodeficiencies are recessive and X-linked (Kinnon and Levinsky 1990, WHO 1992). Remarkably, approximately 30 % of the X-linked genetic defects originate de novo during meiosis and are thus absent in the somatic cells of the mother (Kuijpers and Vossen 2000).
The gene defects may lead to defects in B- and T-lymphocyte development, defects in surface molecules important for the function of these cells, defects in antigen recognition, defects in the communication of these cells with their environment, defects in adhesion, and motility, and defects in effector functions such as killing (Table 6). The relation between genetic defect and phenotype is not always clear. However, some of the disorders illustrate the uniqueness of the immunological requirements for protection against a particular (group of) pathogen(s). Usually mutations that result in truncated proteins or in a total lack of expression of proteins have a more severe phenotype than point mutations. Disease modifying genes may influence the penetrance of disease. For example, the phenotype of X-linked agammaglobulinemia is extremely variable (varying from granulomatous disease to infectious disease) as a result of such modifying genes (Kornfield et al 1997).

When we study these disorders as "experiments of nature", we can learn much of the specific functions of the affected molecules and immune cells. Indeed, often these disturbances were the first indication of the function of immune cells. For example, diseases of antibody production, such as X-linked agammaglobulinemia (XLA), primarily lead to an inability to phagocytose encapsulated bacteria, which are usually opsonized by antibody and complement, and also to a somewhat enhanced susceptibility to viral infections (Bruton 1952). The prevalence of the disease in Caucasians is approximately 1:200,000. Patients with XLA have a defect in the gene encoding Bruton's tyrosine kinase (*btk*) on chromosome Xq21.3-22, which halts the maturation of pre-B cells. This results in a lack of B cells and gammaglobulins in peripheral blood. The *btk* gene is a member of the *src* family of protein-tyrosine kinases. Likely *btk* is involved in the coupling of the pre-B-cell receptor, consisting of heavy chains, surrogate light chains, and Ig-α and Ig-β, to nuclear events that leads to pre-B cell growth and differentiation (Conley 1992, Vetrie et al 1993). A causal therapy is not available, but these patients benefit from treatment with purified IgG and early antibiotic treatment. Because the disease is X-linked, the disease manifests itself almost only in male patients, who become ill before their 6th year. Female carriers of a mutant *btk* gene can be easily identified. Normally female cells randomly inactivate one of their two X-chromosomes. However, in female carriers of mutant *btk* genes, all B cells have the same active X chromosome, thus allowing the detection of carriership.

Rare disorders have been described involving defects in the ability to produce one or more immunoglobulin isotypes, owing to deletions of heavy-chain constant-region genes, or failure to produce κ light-chain genes (Preudhomme and Hanson 1990, Volonakis et al 1992). Patients with X-linked hyper IgM syndrome make only limited IgM antibody responses against T-cell dependent antigens, but they make isotypes other than IgM and IgD only in trace amounts. The defect is in the CD40 ligand (GP39) gene on the X chromosome, which gene is normally expressed on activated T helper cells. The CD40 ligand is a 39 kDa protein, which binds to the CD40 receptor on B cells. The CD40/CD40 ligand binding is important in the T-cell dependent activation of B-cell proliferation and for signaling B cells to switch immunoglobulin isotype from IgM to other isotypes (Allen et al 1993). Because IgM production is undisturbed, these patients have normal or high IgM levels, and undetectable or low levels of IgG, IgA, and IgE.

IgA is the most predominant immunoglobulin isotype in body secretions. Selective IgA deficiency is the most common form of inherited immunodeficiency and is often associated with chronic respiratory and gastrointestinal tract infections, especially with the parasite *Giardia lamblia*. Approximately 1 in 600 individuals in the western world are affected. Great variability in the prevalence is found in different ethnic groups, and a markedly lower frequency has been found in mongoloid populations (Hammarstrom and Smith 1999). Familial clustering and a predominant inheritance pattern further suggest a strong genetic predisposition. IgA deficiency is characterized by a defect of terminal lymphocyte differentiation. The pathogenesis and pattern of inheritance appear complex and involve a particular MHC haplotype, the production of specific autoantibodies, and parental penetrance differences. The recurrence risk of IgA deficiency appears to depend on the sex of parents transmitting the defect. Affected mothers are more likely to produce offspring with IgA deficiency than affected fathers are. This maternal effect is likely produced by the production of anti-IgA antibodies (Vorechovski et al 1999). In approximately two-thirds of the cases, the disease does not lead to an increased occurrence of infections, whereas the remaining patients suffer from bacterial upper and lower respiratory infections (Hammarstrom and Smith 1999). Oxelius et al (1981) reported that IgG2 deficiency in combination with IgA deficiency is critical in determining whether IgA deficient persons have illnesses, such as frequent infections, autoimmune disorders, atopy, and malabsorption syndrome.

The utmost significance of T cells in antimicrobial immunity is illustrated by the very severe and life threatening infections involving almost all pathogens in patients with T-cell defects. Both T cell-mediated immune responses and T-cell-dependent antibody responses are impaired in patients with so called severe combined immune deficiency (SCID). SCID may be based on several genetic defects. A mutation in the γ chain of the IL-2 receptor (which also functions as receptor for IL-4 and IL-7) is the cause of X-linked SCID. In these patients T cells do not develop, indicating the important role of the γ chain of the IL-2 receptor in T cell development (Noguchi et al 1993). Because mice and humans that lack IL-2 have normal T-cell development, this role of the γ chain of the IL-2 receptor must be unrelated to IL-2 binding or IL-2 responses. In unaffected carriers, naive IgM+ cells have inactivated the defective X-chromosome more frequently than the normal one. However, mature memory B cells switched to other isotypes have almost all inactivated the defective X chromosome. This finding indicates that B cell development is partly dependent, but B cell maturation completely dependent on the γ chain of the IL-2 receptor and one or more of the cytokines binding to it.

Severe immunodeficiency is also seen in patients that do not produce IL-2 upon receptor ligation. These IL-2 negative patients may have several defects, among others in the production of other cytokines. However, the patients have increased numbers of phenotypically normal T cells (DiSanto et al 1990, Weinberg and Parker 1990).

SCID is also caused by defects in the recombination activating genes *RAG1* and *RAG2*. These genes take part in the DNA rearrangement of the gene-segments called V, J, and also D at some loci, that together code the variable portion of the antigen receptors on B and T cells. *RAG1* and *RAG2* are involved in the recognition and cleavage of DNA at the border between a conserved recombination signal sequence and the flanking coding DNA segment.

The SCID phenotype is also caused by defects in the genes encoding the enzymes adenosine deaminase (ADA) and purine nucleotide phosphorylase (PNP) (Hirschhorn 1990). These enzymes are involved in nucleotide degradation. Their defects result in the accumulation of toxic nucleotide metabolites that particularly affect developing T cells and to a somewhat lesser degree also B cells. Patients with ADA deficiency may be treated with the enzyme (in the form of irradiated human erythrocytes or as purified bovine enzyme), with bone marrow transplantation, or with gene therapy.

A defect in the signaling pathway in T cells leads to an inherited syndrome known as X-linked lymphoproliferative (XLP) disease. Patients with XLP have mutations in the gene coding for a T cell-specific SLAM (signalling lymphocyte activation molecule)-associated protein (SAP). SAP acts by inhibiting some of the signaling components in NK and T cells. Patients with XLP thus lack an important inhibitor of T cell hyperreactivity. These patients have an abnormal response to the normally benign Epstein-Barr virus, resulting in fatal infectious mononucleosis, hypogammaglobulinemia, virus-associated hemophagocytic syndrome, and malignant lymphoma (Honda et al 2000). Prenatal diagnostics and detection of carriership is possible by RFLP analysis.

In patients with DiGeorge syndrome the thymic epithelium fails to develop normally. As a result, T cells lack the proper environment to mature. The genetic defect has been mapped to chromosome 2q11. These patients have very few T cells and lack T-cell dependent antibody production. These patients have a general increased susceptibility to infectious disease.

Finally, several gene defects have been described that lead to defects in the T cell receptor/CD3 complex and defective signal transduction in T cells (Arnaiz-Villena et al 1992, Chatila et al 1989, Chatila et al 1990, Janeway and Travers 1994, Soudais et al 1993). For example, defects in CD3ε and CD3γ chain expression on T cells result in deficient T cell activation and differentiation.

Inherited immunodeficiency diseases in animals
Several seldom occurring natural immunodeficiency disorders have been described in animals. For example, a recessive autosomal SCID disorder occurs in Arab horses. Horses with SCID have a defective thymocyte differentiation, resulting in a lack of functional T and B cells. Affected foals often acquire an adenoviral pneumonia and die early. The genetic defect has been mapped using linkage analysis and has subse-

Table 6. Inherited immunodeficiency diseases

Syndrome	Gene defect	Immune defect	Disease phenotype	Ref.
DiGeorge syndrome	2q11	thymic aplasia, lack of T cells	increased susceptibility to infections	
Severe combined immune deficiency	*ADA*	lack of T cells	increased susceptibility to infections	Hirschhorn 1990
Severe combined immune deficiency	*PNP*	lack of T cells	increased susceptibility to infections	
Severe combined immune deficiency	*RAG-1* or *RAG-2*	defective rearrangement of TcRα/β; lack of T and B cells		Kuijpers and Vossen 2000
Severe combined immune deficiency	CD3 (γ/δ/ε)	defective expression of TcR		Kuijpers and Vossen 2000
Severe combined immune deficiency	X-linked scid, IL-2Rγ chain deficiency	lack of T cells	increased susceptibility to infections	Noguchi et al 1993
Severe combined immune deficiency	*JAK3*, autosomal recessive	disturbed signalling in T cells	increased susceptibility to infections	
Severe combined immune deficiency	autosomal scid DNA repair defect	lack of T or B cells	increased susceptibility to infections	
Wiskott-Aldrich syndrome	X-linked, *WASP* gene	lack of anti-poly-saccharide antibody responses, defective mobility immune cells	increased susceptibility to encapsulated extracellular bacteria	

continue table 6. Inherited immunodeficiency diseases

Syndrome	Gene defect	Immune defect	Disease phenotype	Ref.
Common variable immunodeficiency	MHC-linked	defective antibody production	increased susceptibility to extracelular bacteria	Volanakis et al 1992
Selective IgA deficiency	MHC-linked	no IgA production	increased susceptibility to respiratory infections	
X-linked agamma-globulinemia	*btk* tyrosine kinase	lack of B cells	increased susceptibility to infections	Bruton 1952
X-linked hyper IgM syndrome	CD40 ligand	no isotype switching	increased susceptibility to encapsulated extracellular bacteria	Allen et al 1993
Selective Ig deficiencies	Ig constant region gene deletions	defect of one or more Ig isotypes	various clinical manifestations	Preudhomme and Hanson 1990
Natural killer cell defect		loss of NK cell function	increased severity of herpesvirus infections	Biron et al 1989
X-linked lymphoproliferative disease	*SAP*	lack of T cell inhibitor activity	abnormal response to Epstein-Barr virus	Honda et al 2000
Hyper IgE syndrome	unknown (chromosome 4q)	unknown	increased frequency *S. aureus* and group A streptococci	Grimbacher et al, 1999 Holland 2000

Adapted from Janeway and Travers (1994).

quently been identified as a 5 bp deletion in the gene encoding the catalytic subunit of the DNA-dependent protein kinase (DNA-PKcs), which is essential for V(D)J gene recombination and DNA repair. The frequency of the affected allele is approximately 4 %. A genetic test is available for identification of affected horses, and for detection of carriers to prevent their participation in breeding (Don-van 't Slot and van der Kolk 2000).

Vitamin D receptor

A role of vitamin D in infectious disease resistance has been suggested previously, because people with vitamin D deficiency appeared more susceptible to tuberculosis and because vitamin D treatment appeared beneficial in tuberculosis patients (Davies 1985). Vitamin D is not only involved in the regulation of calcium metabolism, but the hormone has also immunoregulatory functions. Its active metabolite, 1,25 dihydroxyvitamin D_3 (1,25D_3) binds to the vitamin D receptor, which is present on monocytes and activated B and T cells. As a result, it stimulates monocytes and cell-mediated immunity, and it suppresses lymphocyte proliferation and cytokine synthesis (Rook et al 1986, Tsoukas et al 1984, Provvedini et al 1983). Furthermore it is one of the few mediators that impair the growth of *M. tuberculosis* in macrophages (Rook 1988). 1,25 D_3 also inhibits, in addition to other immunomodulatory effects, the production of IL-12 by activated macrophages and dendritic cells. Because IL-12 is pivotal in the development of Th1 cells, this may have strong effects on the Th1/Th2 balance during inflammatory responses. 1,25 D_3 inhibits IL-12 mRNA expression by acting at the transcriptional level (D'Ambrosio et al 1998).

In several populations the vitamin D receptor shows single-base change polymorphisms at its carboxyterminal site, which are associated with the risk of developing osteoporosis (Eisman 1996, Morrison et al 1994). Persons homozygous for the t allele have increased mRNA levels *in vitro*, and have the highest levels of circulating 1,25D_3 (Howard et al 1995, Morrison et al 1994). These persons are most at risk of osteoporosis, but they appear protected against tuberculosis and persistent hepatitis B infection (Bellamy et al 1999). The finding supports the notion that vitamin D may be immunomodulating and may have beneficial effects during tuberculosis.

In Bengali leprosy patients from Calcutta the vitamin D receptor type further appeared associated with the course of leprosy. The tt genotype was associated with tuberculoid leprosy, and the TT genotype was associated with lepromatous leprosy (Roy et al 1999). These findings further underline the immunoregulatory functions of vitamin D, because lepromatous leprosy, or the multibacillary form of the disease, has been associated with a Th2 response, while tuberculoid leprosy, or the resistant form of the disease, has been associated with a Th1 response (Yamamura et al 1991).

References

Allen, R.C., Armintage, R.J., Conley, M.E., Rosenblatt, H., Jenkins, N.A., Copeland, N.G., Bedell, M.A., Edelhoff, S., Disteche, C.M., Simoneaux, D.K., Fanslow, W.C., Belmont, J., Spriggs, M.K. (1993). CD40 ligand gene defects responsible for X-linked hyper IgM syndrome. Science 259, 990-993.

Arnaiz-Villena, A., Timon, M., Corell, A., Perez-Aciego, P., Martin-Villa, J.M., Regueiro, J.R. (1992). Primary immunodeficiency caused by mutations in the gene encoding the CD3-gamma subunit in the T-lymphocyte receptor. New Engl. J. Med. 327, 529-533.

Bellamy, R.J., Ruwende, C., Corrah, T., McAdam, K.P., Thursz, M., Whittle, H.C. Hill, A.V.S. (1999). Tuberculosis and chronic hepatitis B virus infection in Africans and variation in the vitamin D receptor gene. J. Infect. Dis. 179, 721-724.

Biron, C.A., Byron, K.S., Sullivan, J.L. (1989). Severe herpes virus infection in an adolescent without natural killer cells. New Engl. J. Med. 30, 1731-1734.

Bruton, O.C. (1952). Agammaglobulinemia. Pediatrics 9, 722-728.

Chatila, T., Wong, Young, M., Miller, R., Terhorst, C., Geha, R.S. (1989). An immunodeficiency characterized by defective signal transduction in T lymphocytes. New Engl. J. Med. 320, 696-702.

Chatila, T., Castigli, E., Pahwa, R., Pahwa S., Chirmule, N., Oyaizu, N., Good, R.A., Geha, R.S. (1990). Primary combined immunodeficiency resulting from defective transcription of multiple T-cell lymphokine genes. Proc. Natl. Acad, Sci. USA 87, 10033-10037.

Conley, M.E. (1992). Molecular approaches to analysis of X-linked immunodeficiencies. Annu. Rev. Immunol. 10, 215-238.

D'Ambrosio, D., Cippitelli, M., Cocciolo, M.G., Mazzeo, D., Di Lucia, P., Lang, R., Sinigaglia, F., Panina-Bordignon, P. (1998). Inhibition of IL-12 production by 1,25-dihydroxyvitamin D3. Involvement of NF-kappaB downregulation in transcriptional repression of the p40 gene. J. Clin. Invest. 101, 252-262.

Davies, P.D.O. (1985). A possible link between vitamin D deficiency and impaired host defence to Mycobacterium tuberculosis. Tubercle 66, 301-306.

DiSanto, J.P., Keever, C.A., Small, T.N., Nicols, G.L., O'Reilly, R.J., Flomenberg, N. (1990). Absence of interleukin 2 production in a severe combined immunodeficiency disease syndrome with T cells. J. Exp. Med. 171, 1697-1704.

Don-van 't Slot, H.P., van der Kolk, J.H. (2000). Severe Combined Immunodeficiency Disease (SCID) bij het Arabische paard. Tijdschr. Diergeneesk. 125, 577-581.

Eisman, J.A. (1996). Vitamin D receptor gene variants: implications for therapy. Curr. Opin. Genet. Dev. 6, 361-365.

Grimbacher, B., Schaffer, A.A., Holland, S.M., et al (1999). Genetic linkage of hyper-IgE syndrome to chromosome 4. Am. J. Hum. Genet. 65, 735-744.

Hammarstrom, L., Smith, C.I.E. (1999). Genetic approach to common variable immunodeficiency and IgA deficiency. In: Ochs, H.D., Smith, C.I.E., Puck, J.M. (eds.): Primary immunodeficiency diseases: A molecular and genetic approach. Oxford University Press, New York, pp. 250-262.

Hirschhorn, R. (1990). Adenosine deaminase deficiency. Immunodef. Rev. 2, 175-198.

Holland, S. (2000). Personal communication.

Honda, K., Kanegane, H., Eguchi, M., Kimura, H., Morishima, T., Masaki, K., Tosato, G., Miyawaki, T., Ishii, E. (2000). Large deletion of the X-linked lymphoproliferative disease gene detected by fluorescence in situ hybridization. Am. J. Hematol. 64, 128-132.

Howard, G., Nguyen, T., Morrison, N., Watanabe, T., Sambrook, P., Eisman, J., Kelly, P.J. (1995). Genetic influences on bone density: physiological correlates of Vitamin D receptor gene alleles in premenopausal women. J. Clin. Endocrinol. Metab. 80, 2800-2815.

Janeway, C.A., Travers, P. (1994). Immunobiology, the immune system in health and disease. Current Biology, Garland Publishing, Blackwell Scientific Publications, Oxford.

Kinnon, C., Levinsky, R.J. (1990). Molecular genetics of inherited immunodeficiency diseases. Immunogenetics 135, 894-900.

Kornfield, S.J., Haire, R.N., Strong, S.J., Brigino, E.N., Tang, H., Sung, S.S. et al (1997). Extreme variation in X-linked agammaglobulinemia phenotype in a three generation family. J. Allergy Clin. Immunol. 100, 702-706.

Kuijpers, T.W., Vossen, J.M.J.J. (2000). Immunologie in de medische praktijk. XXXIII. Erfelijke immuundeficiënties: van genotype naar fenotype. Ned. Tijdschr. Geneesk. 37, 1768-1773.

Morrison, N.A., Qi, J.C., Tokita, A., Kelly, P.J., Crofts, L., Nguyen, T.V., Sambrook, P.N., Eisman, J.A. (1994). Prediction of bone density from vitamin D receptor alleles. Nature 367, 284-287.

Noguchi, M., Yi, H., Rosenblatt, H.M., Filipovich, A.H., Adelstein, S., Modi, W.S., McBride, O.W., Leonard, W.J. (1993). Interleukin-2 receptor gamma-chain mutation results in X-linked severe combined immunodeficiency. Cell 73, 147-157.

Oxelius, V.-A., Laurell, A.-B., Lindquist, B., Golebiowska, H., Axelsson, U., Bjorkander, J., Hanson, L.A. (1981). IgG subclasses in selective IgA deficiency: importance of IgG2-IgA deficiency. New Engl. J. Med. 304, 1476-1477.

Preudhomme, J.L., Hanson, L.A. (1990). IgG subclass deficiency. Immunodef. Rev. 2, 129-149.

Provvedini, D.M., Tsoukas, C.D., Deftos, L.J., Manolagas, S.C. (1983). 1,25 dihydroxyvitamin D3 receptors in human leukocytes. Science 221, 1181-1183.

Rook, G.A.W., Steele, J., Frather, L., Barker, S., Karmali, R., O'Riordan, J. (1986). Vitamin D3 gamma interferon, and control of Mycobacterium tuberculosis by human monocytes. Immunology 57, 159-163.

Rook, G.A. (1988). The role of vitamin D in tuberculosis. Am. Rev. Respir. Dis 138, 768-770.

Roy, S., Frodsham, A., Saha, B., Hazra, S.K., Mascie-Taylor, C.G., Hill, A.V. (1999). Association of vitamin D receptor genotype with leprosy type. J. Infect. Dis. 179, 187-191.

Soudais, C., de Villartay, J.P., Le Deist, F., Fischer, A., Lisowska-Grospierre, B. (1993). Independent mutations of the human CD3e gene resulting in a T cell receptor/CD3 complex immunodeficiency. Nature Genet. 3, 77-81.

Tsoukas, C.D., Provvedini, D.M. Manolagas, S.C. (1984). 1,25 dihydroxyvitamin D3: a novel immunoregulatory hormone. Science 224, 1438-1440.

Vetrie, D., Vorechovsky, I., Sideras, P., Holland, J., Davies, A., Flinter, F., Hammarström, L., Kinnon, C., Levinsky, R., Bobrow, M., Edvard Smith, C.I., Bentley, D.R. (1993). The gene involved in X-linked agammaglobulinemia is a member of the src family of protein-tyrosine kinases. Nature 361, 226-233.

Volanakis, J.E., Zhu, Z.B., Schaffer, F.M., Macon, K.J., Palermos, J., Barger, B.O., Go, R., Campbell, R.D., Schroeder, H.W., Cooper, M.D. (1992). Major histocompatibility class III genes and susceptibility to immunoglobulin A deficiency and common variable immunodeficiency. J. Clin. Investig. 89, 1914-1922.

Vorechovski, I., Webster, A.D., Plebani, A., Hammarstrom, L. (1999). Genetic linkage of IgA deficiency to the major histocompatibility complex: evidence for allele segregation distortion, parent-of-origin penetrance differences, and the role of anti-IgA antibodies in disease predispostiion. Am. J. Hum. Genet. 64, 1096-1109.

Weinberg, K., Parkman, R. (1990). Severe combined immunodeficiency due to a specific defect in the production of interleukin-2. New Engl. J. Med. 322, 1718-1723.

WHO Primary immunodeficiency diseases (1992). Immunodef. Ref. 3, 195-236.

Yamamura, M., Uyemura, K., Deans, R.J., Weinberg, K., Rea, T.H., Bloom, B.R., Modlin, R.L. (1991). Defining protective responses to pathogens: cytokine profiles in leprosy lesions. Science 254, 277-279.

Chapter 12. Genes involved in accessory effector cell function

Functional Fc receptor polymorphisms
During the humoral immune response antibodies can activate several accessory effector cells in order to help eliminating micro-organisms from the body, usually by ingesting antibody-coated bacteria and killing them, or by releasing stored inflammatory mediators. These cells include macrophages and polymorphonuclear neutrophilic leukocytes, as well as natural killer cells, eosinophils, and mast cells. The accessory cells are activated when their Fc receptors aggregate upon binding of the constant parts, or Fc domains, of antibody that is bound to a micro-organism. Fc receptors are thus crucial links between the cellular and humoral parts of the immune system. Specific Fc receptors have been identified for IgM, IgA, IgD, IgE, and IgG.

Three main subclasses of the leukocyte receptor for the constant or Fc parts of IgG (FcγR) have been recognized in man: FcγRI (CD64), FcγRII (CD32), and FcγRIII (CD16) (Fig. 9). A broad diversity in FcγR isoforms exists, because each FcγR subclass is encoded by two or three genes, and alternative RNA splicing leads to multiple transcripts. FcγRs are members of the immunoglobulin superfamily. The genes encoding their transcripts are located on chromosome 1 in region 1q21-24. Expression patterns of the FcγRs differ widely between the various effector cells. For example, FcγRII is the most widely distributed receptor, while FcγRI and III exhibit more restricted patterns of expression. The basic architecture of leukocyte FcγR has been reviewed recently (Van der Pol and Van de Winkel 1998). Both deficiencies and polymorphisms of FcγR have been described. Deficiencies are seldom. Remarkably, deficiencies of both FcγRI and FcγRIIIb do not appear to lead to enhanced susceptibility to infectious disease. If true, this is probably due to the take over of their function by other Fc receptors (Ceuppens et al 1988, Haas et al 1995, Van de Winkel 1995).

Several functional allelic variants of FcγRIIa, -IIIa, and -IIIb have been described. Allotyping of these receptors is usually based on immunofluorescence or PCR-based techniques (Pol et al 1998). The allelic variants are expressed codominantly and differ in the function of their extracellular binding domain (for a summary see Table 7). The three leukocyte FcγR classes differ in their affinity for and interaction with IgG subclasses. Genetically determined structural differences, variable glycosylation patterns, and variation in association with signaling subunits are responsible for these differences (Van der Pol and Van de Winkel 1998). Several residues, among others amino acid 131, which can either be an arginine (R) or a histidine (H), influence the binding affinity of FcγRIIa (Hulett et al 1995). The frequencies of these alleles differ among various ethnic populations. Because the FcγRIIa-H131 is the only Fc receptor binding to IgG2 antibodies, which are often directed against the polysaccharides from encapsulated bacteria such as pneumococci and meningococci, several studies examined the influence of the FcγRIIa allotype on susceptibility to these infections. These studies were further initiated because IgG2 deficiency is often associated with increased susceptibility to *Streptococcus pneumoniae* (Jefferis and Kumararatne 1990). Children suffering from recurrent bacterial upper respiratory tract infections, which was often associated with *S. pneumoniae*, were indeed less often homozygous for the FcγRIIa-H131 allotype. In contrast, the frequency of FcγRIIIb allotypes was not different from controls (Sanders et al 1994). Among survivors from septic meningococcal septic shock, the frequency of the homozygous FcγRIIa-R131 allotype was higher (44 %) than in the control population (23 %) (OR = 2.67; 95 % CI: 1.09-6.53). No skewed distribution of the FcγRIIIb allotypes was noted in these children (Bredius et al 1994). These findings indicate an import role for anti-*N. meningitidis* IgG2 and the FcγRIIa polymorphism, which is further supported by the finding that neutrophils with the FcγRIIa-R/R131 allotype phagocytized *N. meningitidis* opsonized with polyclonal IgG2 antibodies less effectively than did IIa-H/H131 neutrophils.

FcγRIIIa bears a (valine to phenylalanine) polymorphism at amino acid position 158 in the 2nd Ig-like domain of the molecule, which results in different binding affinities for IgG1 and 3 (Table 7). This polymorphism appears relevant for the function of natural killer (NK) cells. NK cells from FcγRIIIa-158 V homozygous donors show larger calcium influxes and more rapid induction of apoptosis following incubation with IgG aggre-

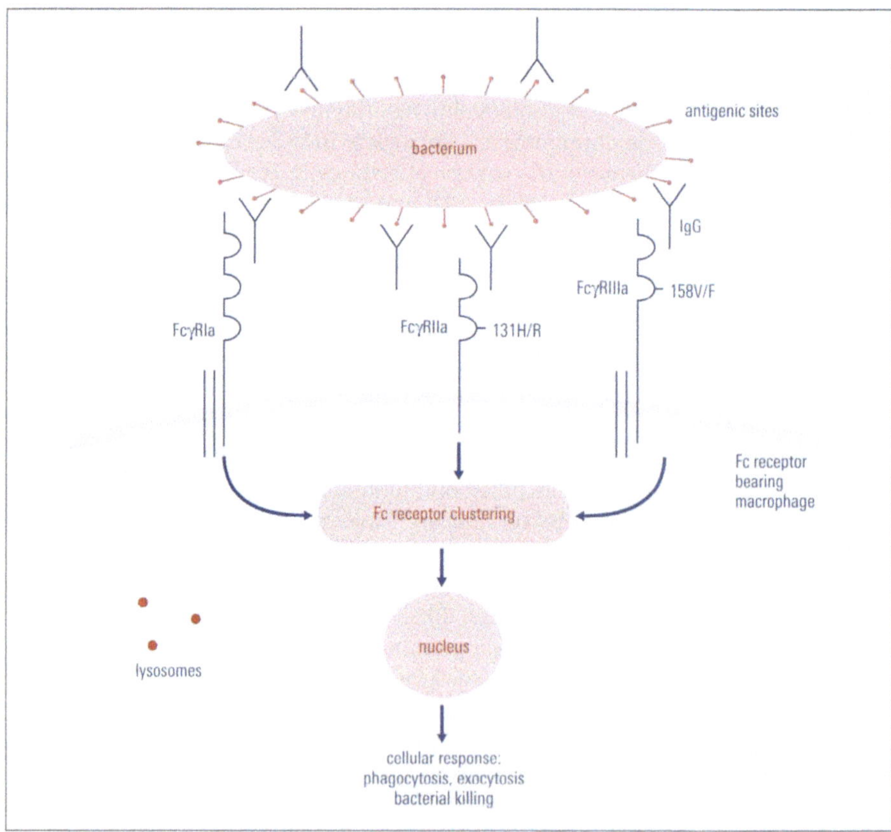

Figure 9. Upon binding of Fc receptors to the constant region of immunoglobulin chains, Fc receptor clustering may occur through binding of multiple antibodies to the same particle. Complex cellular responses follow. The cellular response may lead to phagocytosis, bacterial killing by antibody-dependent cellular cytotoxicity, and release of inflammatory mediators. Expression of different FcγRs is shown. The extracellular parts of the FcγRs consist of two to three Ig-like domains containing disulphide bonds. Amino acid substitutions at positions 131 (FcγRIIa) and 158 (FcγRIIIa) influence affinity for immunoglobulin Fc chains of various isotypes.

gates than NK cells from FcγRIIIa-158 F homozygous donors (Wu et al 1977). FcγRIIIa further bears a non-functional tri-allelic polymorphism resulting in amino acid variation at position 48 in the 1st Ig-like domain. This polymorphism appears closely linked to the FcγRIIIa-158 polymorphism (Koene et al 1997).

FcγRIIIb bears the neutrophil antigen (NA) polymorphism. The FcγRIIIb-NA1 and NA2 isoforms differ by four amino acids in the membrane distal Ig-like domain, resulting in different patterns of glycosylation and ligand affinity (Table 7) (Ravetch and Perussia 1989). The allelic frequencies in Caucasians are 37 % and 63 % for NA1 and NA2 respectively. The influence of this and other polymorphisms on susceptibility to infections has been studied in patients with deficiencies in the late components of the complement system (C5-C9: membrane attack complex), since the complement system of these patients is not able to lyse Gram-negative bacteria through formation of a membrane attack complex. These patients therefore are particularly predisposed to infection with encapsulated bacteria, such as pneumococci, meningococci, and invasive strains of *Haemophilus influenzae*. They depend on efficient phagocytosis for adequate defense against these bacteria. However, combined FcγRIIa-R131 and FcγRIIIb-NA2 homozygosity results in depressed phagocytotic function and is associated with a significantly increased frequency of meningococcal infections in patients with deficiencies in the membrane attack complex (OR = 13.9; 95 % CI: 1.2-478) (Fijen et al 1993). It has been suggested that the low frequency of these allotypes in Japanese people, compared to European Caucasians, is the explanation for the low frequency of infection with encapsulated bacteria in Japan (Winkel and Capel 1993).

FcγR polymorphisms may further be involved in chronic periodontitis. Anaerobic Gram-negative bacteria, such as *Actinobacillus actinomycetemcomitans*, are probably causative agents for various types of periodontitis (Van der Pol and Van de Winkel 1998). Localized juvenile periodontitis, which often occurs in young adult blacks, appears associated with elevated levels of IgG2 production and the FcγRIIa-R131 allotype, confirming the need for IgG2-opsonization and phagocytosis in the clearance of this infection (Wilson and Kalmar 1996). FcγRIIIb polymorphism appears to influence the severity and frequency of the most frequently occurring form of periodontitis in adults. Japanese patients with one or two FcγRIIIb-NA2 alleles have a higher chance of recurrent adult periodontitis after treatment than homozygous FcγRIIIb-NA1 patients (Kobayashi et al 1997).

In conclusion, the examples provide arguments in favor of determining the FcγRII and FcγRIII genotypes to further evaluate their contribution to infection susceptibility.

Deficient oxygen metabolism in phagocytes
Defects in cytochrome b, the *phox*-proteins, or NADPH oxidase lead to deficient oxygen metabolism in phagocytes (see chapter 8). In these patients phagocytosed bacteria are not killed due to a lack of oxygen radicals, and chronic granulomatous disease (X-linked or autosomal recessive) may follow. Most patients are detected before they are 5 years old when they suffer from infections caused by staphylococci, Gram-negative bacteria, and fungi. Causal therapy of this disorder is impossible, but early and prophylactic treatment with antibiotics and antimycotics clearly benefits the prognosis of these patients (Roos et al 1993).

Table 7. Binding affinities and associated effector cell functions of Fcg receptor allotypes

Receptor	Gene frequency	Binding affinity				Effector cell function
		IgG1	IgG2	IgG3	IgG4	
FcγRI (CD64)		+++	-	++++	++	
FcγRIIa-R131	0.47	++	-	++	-	poor phagocytosis of IgG2-coated particles
FcγRIIa-H131 (CD32)	0.53	++	++	++	-	efficient phagocytosis of IgG2-coated particles
FcγRIIIa-F158	0.57	++/+++	-	++/+++	-	
FcγRIIIa-V158	0.43	+++	-	+++	+	
FcγRIIIb-NA1	0.37	+++	-	++	-	efficient phagocytosis of IgG1/3-coated particles
FcγRIIIb-NA2 (CD16)	0.62	+/++	-	+/++	-	poor phagocytosis of IgG1/3-coated particles

Adapted from Pol et al (1998) and De Haas (1996).

FcγRIIa-R131: Arginine at amino acid position 131 in the extracellular binding domain

FcγRIIa-H131: Histidine at amino acid position 131

FcγRIIIa-F158: Phenylalanine at amino acid position 158 in the extracellular binding domain

FcγRIIIa-V158: Valine at amino acid position 158

FcγRIIIb-NA1 and FcγRIIIb-NA2 differ at several amino acids resulting in different glycosylation and neutrophil antigen (NA) polymorphism.

Note: FcγRIIa is expressed by neutrophilic granulocytes, monocytes/macrophages and platelets. FcγRIIIa is expressed by natural killer cells and monocytes/macrophages. FcγRIIIb is expressed by neutrophilic granulocytes.

References

Bredius, R.G., Derkx, B.H., Fijen, C.A., Wit, T.P. de, Haas, M. de, Weening, R.S., van de Winkel, J.G., Out, T.A. (1994). Fc gamma receptor IIa (CD32) polymorphism in fulminant meningococcal septic shock in children. J. Infect. Dis. 170, 848-853.

Ceuppens, J.L., Baroja, M.L., Vaeck, F. van, Anderson, C.L. (1988). Defect in the membrane expression of high affinity 72-kD Fc gamma receptors on phagocytic cells in four healthy subjects. J. Clin. Invest. 82, 571-578.

Fijen, C.A., Bredius, R.G., Kuijper, E.J. (1993). Polymorphism of IgG Fc receptors in meningococcal disease. Ann. Intern. Med. 119, 636.

Haas, M. de, Kleijer, M., Zwieten, R. van Roos, D., Von dem Borne, A.E. (1995). Neutrophil Fc gamma RIIIb deficiency, nature, and clinical consequences: a study of 21 individuals from 14 families. Blood 86, 2403-2413.

Haas, M. de (1996). Bepaling van IgG-Fc-receptor polymorfismen vergroot het inzicht in ziekterisico. CLB Bulletin 96-3, 4-6.

Hulett, M.D., Witort, E., Brinkworth, R.I., McKenzie, I.F.C., Hogarth, P.M. (1995). Multiple regions of human FcgRII (CD32) contribute to the binding of IgG. J. Biol. Chem. 269, 15287-15293.

Jefferis, R., Kumararatne, D.S. (1990). Selective IgG subclass deficiency: quantification and clinical relevance. Clin. Exp. Immunol. 81, 357-367.

Kobayashi, T., Westerdaal, N.A., Miyazaki, A., Pol, W.L., van der, Suzuki, T., Yoshie, H., van de Winkel, J.G.J., Hara, K. (1997). Relevance of immunoglobulin G Fc receptor polymorphism for recurrence of adult periodontitis in Japanse patients. Infect. Immun. 65, 3556-3560.

Koene, H.R., Kleijer, M., Algra, J., Roos, D., von dem Borne, A.E.G.Kr., de Haas, M. (1997). FcgRIIIa-158 V/F polymorphism influences the binding of IgG by NK cell FcgRIIIa, independently of the FcgRIIIa-48L/R/H phenotype. Blood 90, 1109-1114.

Pol, W.-L. van der, Kuijper, E.J., Koene, H.R., Haas, M. de, Sanders, E.A.M., Winkel, J.G.J. (1998). Immunologie in de medische praktijk. XI. IgG-receptoren: de rol van polymorfismen in auto-immuun- en infectieziekten. Ned. Tijdschr. Geneesk. 142, 340-345.

Pol, W..-L. van der, Winkel, J.G.J. (1998). IgG receptor polymorphisms: risk factors for disease. Immunogenetics 48, 222-232.

Ravetch, J.V., Perussia, B. (1989). Alternative membrane forms of FcgRIII (CD16) on human natural killer cells and neutrophils. Cell type-specific expression of two genes that differ in single nucleotide substitutions. J. Exp. Med. 170, 481-497.

Roos, D., De Boer, M., De Klein, A., Bolscher, B.G., Weening, R.S. (1993). Chronic granulomatous disease: mutations in cytochrome b558. Immunodeficiency 4, 289-301.

Sanders, L.A., Winkel, J.G.J. van de, Rijkers, G.T. Voorhoorst-Ogink, M.M., Haas, M. de, Capel, P.J., Zegers, B.J.M. (1994). Fc gamma receptor IIa (CD32) heterogeneity in patients with recurrent bacterial respiratory tract infections. J. Infect. Dis. 170, 854-861.

Wilson, M.E., Kalmar, J.L. (1996). FcgRIIa (CD32): a potential marker defining susceptibility to localized juvenile periodontitis. J. Periodontol. 67, 323-331.

Winkel, J.G.J. van de, Capel, P.J.A. (1993). Human IgG Fc receptor heterogeneity: molecular aspects and clinical implications. Immunol. Today 14, 215-221.

Winkel, J.G.J. van de, Wit, T.P.M. de, Ernst, L.K., Capel, P.J.A., Ceuppens, J.L. (1995). Molecular basis for a familial defect in phagocyte expression of IgG receptor I (CD64). J. Immunol. 154, 2896-2903.

Wu, J., Edberg, J.C., Redecha, P.B., Bansal, V., Guyre, P.M., Coleman, K., Salmon, J.E., Kimberley, R.P. (1997). A novel polymorphism of FcgRIIIa (CD16) alters receptor function and predisposes to autoimmune disease. J. Clin. Invest. 100, 1059-1070.

Chapter 13. Genes involved in tissue function and integrity

A normal function of tissues and organs is apparently required to provide resistance against infection. Obviously this is most apparent in those tissues that are exposed to the environment, such as the skin and the mucosae, and that have a function as barrier. Defects in genes that in themselves do not have a function in disease pathogenesis, inflammation, or recovery may thus lead to enhanced susceptibility to infection. Usually the causative organisms, such as the skin bacterium *Staphylococcus epidermidis* and the aquatic bacterium *Pseudomonas aeruginosa*, are ubiquitous and only pathogenic for a particular set of compromised individuals. A striking example is cystic fibrosis (CF). CF is the most common life-shortening autosomal recessive disorder in populations of European origin (Rosenstein and Zeitlin 1998). Approximately 1 in 3,000 white persons are affected. African Americans and Asian people are affected much less frequently. The genetic defects in CF lead to the loss of the CF transmembrane regulator (CFTR) chloride channel in the apical membranes of epithelial cells (Costerton et al 1999, Welsh and Smith 1993). The gene encoding CFTR is located on chromosome 7, band q31.2. The gene has 27 exons and encodes a protein of 1,480 amino acids. Over 600 different mutations leading to CF are known. The commonest of them is the ΔF508 mutation, which has an allele frequency in Europeans of 4–5 % and which is present on 70% of CF chromosomes. The different mutations may lead to a lack of synthesis, a block in processing, a block in regulation, an altered conductance, or a reduced synthesis of the CFTR protein (Rosenstein and Zeitlin 1998). The ΔF508 mutation leads to retention of the protein in the endoplasmic reticulum and subsequent degradation by a proteasome-dependent pathway. There is a striking variability in severity of disease, which is only partly explained by the underlying genotype. The genetic defect leads to slowly developing, but persistent bacterial infections in the lungs. Most of these patients are persistently colonized by mucoid-producing strains of *P. aeruginosa*, which may lead to chronic lung damage and early death.

A number of pathogenic pathways appear at work in maintaining the persistent infection in CF patients. Among others the absence of a functional chloride channel leads to an elevated salt content in the airway surface fluid, a very thin layer covering the mucosal surface of airway epithelia (Goldman et al 1997, Joris et al 1993, Zabner et al 1998). This alteration may be enough to allow the colonization of *P. aeruginosa* as a biofilm (a slimy conglomeration of bacteria) on the airway epithelium (Costerton et al 1999). But, in addition, the normal airway surface fluid contains an antibacterial protein, known as human β-defensin-1, which activity is salt-sensitive. Airway surface fluid from CF patients fails to kill *P. aeruginosa*, because its β-defensin-1 is inactivated by the elevated salt concentration (Goldman et al 1997, Smith et al 1996). The *P. aeruginosa* biofilm in the airways of CF patients is a developed community in which the individual bacterial cells are embedded in an extracellular polysaccharide matrix. High concentrations of bactericidal and opsonizing antibodies to *P. aeruginosa* are released in blood gend in the lungs, but these antibodies (just like antibiotics) fail to penetrate deep in the bacterial colonies and to clear the infection. On the contrary, circulating antibodies to *P. aeruginosa* appear to promote inflammation and pulmonary tissue damage. In CF patients, a high concentration of antibodies to *P. aeruginosa* correlates with a poor clinical outcome and immune suppression may be part of the treatment of CF patients. The CFTR further functions as a cellular receptor for binding, endocytosing, and clearing *P. aeruginosa* from the lung. This function is severely impaired by the ΔF508 mutation (Pier et al 1997). Other pathogenic mechanisms enhancing lung damage and bacterial colonization in CF patients may be reduced production by epithelial cells of the anti-inflammatory cytokine IL-10, and release of elastase produced by neutrophils. Neutrophil elastase disrupts epithelial cells and destroys the matrix supporting the epithelium. Elastase further digests the C3b receptor on neutrophils, which results in limited phagocytosis of pathogens. Elastase also interferes with the normal binding by neutrophils of antibodies to *P. aeruginosa* by cleaving at the Fc region (Greenberger 1997). The mechanism by which the CFTR mutations may have been maintained in the population is discussed in chapter 7.

References

Costerton, J.W., Stewart, P.S., Greenberg, E.P. (1999). Bacterial biofilms: A common cause of persistent infections. Science 284, 1318-1322.

Greenberger, P.A. (1997). Immunologic aspects of lung diseases and cystic fibrosis. J. Am. Med. Ass. 278, 1924-1930.

Goldman, M.J., Anderson, G.M. Stolzenberg, E.D., Kari, U.P., Zasloff, M., Wilson, J.M. (1997). Human beta-defensin-1 is a salt-sensitive antibiotic in lung that is inactivated in cystic fibrosis. Cell 88, 553-560.

Joris, L., Dab, I., Quinton, P.M. (1993). Elemental composition of human airway surface fluid in healthy and diseased airways. Am. Rev. Respir. Dis. 148, 1633-1637.

Pier, G.B., Grout, M., Zaidi, T.S. (1997). Cystic fibrosis transmembrane conductance regulator is an epithelial cell receptor for clearance of Pseudomonas aeruginosa from the lung. Proc. Natl. Acad. Sci. USA 94, 12088-12093.

Rosenstein, B.J., Zeitlin, P.L. (1998). Cystic fibrosis. Lancet 351, 277-282.

Smith, J.J., Travis, S.M., Greenberg, E.P., Welsh, M.J. (1996). Cystic fibrosis airway epithelia fail to kill bacteria because of abnormal airway surface fluid. Cell 85, 229-236.

Welsh, M.J., Smith, A.E. (1993). Molecular mechanisms of CFTR chloride channel dysfunction in cystic fibrosis. Cell 73, 1251-1254.

Zabner, J., Smith, J.J., Karp, P.H., Widdicombe, J.H., Welsh, M.J. (1998). Loss of CFTR chloride channels alters salt absorption by cystic fibrosis airway epithelia in vitro. Mol. Cell 2, 397-403.

Chapter 14. Infectious diseases, blood groups, and secretor status

Blood groups and infectious diseases

Twenty-three human blood group systems have been identified, and genes for 20 of them have now been cloned. The molecular bases of the important polymorphisms within most of these systems are known. Little is known, however, about the biological significance of the blood group polymorphisms. Several blood group antigens are expressed at mucosal surfaces in the respiratory and gastrointestinal tracts, the most frequent routes of entry of pathogenic micro-organisms. Many blood group antigens are receptors for pathogenic micro-organisms and these pathogens might have played an important part in the evolution of blood group polymorphism. The distribution of human blood groups in different parts of the world therefore likely reflects different evolutionary forces important for survival of infectious diseases.

Most blood group polymorphisms are due to missense mutations, resulting in amino acid substitutions, but other types of mutations, including gene deletion, single base deletion, homologous recombination, are also involved (Daniels 1997). The best-known blood group antigens are the A, B, and AB antigens on the surface of human erythrocytes. In fact these blood group antigens functioned as the first identified chromosome markers. Blood group O erythrocytes do not possess these antigens, but blood of such people have antibodies to A and B antigens. Persons with blood group A have antibodies to antigen B and vice versa. The ABH and Lewis blood group antigens are not limited to erythrocytes, but are found in secretions and cell membranes throughout the body. Specific antigens consist of glycoproteins, which contain varying percentages of galactose, fucose, hexosamine, N-acetyl-glucosamine, and N-acetyl-galactosamine (Berger et al 1989). The A and B alleles determine a specific glycosyltransferring enzyme. A and B differ in 7 single nucleotides resulting in 4 amino acid substitutions, while a single base deletion in the O allele causes a shifted reading frame resulting in translation of an entirely different protein (Yamamoto et al 1990). The literature on the relationship between blood groups and susceptibility to infectious disease is large, but contains many conflicting findings and speculation. These are probably based on errors of data stratification and sampling bias. Mechanistic explanations for blood group-disease associations are therefore urgently needed. For example, the influence of erythrocyte membrane on susceptibility to infection may reflect antigenic similarity, receptor function, secretor status influencing microbial adhesion through specific receptors, autoimmune reactions, or anthropological and geographical influences (Berger et al 1989). Moreover, real relationships between blood groups and infections may be difficult to detect because of the large diversity of infectious diseases on the one hand, and the many other selective mechanisms acting on blood group distributions on the other (Weatherall et al 1988).

Some possible relationships between blood type and infectious disease that have been suggested may be reflected in geographic differences in the occurrence of blood groups. For example the high blood group O frequency in Central and South America may have been due to a selective advantage of blood group O against syphilis, but other possible explanations for the high blood group O frequency in this region, including even a founder effect, cannot be excluded (see chapter 3). It has been suggested that the high prevalence of blood group O among populations in Latin America may have intensified the cholera epidemic in that region (Swerdlow et al 1994). The higher frequency of blood group O in marginal European populations may have been caused by lower selection against O carriers by plague. In contrast, selection by smallpox might have caused the relatively low frequency of blood group A in central and southern Asia. In these areas blood group O is not frequent as well, giving a relatively high frequency of blood group B. Thus this high group B frequency might be due to long-standing selection against group A by smallpox, as well as against group O by plague (Weatherall et al 1988).

In addition, both cholera and enterotoxigenic *E. coli* may have contributed to selection against blood group O in central and southern Asia (Glass et al 1985, Levine et al 1997, Black et al 1987). The lowest blood group O prevalence is found in people living in the Ganges River delta, among whom 12 – 20 % have blood group O. This low prevalence has been attributed to selective pressure exerted by cholera over many generations. Indeed several studies found that individuals with blood group O are at higher risk of acquiring cholera due to *Vibrio cholerae* 01 (including the E1 Tor biotype), than those with other blood groups. Persons with blood group AB are relatively resistant to cholera (Faruque et al 1994, Glass et al 1985). Clemens et al (1989) found that vaccine efficacy is also modulated by blood group type. During a large field trial in Bangladesh of killed cholera vaccine, persons with blood group O were significantly less protected against severe cholera by the vaccine than were persons with the AB blood group.

Erythrocyte receptors

Perhaps better documented than some of the examples mentioned above is the relationship between the lack of the Duffy blood group antigen (FyFy) and resistance of human erythrocytes against invasion by merozoites of the human malarial parasite *P. vivax*, and *P. knowlesi*, a parasite resembling *P. vivax*. The Duffy glycoprotein functions as chemokine receptor, but also as a receptor for *P. vivax*. In contrast to the rest of the world population, West Africans are almost entirely Duffy negative. Duffy-negative volunteers experimentally inoculated with *P. vivax* were completely refractory to infection. Selection by *P.vivax* apparently resulted in fixation of the Duffy-negative gene, in contrast to selective forces resulting in balanced polymorphisms. This may have happened because Duffy-negatives are hematologically completely normal (Miller et al 1975, Weatherall et al 1988). As described in chapter 7, the absence of the Duffy glycoprotein from red cells results from a mutation within an erythroid-specific transcription factor binding site within the FY gene. The glycoprotein is normally present in other tissues, therewith further reducing any potential selective disadvantage.

The blood group P antigen, also known as globoside (Gb4, a tetrahexose ceramide), is the receptor for human parvovirus B19. This virus, discovered in 1974, is a single-stranded DNA virus, which causes erythema infectiosum, arthralgia (mainly in women), aplastic crisis in patients with red cell defects, chronic anemia in immunocompromised patients, and fetal hydrops. Seroprevalence in developed countries increases with age up to 50 % by the age of 15 and up to 85 % or more in those over 70 years. The virus may be transmitted by the respiratory route and by transfusion of infected blood and blood products. After an incubation period of six to eight days, viremia occurs and the virus enters its target cells in the bone marrow, proliferating erythroid precursor cells in the late S phase of the cell cycle, through the blood group

P antigen. Reticulocyte numbers subsequently fall dramatically, resulting in a temporary drop in hemoglobin levels. In immunologically normal individuals the infection is cleared by the humoral immune system. In patients with absent or dysfunctional humoral immunity, persistent infection can occur, which results in chronic suppression of erythropoiesis with chronic anemia (Van Elsacker-Niele and Kroes 1999, Kerr 1996). P antigen is also present on endothelial cells and fetal myocardial cells, which may be targets of viral infection in the fetus after transplacental transmission. The tissue distribution of the P antigen is therewith consistent with the known tropism of parvovirus B19. The blood group P system consists of three antigens: P antigen, P^k antigen and P_1 antigen. Almost all humans have either the P1 phenotype (both P and P_1 antigen present) or the P2 phenotype (only P antigen present). Approximately 1:200 000 individuals either lack all three antigens or have the P_1^k phenotype (both P_1 and P^k antigen present). These rare individuals lacking the P antigen are resistant to B19 infection and their bone marrow cells cannot be infected with B19 *in vitro* (Brown et al 1994, Van Elsacker-Niele and Kroes 1999).

Antigenic relatedness
Antigens cross-reacting with blood group antigens are present on *Enterobacteriaceae* and have led to the speculation that such strains cannot be recognized as foreign and may thus evade the immune response. For example, it has been hypothesized that the lack of anti-B isoagglutinin results in an increased incidence of upper urinary tract infection as well as septicemia caused by *Escherichia coli* in persons with type B blood (Blackwell et al 1984, Kinane et al 1982, Ratner et al 1986, Wittels and Lichtman 1986). Unfortunately the data appear conflicting, thus making the hypothesis controversial. Otherwise *Streptococcus pneumoniae* is sensitive to the anti-A antibody in group B individuals, which results in a decreased disease incidence in these persons (Reed et al 1974).
Helicobacter pylori lipopolysaccharide (LPS) expresses Lewis x or y, or both, blood group antigens, which mimic those occurring on human gastric epithelial cells. The mimicry may have diverging roles during infection. Infection may break tolerance leading to anti-Lewis antibodies that bind gastric mucosa and cause damage. Otherwise mimicry may cause "invisibility" of the pathogen, thus aiding in persistence of the infection (Vandenbroucke-Grauls and Appelman 1998).

Autoimmune reactions
Antibodies crossreacting with the I or i blood group antigens that are induced after infection with *Mycoplasma pneumoniae* may cause severe anemia in cold hemagglutinin disease. These antibodies are primarily of the IgM isotype and react mainly at temperatures below 37 ^{0}C. Similar mechanisms may be caused by antibodies specific for blood group P. Cold agglutinin disease may also occur after a variety of other infectious and non-infectious disease, including infectious mononucleosis, cytomegalovirus infection, listeriosis, mumps orchitis, syphilis, endocarditis, trypanosomiasis, and malaria (Berger et al 1989, Pruzanski and Shumak 1977).

Secretor status
Most people secrete the water-soluble forms of the ABO blood group antigens into saliva and other body fluids, but approximately 20 % of most human populations have a genetically determined inability to do so (Hill 1996). When blood group anti-

gens are present in mucus they may bind micro-organisms, aid in the elimination of the pathogen, and minimize colonization of tissues. However, exceptions illustrate that other outcomes may be possible (see below).

Non-secretors are homozygous for a nonsense mutation at position 143 (i.e. encoding a stop codon resulting in premature termination) of the *Sec2* gene, that encodes an $\alpha(1,2)$fucosyltransferase (Kelly et al 1995). This gene is in close proximation to the H blood group $\alpha(1,2)$fucosyltransferase gene on chromosome 19. The same enzyme-inactivating mutation has been found in different populations, i.e. Americans and sub-Saharan Africans (Hill 1996). The mutation may have been maintained by a balanced polymorphism based on a differential resistance to various microbial pathogens. For example, non-secretors have been reported to be at increased risk of invasive meningococcal disease, *Candida* infections, and recurrent urinary tract infections. However, they appear to have a lower risk to various respiratory viral infections, such as influenza virus and respiratory syncytial virus (Blackwell et al 1986, Kinane et al 1983, Raza et al 1991, Thom et al 1989). Non-secretors also appear to be less susceptible (OR = 0.46, 95 % CI: 0.20 − 1.09) to acquiring HIV infection through heterosexual intercourse, while they have an equal chance of acquiring the virus through intravenous drug use (Blackwell et al 1991).

The lower frequency of urinary tract infection in secretors may be caused by secretory A, B, or H substances covering receptor oligosaccharides, thereby preventing access of adhesins on bacterial pili (from *E. coli*) to receptors of the urinary epithelium (Lomberg et al 1986, Scheinfeld et al 1989). These *E.coli* receptors on human uroepithelial cells are glycosphingolipids from the P blood group antigen system (Korhonon et al 1982, Mulholland et al 1984). These receptors are found on epithelial cells of the renal collecting ducts, pelvis, ureter, bladder, vagina, as well as vaginal secretions. Bacterial pili binding to such substances are termed "P fimbriae". These bacterial pili bind to the Gal(α1-4)Gal moiety of the glycolipid part of blood group substance P expressed on uroepithelial cells. Interestingly, an antigen similar to P1 is found in *E.coli* itself (Berger et al 1989, Roland 1973). Svanborg Eden et al (1983) subsequently presented evidence that the P1 blood group phenotype is also a factor in susceptibility to urinary tract infection. The P1 phenotype was found in 97% of patients with recurrent pyelonephritis, as compared to 75 % in healthy children. A mechanistic explanation may be that persons of blood group P1 have a higher density of receptor glycolipids in their red cell membrane than do persons of the P2 phenotype.

Secretors also less often experience rheumatic fever than non-secretors. Rheumatic fever is a complication of a *Streptococcus pyogenes* throat infection and has a clear familial predisposition. Antibodies are formed to antigens in the streptococcal cell wall, notably streptolysin O, that cross-react with the sarcolemma of the human heart and with tissues elsewhere. As a result, inflammatory lesions may be found in the heart, skin, joints, and rarely the central nervous system. In non-secretors the streptococci appear to persist longer leading to higher titers of antibacterial antibodies, among which are antibodies directed against streptolysin O.

References

Berger, S.A., Young, N.A., Edberg, S.C. (1989). Relationship between infectious diseases and human blood type. Eur. J. Clin. Microbiol. Infect. Dis. 8, 681-689.

Black, R.E., Levine, M.M., Clements, M.L., Hughes, T., O'Donnell, S. (1987). Association between O blood group and occurrence and severity of diarrhea due to Escherichia coli. Trans. R. Soc. Trop. Med. Hyg. 81, 120-123.

Blackwell, C.C., Andrew, S., May, S.J., Weir, D.M., MacCallum, C., Brettle, R.P. (1984). ABO blood group and susceptibility to urinary tract infection: no evidence for involvement of isohaemagglutinins. J. Clin. Laborat. Immunol. 15, 191-194.

Blackwell, C.C., Jonsdottir, K., Hanson, M., Todd, W.T., Chaudhuri, A.K., Mathew, B., Brettle, R.P., Wei, D.M. (1986). Non-secretion of ABO antigens predisposing to infection by Neisseria meningitidis and Streptococcus pneumoniae. Lancet 2, 284-285.

Blackwell, C.C., James, V.S., Davidson, S., Wyld, R., Brettle, R.P., Robertson, R.J., Weir, D.M. (1991). Secretor status and heterosexual transmission of HIV. Br. Med. J. 303, 825-826.

Brown, K.E., Hibbs, J.R., Gallinella, G., Anderson, S.M., Lehman, E.D., McCarthy, P., Young, N.S. (1994). Resistance to parvovirus B19 infection due to lack of virus receptor (erythrocyte P antigen). New Engl. J. Med. 330, 1192-1196.

Clemens, J.D., Sack, D.A., Harris, J.R., Chakraborty, J., Khan, M.R., Huda, S., Ahmed, F., Gomes, J., Rao, M.R., Svennerholm, A.-M., Holmgren, J. (1989). ABO blood groups and cholera: new observations on specificity of risk and modification of vaccine efficacy. J. Infect. Dis. 159, 770-773.

Daniels, G. (1997). Blood group polymorphisms: molecular approach and biological significance. Transfus. Clin. Biol. 4, 383-390.

Faruque, A.S.G., Mahalanabis, D., Hoque, S.S., Albert, M.J. (1994). The relationship between ABO blood groups and susceptibility to diarrhea due to Vibrio cholerae O139. Clin. Infect. Dis. 18, 827-828.

Glass, R.I., Holmgren, J., Haley, C.E., Khan, M.R., Svennerholm, A.M., Stoll, B.J., Belayet-Hossain, K., Black, R.E., Yunus, M., Barua, D. (1985). Predisposition for cholera of individuals with O blood group: possible evolutionary significance. Am. J. Epidemiol. 121, 791-796.

Hill, A.V.S. (1996). Genetics of infectious disease resistance. Curr. Opin. Genet. Developm. 6, 348-353.

Kelly, R.J., Rouquier, S., Giogi, D., Lennon, G.G., Lowe, J.B. (1995). Sequence and expression of a candidate for the human Secretor blood group alpha(1,2) fucosyltransferase gene (FUT2). Homozygosity for an enzyme-inactivating nonsense mutation commonly correlates with the non-secretor phenotype. J. Biol. Chem. 270, 4640-4649.

Kerr, J.R. (1996). Parvovirus B19 infection. Eur. J. Infect. Dis. 15, 10-29.

Kinane, D.F., Blackwell, C.C., Winstaney, F.P., Weir, D.M. (1983). Blood group, secretor status, and susceptibility to infection by Neisseria gonorrhoeae. Br. J. Vener. Dis. 59, 44-46.

Kinane, D.F., Blackwell, C.C., Brettle, R.P., Weir, D.M., Winstaney, F.P., Elton, R.A. (1982). ABO blood group, secretor status, and susceptibility to recurrent urinary tract infection in women. Brit. Med. J. 285, 7-9.

Korhonon, T.K., Vaisanen, V., Saxen, H., Hultberg, H., Svenson, S.B. (1982). P-antigen-recognizing fimbriae from human uropathogenic Escherichia coli strains. Infect. Immun. 37, 286-291.

Levine, M.M., Nalin, D.R., Rennels, M.B, Hornick, R.B., Sotman, S., Van Blerk, G., Hughes, T.P., O'Donnell, S., Barua, D. (1979). Genetic susceptibility to cholera. Ann. Human Biol. 6, 369-374.

Lomberg, H., Cedergen, B., Leffler, H., Nilsson, B., Carlstrom, A., Svanborg-Eden, C. (1986). Influence of blood group on the availability of receptors for attachment or uropathogenic Escherichia coli. Infection and Immunity 51, 919-926.

Miller, L.H., Mason, S.J., Dvoral, J.A., McGinnes, M.H., Rothman, I.K. (1975). Erythrocyte receptors for (Plasmodium knowlesi) malaria: Duffy blood group determinants. Science 189, 561-563.

Mulholland, S.G., Mooreville, M., Parsons, C.L. (1984). Urinary tract infections and P blood group antigens. Urology 25, 232-235.

Pruzanski, W., Shumak, K.H. (1977). Biologic activity of cold-reacting autoantibodies. New Engl. J. Med. 297, 583-589.

Ratner, J.J., Thomas, V.L., Forland, M. (1986). Relationship between human blood groups, bacterial pathogens, and urinary tract infections. Am. J. Med. Sci. 292, 86-91.

Raza, M.W., Blackwell, C.C., Molyneaux, P., James, V.S., Ogilvie, M.M., Inglis, J.M., Weir, D.M. (1991). Association between secretor status and respiratory illness. Brit. Med. J. 303, 815-818.

Reed, W.P., Drach, G.W., Williams, R.C. (1974). Antigens common to human and bacterial cells. IV: Studies of human pneumococcal disease. J. Clin. Laborat. Med. 83, 599-610.

Roland, F. (1973). Presence of human blood group substance P1 in gram negative bacilli. Ann. Microbiol. 124A, 375.

Scheinfeld, J., Schaeffer, A.J., Condon-Cardo, C., Rogatko, A., Fair, W.R. (1989). Association of Lewis blood group phenotype with recurrent urinary tract infections in women. New Engl. J. Med. 320, 273-277.

Svanborg Eden, C., Hagberg, L., Hanson, L.A., Hull, S., Hull, R., Jodal, U., Leffler, H., Lomberg, H., Straube, E. (1983). Bacterial adherence-a pathogenetic mechanism in urinary tract infections caused by Escherichia coli. Prog. Allergy 33, 175-188.

Swerdlow, D.L., Mintz, E.D., Rodriguez, M., Tejada, E., Ocampo, C., et al. (1994). Severe life-threatening cholera associated with blood group O in Peru: implications for the Latin American epidemic. J. Infect. Dis. 170, 468-472.

Thom, S.M., Blackwell, C.C., MacCallum, C.J., Weir, D.M., Brettle, R.P., Kinane, D.F., Wray, D. (1989). Non-secretion of blood group antigens and susceptibility to infection by Candida species FEMS Microbiol. Immunol. 1, 401-405.

Vandenbroucke-Grauls, C.M., Appelmelk, BJ. (1998). *Helicobacter pylori* LPS: molecular mimicry with the host and role in autoimmunity. Ital. J. Gastroenterol. Hepatol. 30, Suppl 3, S259-20.

Van Elsacker-Niele, A.M., Kroes, A.C. (1999). Human parvovirus B19: relevance in internal medicine. Neth. J. Med. 54, 221-230.

Weatherall, D.J., Bell, J.I., Clegg, J.B., Flint, J., Higgs, D.R., Hill, A.V.S., Pasvol, G., Thein, S.L. (1988). Genetic factors as determinants of infectious disease transmission in human communities. Phil. Trans. Roy. Soc. London B 321, 327-348.

Wittels, E.G., Lichtman, H.C. (1986). Blood group incidence and *Escherichia coli* bacterial sepsis. Transfusion 26, 533-535.

Yamamoto, F., Clausen, H., White, T., Marken, J., Hakomori, S. (1990). Molecular genetic basis of the histo-blood group ABO system. Nature 345, 229-233.

Part III   Genetics of specific infectious and multicausal diseases

Chapter 15. *Helicobacter pylori*

Helicobacter pylori is a Gram-negative, microaerophilic bacterial rod which chronically infects up to 50 % of the world's human population. Yet only a minority of the infected individuals develop disease, albeit often severe and chronic. *H. pylori* has recently been demonstrated to be involved in the pathogenesis of gastritis, peptic ulcers, and gastric cancer. In addition to host genetics, the differences in clinical outcome appear to be partly explained by the strain of the infecting bacterium, in particular whether they produce *cagA* or *picB*, or vacuolating cytotoxins. Infection is often acquired in childhood and the bacterium may then live for decades, if not for a lifetime, in the human stomach. Although *H. pylori* infection induces a Th1-mediated immune response, the response is apparently unable to eliminate the bacteria (Covacci et al 1999). Bacteria isolated from the same patient at intervals of several years have identical DNA fingerprints, and mixed infections are uncommon. Often children are infected by a strain acquired from one of their parents. Hence transmission is mainly frequent within a family or among infants within a community. Although the DNA fingerprints remain similar, the bacterium accumulates much genetic variability during infection. In 20 to 30 % of the infected persons, the infection may subsequently give rise to severe and life-threatening disease, finally resulting in hundreds of thousands of deaths each year worldwide (Covacci et al 1999, Miehlke et al 1999, Marshall et al 1998, Blaser et al 1998).

Co-evolution between man and H. pylori
Based on differences in nucleotide sequences, Asian strains of *H. pylori* can be discriminated from those isolated from the white population in Europe, North America, and South Africa. Furthermore, the m1 form of the bacterial toxin VacA is predominant in Western, Korean and Japanese isolates, whereas the m2 form is found in most Chinese isolates. Strains isolated from the indigenous Maori population of New Zealand differed from strains isolated from late colonizers of that country. These regional differences in genetically distinct bacterial strains coincide with regional differences in genetically distinct human populations, and thus suggest a co-evolution between the two species that has started at least 100,000 years ago (Achtman et al 1999, Covacci et al 1999, Ito et al 1997, Yang et al 1998, Pan et al 1998, Suerbaum et al 1998, Kersulyte et al 1999, Ende et al 1998).

Evidence for host genetic factors in susceptibility
The prevalence of *H. pylori* infection varies considerably between geographical areas and ethnic groups, very likely indicative for a contribution of genetic factors predisposing to infection. For example, in the USA the rate of acquisition of *H. pylori* infection in blacks and hispanics is twice that seen in whites. This difference remains after adjustment for age and socioeconomic factors. High prevalences of infection are found in Central and South America, Saudi Arabia, and India (Graham et al 1994).

A contribution of host genetics for acquiring *H. pylori* infection has also been clearly demonstrated in twin studies. Twin studies indicate that the correlation coefficient for the relative importance of genetic effects (heritability) on acquisition of the infection is approximately 0.6 – 0.7. Shared and non-shared environmental and bacterial factors account for the remaining variance (Graham et al 1994, Malaty et al 1994). Genetic influences for liability for peptic ulcer disease appear to be independent of genetic influences important for acquiring *H. pylori* infection (Malaty et al 2000). Other risk factors for infection are low socioeconomic status, a high population density, sharing an infected environment, and gender. Males in many populations appear to have 20 to 30 % higher rates of infection than females (Replogle et al 1995).

This sexual influence appears to interact with genetic polymorphisms at the TNF locus, perhaps because in females the expression of TNF-α is modulated by estrogen and progesterone. Kunstmann et al (1999) examined the relationship of polymorphisms in this region and *H. pylori* infection. They examined the TNF-308 polymorphism and one of five microsatellites (TNFa) in this region. Of these microsatellites TNFa is the most polymorphic, exhibiting 14 different alleles. The genotype TNF1/TNF1 of the TNF-308 polymorphism (associated with lower TNF-α expression) appeared a clear risk factor for duodenal ulcer in infected women (OR = 10.7). The frequency of the microsatellite allele TNFa6 was lower in infected females (OR = 0.51), as well as in infected females with gastric ulcer (OR = 0.13) compared to uninfected controls. Infected men with duodenal ulcer had a decreased frequency of allele TNFa10 (OR = 0.19). However, it remains to be established whether these findings represent real causal relationships between levels of TNF-α expression or linkage disequilibrium with other polymorphisms in the region (notably HLA) that cause susceptibility or resistance.

HLA association studies in patients with gastric cancer revealed that the HLA-DQA1*0102 allele was inversely associated with *H. pylori* seropositivity, but not with a reduced risk of gastric cancer. In contrast, the HLA-DRB1*1601 allele was associated with an increased gastric cancer risk. This effect was more pronounced among *H. pylori*-negative subjects. Because none of the HLA alleles were associated with both *H. pylori* infection and gastric cancer, the HLA DR-DQ alleles are linked with gastric cancer risk through other mechanisms than an increased susceptibility to *H. pylori* infection (Magnusson et al 2001).

H. pylori infection and blood groups

Peptic ulcer is a disease, which already previously, before the recognition of *H. pylori*, was known to be associated with blood group O in certain populations. This observation might be one explanation for the large geographical differences in the incidence of infection. For example, the incidence of the disease is very high in Peru, perhaps because native Americans are all of blood type O in that country (Wills 1996). In Peru the bacterium may also easily spread through the water supply. Blood group O is usually frequent in populations that have been isolated for many generations, such as in Central and South American Indians. These populations also differ in other genetic markers, such as the R blood groups. Although it has been suggested that natural selection might be responsible for the high blood group O frequencies, a founder effect cannot be excluded (Hill and Motulsky 1999).

An explanation for these observations has recently been suggested, but not yet firmly proven. *H. pylori* binds strongly to carbohydrate chains on the surface of gastric epithelial cells of people with blood group O, and less strongly to those of people with blood groups A and B (Boren et al 1993). Other epithelial structures that have been implicated in adhesion of the bacterium are lipids, gangliosides, and sulfated carbohydrates. The adhesins on the bacterial surface that bind to the epithelium are poorly understood. The best characterized of them is the BabA protein that binds the Lewis b blood group antigen on the gastric epithelium (Ilver et al 1998). The significance of this interaction is illustrated by the finding that soluble glycoproteins presenting the Lewis b antigen or antibodies directed to the Lewis b antigen inhibit bacterial binding. Furthermore, gastric tissue lacking Lewis b expression does not bind *H. pylori* (Boren et al 1993). Individuals with blood group O have more Lewis b expression as compared to those with blood groups A and B (Nguyen et al 1999). The ability of the Lewis b blood group antigen to act as receptor for *H. pylori* might thus explain the higher prevalence of ulcers associated with *H. pylori* infection in populations with high blood group O frequencies.
However, despite the suggested relation between *H. pylori* infection or gastric disease and blood groups antigens, a recent study shed doubt on this interaction. Umlauft et al (1996) failed to find an *in vivo* correlation between them in a white population. In their study, patients with *H. pylori* infection and disease had a distribution of blood group antigens similar to the control population.

Interestingly, the bacterial isolates from Le (a+b-) hosts express Lewis x more strongly than Lewis y, isolates from Le (a-b+) hosts express more Lewis y than Lewis x, while isolates from Le (a-b-) hosts express Lewis x and Lewis y approximately equal (Dunn et al 1997). These findings suggest selection of *H. pylori* strains, or phenotypic regulation, or both, by the host Lewis phenotype. Moreover, when such bacterial-host interactions are not taken into account, they may hamper the identification of important host susceptibility genes.

The Lewis blood group antigens have also been implicated in another aspect of the pathogenesis. The lipopolysaccharide of some *H. pylori* strains contains structures, which mimic blood group antigens expressed on the gastric mucosa, i.e. fucosylated Lewis x and Lewis y. This antigenic mimicry may result in immune tolerance directed against the bacterial antigens or in induction of autoantibodies directed against gastric epithelial cells, which are frequently found in patients with chronic active gastritis (Appelmelk et al 1997).

Recently more direct experimental evidence linked gastric ulcers with *H. pylori* and blood group secretor status. Secretors have a lower risk of ulceration, probably because *H. pylori* binding to secreted, soluble Lewis b antigens can be easily removed by peristalsis and ciliary activity (Anstee 2000, Nguyen et al 1999).

In conclusion, although the findings so far do not exclude that bacterial colonization, the inflammatory response, and disease expression may be influenced partly by host expression of ABO and Lewis blood group antigens, their precise relation remains to be defined. Moreover, there are likely many other and different aspects of genetic susceptibility to *H. pylori* that are still unknown.

References

Achtman, M., Azuma, T., Berg, D.E., Ito, Y., Morelli, G., Pan, Z.J., Suerbaum, S., Thompson, S.A., van der Ende, A., van Doorn, L.J. (1999). Recombination and clonal groupings within *Helicobacter pylori* from different geographical regions. Mol. Microbiol. 32, 459-470.

Anstee, D. (2000). Personal communication.

Appelmelk, B.J., Negrini, R., Moran, A.P., Kuipers, E.J. (1997). Molecular mimicry between *Helicobacter pylori* and the host. Trends Microbiol. 5, 70-73.

Blaser, M.J. (1998). Helicobacter pylori and gastric diseases. Br. Med. J. 316, 1507-1510.

Boren, T., Falk, P., Roth, K.A., Larson, G., Normark, S. (1993). Attachment of *Helicobacter pylori* to human gastric epithelium mediated by blood group antigens. Science 262, 1892-1895.

Covacci, A., Telford, J.L., Del Giudice, G., Parsonnet, J., Rappuoli, R. (1999). *Helicobacter pylori* virulence and genetic geography. Science 284, 1328-1333.

Dunn, B.E., Cohen, H., Blaser, M.J. (1997). *Helicobacter pylori*. Clin. Microbiol. Rev. 10, 720-741.

Ende, A. van der, Pan, Z.J., Bart, A., van der Hulst, R.W., Feller, M., Xiao, S.D., Tytgat, G.N., Dankert, J. (1998). CagA-positive *Helicobacter pylori* populations in China and The Netherlands are distinct. Infect. Immun. 66, 1822-1826.

Graham, D.Y., Malaty, H.M., Go, M.F. (1994). Are there susceptible hosts to *Helicobacter pylori* infection? Scand. J. Gastroenterol. Suppl. 205, 6-10.

Hill, A.V.S., Motulsky, A.G. (1999). Genetic variation and human disease: The role of natural selection. In: Evolution in health and disease, Ed. by S.C. Stearns, Oxford University press, pp. 50-61.

Ilver, D., Arnvist, A., Ogren, J., Frick, I.M., Kersulyte, D., Incecik, E.T., Berg, D.E., Covacci, A., Engstrand, L., Boren, T. (1998). *Helicobacter pylori* adhesin binding fucosylated histo-blood group antigens revealed by retagging. Science 279, 373-377.

Ito, Y., Azuma, T., Ito, S., Miyaji, H., Hirai, M., Yamazaki, Y., Sato, F., Kato, T., Kohli, Y., Kuriyama, M. (1997). Analysis and typing of the vacA gene from cagA-positive strains of *Helicobacter pylori* isolated in Japan. J. Clin. Microbiol. 35, 1710-1714.

Kersulyte, D., Chalkauskas, H., Berg, D.E. (1999). Emergence of recombinant strains of *Helicobacter pylori* during human infection. Mol. Microbiol. 31, 31-43.

Kunstmann, E., Epplen, C., Elitok, E., Harder, M., Suerbaum, S., Peitz, U., Schmiegel, W., Epplen, J.T. (1999). *Helicobacter pylori* infection and polymorphisms in the tumor necrosis factor region. Electrophoresis 20, 1756-1761.

Magnusson, P.K.E., Enroth, H., Eriksson, I., Held, M., Nyren, O., Engstrand, L., Hansson, L.E., Gyllensten, U.B. (2001). Gastric cancer and human leukocyte antigen: distinct DQ and DR alleles are associated with development of gastric cancer and infection by *Helicobacter pylori*. Cancer Res. 61, 2684-2689.

Malaty, H.M., Engstrand, L., Pedersen, N.L., Graham, D.Y. (1994). *Helicobacter pylori* infection: genetic and environmental influences, a study of twins. Ann. Intern. Med. 120, 982-986.

Malaty, H.M., Graham, D.Y., Isaksson, I., Engstrand, L., Pedersen, N.L. (2000). Are genetic influences on peptic ulcer dependent or independent of genetic influences for *Helicobacter pylori* infection? Arch. Intern. Med. 160, 105-109.

Marshall, D.G., Dundon, W.G., Beesely, S.M., Smyth, C.J. (1998). *Helicobacter pylori*-a conundrum of genetic diversity. Microbiology 144, 2925-2939.

Miehlke, S., Thomas, R., Gutierrez, O., Graham, D.Y., Go, M.F. (1999). DNA fingerprinting of single colonies of *Helicobacter pylori* from gastric cancer patients suggests infection with a single predominant strain. J. Clin. Microbiol. 245-247.

Nguyen, T., N., Barkun, A.N., Fallone, C.A. (1999). Host determinants of *Helicobacter pylori* infection and its clinical outcome. Helicobacter 4, 185-197.

Pan, Z.J., Berg, D.E., van der Hulst, R.W., Su, W.W., Raudonikiene, A., Xiao, S.D., Dankert, J., Tytgat, G.N., van der Ende, A. (1998). Prevalence of vacuolating cytotoxin production and distribution of distinct vacA alleles in *Helicobacter pylori* from China. J. Infect. Dis. 178, 220-226.

Replogle, M.L., Glaser, S.L., Hiatt, R.A., Parsonnet, J. (1995). Biologic sex as a risk factor for *Helicobacter pylori* infection in young healthy young adults. Am. J. Hum. Epidemiol. 142, 856-863.

Suerbaum, S., Smith, J.M., Bapumia, K., Morelli, G., Smith, N.H., Kunstmann, E., Dyrek, I., Achtman, M. (1998). Free recombination within *Helicobacter pylori*. Proc. Natl. Acad. Sci. USA 95, 1261912624.

Umlauft, F., Keeffe, E.B., Offner, F., Weiss, G., Feichtinger, H., Lehmann, E., Kilga-Nogler, S., Schwab, G., Propst, A., Grussnewald, K., Judmaier, G. (1996). *Helicobacter pylori* infection and blood group antigens: Lack of clinical association. Am. J. Gastroenterol. 91, 2135-2138.

Wills, C. 1996. Plagues. Harper Collins Publishers, London.

Yang, J.C., Kuo, C.H., Wang, H.J., Wang, T.C., Chang, C.S., Wang, W.C. (1998). Vacuolating toxin gene polymorphism among *Helicobacter pylori* clinical isolates and its association with m1, m2, or chimeric vacA middle types. Scand. J. Gastroenterol. 33, 1152-1157.

Chapter 16. Tuberculosis

Tuberculosis, the white plague, represents a complex interaction between human and bacterial genomes. Its spread throughout the world is enormous with approximately 30 million active cases. Indeed it is estimated that one-third of the World's population is infected with *Mycobacterium tuberculosis*, of which 10 % will ever develop overt clinical disease (Murray et al 1990). Annually tuberculosis accounts for more than 3 million deaths. Moreover, co-infection with HIV is a major cause for the increased incidence that has occurred during the past years.

Most infections run silently and there is no comprehensive understanding of the factors that determine clinical disease, reactivation, and reinfection. Host genetics clearly have a profound impact on the course of infection as shown in twin studies and in different racial groups. Apparently the majority of infected persons is resistant to developing clinical tuberculosis. Evidence for a contribution of host genetics to the development of tuberculosis was already obtained in the 1930's and 1940's, when Diehl and von Verscheur (1936) and Kallmann and Reisner (1942) found higher concordance rates among monozygous (65 – 88 %) than among dizygous (25 – 28 %) twin pairs. Further evidence for host genetic factors in tuberculosis came from studies in mice and rabbits (Lurie 1941, Blackwell 1989).

Human genetic studies
Bellamy and Hill (1998) used the affected sib-pair method to carry out a genome screen on 92 sib pair families from The Gambia and South Africa using 282 highly informative microsatellite markers. Five of the markers, d3s1262, d15s128, d6s276, d8s272, dxs984, showed significant evidence of co-segregation with tuberculosis, clearly indicating the multigenic nature of tuberculosis susceptibility. One of the markers tested, d6s276 is located in the MHC complex on chromosome 6p. In a second set of 81 sibpairs from the same countries, 22 markers were genotyped to identify whether any of the provisionally identified regions contained a potential tuberculosis-susceptibility gene. Markers on chromosomes 15q and Xq showed suggestive evidence of linkage to tuberculosis (LOD = 2.00 and 1.77 respectively). Such an X chromosome susceptibility gene may contribute to the excess of males with tuberculosis that is observed in many different populations. In some countries twice as many males compared with females are affected by tuberculosis. An independent analysis, designated common ancestry using microsatellite (CAM) mapping, supported the potential identification of these susceptibility loci. CAM mapping is an extended form of homozygosity mapping that looks for an association between disease incidence and regions of high homozygosity/heterozygosity (Bellamy et al 2000).

Several studies addressed the association of MHC with tuberculosis. Reduced prevalences of HLA-DR3, HLA-DQw3, HLA-DQB1*0402, HLA-DR4 or HLA-DR8 have been observed in different studies that compared tuberculosis patients with controls. Otherwise, Goldfeld et al (1998) found evidence for a significant association between the HLA-DQB1*0503 allele and increased susceptibility to tuberculosis in Cambodian patients.
In South India, 29 % of the total risk of developing pulmonary tuberculosis was attributable to HLA-DR2 (subtypes DRB1*1501-02) (McNicholl et al 2000). The latter asso-

ciation has been suggested to reflect the role of HLA-DR2 in the antigen presentation of the 38-kDA protein of *M. tuberculosis* to T cells. This protein may therefore have an important role in the development of pulmonary tuberculosis. Other suggested explanations for the susceptibility to tuberculosis in DR2-positive patients include reduced cell-mediated responses, low plasma lysozyme levels, and failure of anti-tubercular drugs. The effect of HLA-DR2 in immunity to tuberculosis therefore appears complex. In an association study with sputum smear-positive tuberculosis from Surabaya, Indonesia, the attributable risk for tuberculosis associated with HLA-DR2 was 36 % (confirming the findings from South India), and with HLA-DQw1 39 %, while HLA-DQw3 had a preventive fraction of 57 %. This latter finding means that HLA-DQw3 could prevent 57 % of clinical tuberculosis cases (McNicholl et al 2000).

In contrast to these HLA-related findings, a Brazilian family study failed to find linkage of tuberculosis susceptibility to HLA genes (McNicholl et al 2000). The reported findings, therefore, appear not generally applicable in different continents. As indicated throughout this text, such interpopulation heterogeneity in HLA associations appears to be very common in other diseases as well, and may result from complex interactions between HLA, background genetic make-up, and polymorphisms in immunodominant antigens from the pathogen.

In a case control study in an African population in The Gambia, involving many candidate genes, Bellamy and Hill (1998) found persons with the so called homozygous tt genotype (referring to a postition-352 Taq1 restriction enzyme polymorphism) for the vitamin D receptor gene significantly underrepresented among tuberculosis cases compared to controls. The finding suggests that people with this genotype are protected against tuberculosis and supports the suggestion that vitamin D may protect against tuberculosis (see chapter 11). Furthermore, four out of at least 11 *NRAMP1* gene variants were all strongly associated with increased susceptibility to clinical tuberculosis. These included a microsatellite 5' to the transcription start site, a TGTG deletion in the 3' untranslated region, a G to C transversion in intron 4, and a non-conservative amino acid substitution at codon 543 (5'(CA)n, 3'UTR, INTR4, and D543N). Because the microsatellite and the G to C conversion, as well as the TGTG deletion and the codon 543 mutation are in linkage disequilibrium, the results are not independent. Persons who were heterozygous carriers of both the INTR4 and 3'UTR *NRAMP1* gene variants were significantly overrepresented among the tuberculosis cases compared to persons who did not possess either of the variant alleles (OR = 4.07; 95 % CI = 1.86-9.12). The differences between the groups were nonetheless small and people with any *NRAMP1* genotype may develop tuberculosis. Unfortunately none of these polymorphisms have been shown to be functional so far.

Although these studies clearly appear to indicate the role of the vitamin D receptor and the *NRAMP1* gene in tuberculosis in The Gambia, it remains to be elucidated whether they are important in other populations as well. So far, linkage analysis between the phenotype of disease during a tuberculosis outbreak in an extended multisib Canadian Indian family and allelic variants in the region of chromosome 2 encoding *NRAMP1* has revealed a significant LOD score (Skamene 1994). Analysis of multicase tuberculosis families in Brazil also showed that gene markers tightly linked to *NRAMP1*, such as IL8RB and D2S1471, were associated with susceptibility to clinical tuberculosis. In contrast, *NRAMP1* itself was not, suggesting that the effects of tuberculosis susceptibility genes may indeed differ in different populations (McNicholl et al

2000). In Koreans a significant association was found between tuberculosis patients and the 3'UTR variant allele of the *NRAMP1* gene (OR = 1.845; 95 % CI =1.097 – 3.104; *P* = 0.02). This variant is very uncommon in Caucasians, but is present in Koreans and West Africans (Ryu et al 2000). These observations may therefore explain in part why results differ in different populations, as well why Africans and Koreans have greater susceptibility to tuberculosis than Caucasians.

Haptoglobin is an antioxidant and hemoglobin-binding protein, which is expressed by a genetic polymorphism as three major phenotypes: 1-1, 2-1, and 2-2. Functional differences between these phenotypes are explained by modulation of oxidative stress, prostaglandin synthesis, and immune response (see chapter 28). In a Zimbabwean study, the haptoglobin (Hp) phenotype distributions did not differ between tuberculosis patients and controls. However, mortality during an 18-month follow-up period among cases with Hp phenotype 2-2 was greater (33 %) compared with cases with Hp 2-1 (19 %) and Hp 1-1 (10 %). The odds of dying were 6.1-fold greater with Hp 2-2 than with 1-1 (95 % CI = 1.04-35.1, *P* = 0.04) (Kasvosve et al 2000).

Several studies have revealed that rare immunodeficiency disorders, such as adenosine deaminase deficiency, chronic granulomatous disease, and disorders caused by mutations in the genes encoding IL-12, the IL-12 receptor, and the IFN-γ receptor predispose to severe disseminated infections with the attenuated Bacillus Calmette-Guérin (BCG) vaccine strain or with atypical mycobacteria as *Mycobacterium fortuitum*, *Mycobacterium chelonei* and *Mycobacterium avium* (Casanova et al 1995, Jouanguy et al 1996, Jouanguy et al 1999, Newport et al 1996). The IL-12 and IFN-γ pathways are thus important in clearing infections with intracellular bacteria. Although these disorders may point to interesting candidate genes, it remains to be determined whether common polymorphisms in these genes might influence tuberculosis susceptibility in the general population or in different ethnic groups (Bellamy and Hill 1998).

Animal genetic studies
Inbred families of rabbits can be divided into two distinct groups based on their susceptibility to an experimental inoculation with virulent strains of *Mycobacterium bovis* and *M. tuberculosis*. These differences are explained by the ability of the resistant rabbits to inactivate more tubercle bacilli than the susceptible rabbits (Lurie 1941, Lurie et al 1952).
Similarly, inbred strains of mice were divided in resistant and susceptible to BCG vaccine strains and to other mycobacteria, as well as to other intracellular bacteria such as *Leishmania* and *Salmonella*. The candidate gene responsible for this different susceptibility was isolated by positional cloning and was designated natural resistance-associated macrophage protein 1 (*Nramp1*) (Blackwell 1989, Vidal et al 1993). A single non-conservative amino acid substitution of glycine by aspartic acid at position 169 correlated with susceptibility (Malo et al 1994). Proof that *Nramp1,* and not a closely linked gene, is responsible for the different susceptibility to mycobacteria was obtained by the development of knockout mice, which displayed a similar susceptible phenotype as homozygous *Nramp1*^{D169} mice and by transgenic expression of the *Nramp1*^{G169} allele (Govoni et al 1996, Vidal et al 1995). After the work in mice, the human homolog of the *Nramp1* gene, designated *NRAMP1*, has been cloned and located on human chromosome 2q35 (Cellier et al 1994). Several polymorphisms in the

human *NRAMP1* gene have subsequently been described (Liu et al 1995). As indicated above, in West Africans from The Gambia, allelic variation in the 3' untranslated region and in the promoter region of the human *NRAMP1* gene was associated with susceptibility to sputum-positive tuberculosis (Bellamy et al 1998).

Multigenic non-Mendelian control of susceptibility to tuberculosis was examined in recombinant congenic mouse strains (Kramnik et al 1998, 2000). Kramnik et al (2000) established that resistance and susceptibility to virulent *M. tuberculosis* is indeed a complex genetic trait. They identified a new locus, designated *sst1* (susceptibility to tuberculosis 1), on mouse chromosome 1, which is located 10-19 cM distal to *Nramp1*. Mice segregating at the *sst1* locus exhibit marked differences in the growth rates of virulent tubercle bacilli, interstitial granulomas, and necrosis in their lungs (see below). However, the resistant allele of *sst1* is not sufficient to confer full protection against virulent *M. tuberculosis*, further underlining that other genes located outside of the *sst1* locus also contribute to controlling tuberculosis infection (Kramnik et al 2000).

Comparative pathogenesis in resistant and susceptible murine strains: role of Nramp1.
As mentioned above, several mice strains differ in susceptibility to *M. tuberculosis* infection. To elucidate some of the mechanisms behind these differences, Kramnik et al (1998) compared the pathogenesis of infection in C57BL/6J (B6) and C3HeB/FeJ (C3H) mice, as a resistant and susceptible strain respectively. Early growth of virulent *M. tuberculosis* was equal in both strains, but differences in bacterial growth appeared at the time the immune response developed, i.e. at two weeks after infection. The inability to control bacterial growth in the susceptible mice is reflected in disorganized granuloma formation. Finally the resistant mice developed a slowly progressing chronic lung disease, while the susceptible mice succumbed from disseminated tuberculosis. Together these data indicate differences in the ability to control disease progression, rather than differences in resisting early infection. Genetic analysis of F2 hybrids of these strains failed to show a protective effect of the resistant allele of *Nramp1*, but, as mentioned above, identified the susceptibility locus *sst1* on chromosome 1. The incomplete penetrance of this locus further indicated that susceptibility is a multigenic trait.

Further evidence for multigenic control and a possible confirmation of a small or negligible effect of *Nramp1* comes from the observation that the resistant allele of *Nramp1* appears to control bacterial multiplication during the early stage of infection only. In mice infection with the non-pathogenic *M. bovis* strain BCG is biphasic, with an early phase, 0-3 weeks post-infection, in which there is rapid proliferation of bacteria in mice bearing the susceptible allele of *Nramp1*, but not in mice bearing the resistant allele (Gros et al 1981, Skamene et al 1998). *Nramp1* also does not affect infection during the chronic stages of infection (reactivation disease). Another important finding, explaining several of the observations made, is that *Nramp1* appears to act differently on different mycobacterial strains. For example, North and Medina (1996) have found that the gene affects the growth of the attenuated BCG vaccine strain, but does not influence the growth of virulent *M. tuberculosis*.

The contribution of knockout mice in understanding tuberculosis
Mice with targeted gene disruptions are very powerful tools to demonstrate the involvement of genes in the pathogenesis of disease and to point to candidate genes for subsequent studies in humans. Mice deficient for the genes encoding interferon-γ, the interferon-γ receptor, IL-12, inducible nitric oxide synthase (iNOS), and $\beta_2$microglobulin are all very susceptible to mycobacterial infection (Cooper et al 1993, Flynn et al 1992, Kamijo et al 1993, MacMicking et al 1997). These findings thus underline the important role of cellular immunity, especially MHC class-I restricted T cell immunity, in limiting infections with mycobacteria.

References

Bellamy, R.J., Hill, A.V.S. (1998). Host genetic susceptibility to human tuberculosis. In: Genetics and tuberculosis, Wiley, Chichester (Novartis Foundation Symposium 217), 3-23.

Bellamy, R., Ruwende, C., Corrah, T., McAdam, K.P., Whittle, H.C., Hill, A.V.S. (1998). Variations in the *NRAMP1* gene and susceptiblity to tuberculosis in West Africans. N. Engl. J. Med. 338, 640-644.

Bellamy, R., Beyers, N., McAdam, K.P.W.J., Ruwende, C., Gie, R., Samaai, P., Bester, D., Meyer, M., et al. (2000). Genetic susceptibility to tuberculosis in Africans: A genome-wide scan. Proc. Natl. Adac. Sci. USA 97, 8005-8009.

Blackwell, J. (1989). The macrophage resistance gene *Lsh/Ity/Bcg*. Res. Immunol. 140, 767-828.

Casanova, J.-L., Jouanguy, E., Lamhamedi, S., Blanche, S., Fischer, A. (1995). Immunological conditions of children with BCG disseminated infection. Lancet 346, 581.

Cellier, M., Govoni, G., Vidal, S., Kwan, T., Groulx, N., Liu, J., Sanchez, F., Skamene, E., Schurr, E., Gros, P. (1994). Human natural resistance-associated macrophage protein: cDNA cloning, chromosomal mapping, genome organization, and tissue-specific expression. J. Exp. Med. 180, 1741-1752.

Cooper, A.M., Dalton, D.K., Stewart, T.A., Griffin, J.P., Russell, D.G., Orme, I.M. (1993). Disseminated tuberculosis in interferon-γ gene-disrupted mice. J. Exp. Med. 178, 2242-2247.

Diehl, K., von Verscheur, O. (1936). Der Erbeinfluss bei den Tuberculose. Gustav Fischer Jena.

Flynn, J.L., Goldstein, M.M., Triebold, K.J., Koller, B., Bloom, B.R. (1992). Major histocompatibility complex class-I restricted T cells are required for resistance to *Mycobacterium tuberculosis*. Proc. Natl. Acad. Sci. USA 89, 12013-12017.

Goldfeld, A.E., Delgado, J.C., Thim, S., Bozon, Ugilaloro, A.M., Turbay, D., Cohen, C., Yunis, E.J. (1998). Association of an HLA-DQ allele with clinical tuberculosis. JAMA 279, 226-228.

Govoni, G., Vidal, S., Gauthier, S., Skamene, E., Malo, D., Gros, P. (1996). The *Bcg/Ity/Lsh* locus: genetic transfer of resistance to infections in C57BL/6J mice transgenic for the *Nramp1*^{G169} allele. Infect. Immun. 64, 2923-2929.

Gros, P., Skamene, E., Forget, A. (1981). Genetic control of natural resistance to *Mycobacterium bovis* (BCG) in mice. J. Immunol. 127, 2417-2421.

Jouanguy, E., Altare, F., Lamhamedi, S., Revy, P., Emile, J.F., Newport, M., Levin, M., Blanche, S., Seboun, E., Fischer, A., Casanova, J.L. (1996). Interferon-γ-receptor deficiency in an infant with fatal Bacille Calmette-Guérin infection. N. Engl. J. Med. 335, 1956-1961.

Jouangu, E., Lamhamedi-Cherradi, S., Lammas, D., Dorman, S.E., Fondaneche, M.C., et al (1999). A human IFNγR1 small deletion hot spot associated with dominant susceptibility to mycobacterial infection. Nat. Gent. 21, 370-378.

Kallmann, F.J., Reisner, D. (1942). Twin studies on the significance of genetic factors in tuberculosis. Am. Rev. Tuberc. 47, 549-574.

Kamaijo, R., Le, J., Shapiro, D., Havell, E.A., Huang, S., Aguet, M., Bosland, M., Vilcek, J. (1993). Mice that lack the interferon-γ receptor have profoundly altered responses to infection with Bacillus Calmette-Guérin and subsequent challenge with lipopolysaccharide. J. Exp. Med. 178, 1435-1440.

Kasvosve, I., Gomo, Z.A., Mvundura, E., Moyo, V.M., Saungweme, T., Khumalo, H., Gordeuk, V.R., Boelaert, J.R., Delanghe, J.R., De Bacquer, D., Gangaidzo, I.T. (2000). Haptoglobin polymorphism and mortality in patients with tuberculosis. Int. J. Tuberc. Lung Dis. 4, 771-775.

Kramnik, I., Demant, P., Bloom, B.B. (1998). Susceptibility to tuberculosis as a complex genetic trait: analysis using recombinant congenic strains of mice. In: Genetics and tuberculosis, Wiley, Chichester (Novartis Foundation Symposium 217), 120-137.

Kramnik, I., Dietrich, W.F., Demant, P., Bloom, B.R. (2000). Genetic control of resistance to experimental infection with virulent *Mycobacterium tuberculosis*. Proc. Natl. Acad. Sci USA 97, 8560-8565.

Liu, J., Fujiwara, M., Buu, N.T., Sanchez, F.O., Cellier, M., Paradis, A.J., Frappier, D., Skamene, E., Gros, P., Morgan, K. et al. (1995). Identification of polymorphisms and sequence variants in the human homologue of the mouse natural resistance-associated macrophage protein gene. Am. J. Hum. Genet. 56, 845-853.

Lurie, M.B. (1941). Hereditary, constitution and tuberculosis. An experimental study. Am. Rev. Tuberc. 44, S1-S125.

Lurie, M.B., Abramson, S., Heppleston, A.G. (1952). On the response of genetically resistant and susceptible rabbits to the quantitative inhalation of human-type tubercle bacilli and the nature of resistance to tuberculosis. J. Exp. Med. 95, 119-134.

MacMicking, J.D., North, R.J., La Course, R., Mudget, J.S., Shah, S.K., Nathan, C.F., (1997). Identification of nitric oxide synthase as a protective locus against tuberculosis. Proc. Natl. Acad. Sci. USA 94, 5243-5248.

Malo, D., Vogan, K., Vidal, S., Hu, J., Cellier, M., Schurr, E., Fuks, A., Bumstead, N., Morgan, K., Gros, P. (1994). Haplotype mapping and sequence analysis of the mouse *Nramp* gene predict susceptibility to infection with intracellular parasites. Genomics 23, 51-61.

McNicholl, J.M., Downer, M.V., Udhayakumar, V., Alper, C.A., Swerdlow, D.L. (2000). Host-pathogen interactions in emerging and re-emerging diseases: A genomic perspective of tuberculosis, malaria, human immunodeficiency virus infection, hepatitis B, and cholera. Annu. Rev. Public Health 21, 15-46.

Murray, C.J.L., Styblo, K., Rouillon, A. (1990). Tuberculosis in developing countries: burden, intervention and cost. Bull. Int. Union Tuberc. Lung Dis. 65, 6-24.

Newport, M.J., Huxley, C.M., Huston, S., Hawrylowicz, C.M., Oostra, B.A., Williamson, R., Levin, M. (1996). A mutation in the interferon-γ-receptor gene and susceptibility to mycobacterial infection. N. Engl. J. Med. 335, 1941-1949.

North, R.J., Medina, E. (1996). Significance of the antimicrobial resistance gene, Nramp1, in resistance to virulent *Mycobacterium tuberculosis* infection. Res. Immunol. 147, 493-499.

Ryu, S., Park, Y.K., Bai, G.H., Kim, S.J., Park, S.N., Kang, S (2000). 3'UTR polymorphisms in the *NRAMP1* gene are associated with susceptibility to tuberculosis in Koreans. Int. J. Tuberc. Lung Dis. 4, 577-580.

Skamene, E. (1994). The *Bcg* gene story. Immunobiology 191, 451-460.

Skamene, E., Schurr, E., Gros, P. (1998). Infection genomics: *Nramp1* as a major determinant of natural resistance to intracellular infections. Annu. Rev. Med. 49, 275-287.

Vidal, S.M., Malo, D., Vogan, K., Skamene, E., Gros, P. (1993). Natural resistance to infection with intracellular parasites: isolation of a candidate gene for Bcg. Cell 73, 4469-485.

Vidal, S., Tremblay, M.L., Govoni, G., Gauthier, S., Sebastiani, G., Malo, D., Skamene, E., Olivier, M. Jothy, S., Gros, P. (1995). The *Ity/Lsh/Bcg* locus: natural resistance to infection with intracellular parasites is abrogated by disruption of the *Nramp1* gene. J. Exp. Med. 182, 655-666.

Chapter 17. Bacterial meningitis and sepsis

Bacterial meningitis and septicemia are life-threatening diseases of young children. The most common causative pathogens are *Neisseria meningitidis, Streptococcus pneumoniae* and *Haemophilus influenzae*. These micro-organisms colonize the nasopharynx and are transmitted between individuals by the inhalation of droplets (Cartwright et al 1987). Most infected people are asymptomatic carriers. For example, approximately one in every 15 individuals carries the bacterium *N. meningitidis* in his or her nasopharynx, but only about 1 of 2,000 of these carriers develops meningococcal disease. Virulent strains may become invasive by passing the nasopharyngeal mucosa. Depending on the virulence of the infecting strain and the host response, microbial invasion of the bloodstream and the cerebrospinal fluid may subsequently occur. The host inflammatory response to infection contributes significantly to the clinical outcome of the infection. Pathophysiological changes in the course of meningitis are caused by bacterial cell-wall components and host factors, such as cytokines, arachidon acid metabolites, platelet activating factors, granulocytes, and reactive oxygen intermediates (Quagliarello and Scheld 1992). The clinical picture may be very fulminant leading to death within one to two days. Early observations indicated that racial differences in the incidence of bacterial meningitis occurred, thus pointing to a role of genetic factors in the disease. Black and Hispanic populations are at 2-4 times greater risk than Caucasians (Feldman et al 1976, Floyd et al 1974, Fraser et al 1974, Wenger et al 1990).

Pathogenesis
Excellent reviews on the pathogenesis have been written (Kornelisse et al 1995). Briefly, after adherence to epithelial cells invasion across the nasopharyngeal mucosa takes place by an endocytotic process (*N. meningitidis*), or through an intercellular route (*H. influenzae*). The intercellular route occurs through separations in the apical tight junctions of the columnar epithelial cells. After crossing the mucosal barrier, bacteria must overcome host defenses to survive in the bloodstream and to invade the meninges. Encapsulation is therefore an important virulence factor that inhibits neutrophil phagocytosis and lysis by classical complement activation. Bacterial cell-wall components are responsible for initiating the meningeal inflammation. A number of inflammatory cytokines, including TNF-α, IL-1β, and IL-6 also play a pivotal role in triggering the cascades of events leading to meningeal inflammation and sepsis.
TNF-α and IL-1β induce phospholipase A_2 activity, which, together with bacterial antigens, triggers the production of proinflammatory arachidonic acid metabolites from a variety of cells. Bacterial LPS and a variety of cytokines induce platelet-activating factor (PAF), which further recruits and activates polymorphonuclear leukocytes and monocytes. PAF together with LPS and TNF-α induce microvascular damage (Cabellos et al 1992). In septicemia, bacterial toxins activate plasma factors, such as complement and clotting cascades, and cells as neutrophils, macrophages, platelets, and endothelial cells. These activated cells may subsequently produce biologically active substances as cytokines, kinins, eicosinoids, PAF, and nitric oxide (NO) (Brandtzaeg 1996).
Sepsis is associated with excessive complement activation. Complement plays an important role in non-specific host defense mechanism, leading to lysis of bacteria, enhancement of phagocytosis, and neutralization of endotoxin. However, overstimu-

lation may lead to tissue injury, because complement peptides generated during activation have several pro-inflammatory effects (Frank and Fries 1991).

Coagulation disorders and abnormalities of fibrinolysis are also common during bacterial sepsis and they may result in disseminated intravascular coagulation, vascular disruption, necrosis, and outgrowth of meningococci. Levels of natural inhibitors of coagulation are decreased. The fibrinolytic system initially becomes activated by tissue plasminogen activator (t-PA). Subsequently, increased levels of plasminogen activator inhibitor (PAI)-1 inhibit fibrinolysis. Decreased α-2-antiplasmin levels and a high ratio PAI-1/t-PA result in an ineffective fibrinolysis and are associated with a worse outcome in patients with meningococcal septic shock (Kornelisse et al 1996, Brandtzaeg et al 1990). Ultimately the combined actions of bacterial and host factors may lead to shock and multiple organ failure. Up to 50 % of patients with septic shock die from refractory shock, which is characterized by massive intravascular fibrin deposition and formation of microthrombi in various organs.

Genetic factors

Because so many host proteins are involved in the causation or modulation of disease symptoms during bacterial meningitis and sepsis, it is no surprise that several genetic factors have been shown to be associated with the severity of disease. Defects in the complement system, notably in the early components of the alternative pathway, C3, and in the terminal components C5-C9 are associated with an increased risk for meningococcal disease (Figueroa and Densen 1991). Complement deficiencies are predominantly associated with infection by *N. meningitidis* serogroups X, Y, Z, W135, 29E, and to a lesser extent C. For a further discussion on the role of complement factors in bacterial meningitis see chapter 8. As discussed in that chapter, binding of mannose-binding protein (MBP) to bacteria is another route for activating complement. In contrast to deficiencies in other complement factors, MBP-deficiencies may occur frequently, i.e. in up to 5 to 15 % of the Caucasians and in up to 30 % of the Africans. In England variants of the gene for MBP have been found to account for 32 % of the cases of meningococcal disease (Hibberd et al 1999). Recently, Bax et al (1999) described a family with MBP deficiency in which members of three generations experienced meningococcal meningitis.

The role of antibodies and opsonization in host defense against *N. meningitidis* is illustrated by polymorphisms of the Fcγ receptor IIa (CD32) on phagocytes. Genetic polymorphic forms of FcγRIIa have different functions. FcγRIIa-R/R131 neutrophils phagocytose IgG2-opsonized *N. meningitidis* less effectively than FcγRIIa-H/H131 neutrophils. This is a relevant finding, because IgG2 is the major isotype produced in response to bacterial capsular polysaccharides. In a retrospective study it was found that the homozygous FcγRIIa-H/H131 allotype is significantly more present in children who survived fulminant meningococcal septic shock than found in a healthy white population (OR = 2.67; 95% CI = 1.09-6.53). Thus the poor IgG2 binding R131 allotype of FcγRIIa is a risk factor for a poor outcome of meningococcal invasive infections (Bredius et al 1994). Notably, the Japanese population is in majority FcγRIIa-H/H131, which perhaps explains the low incidence of infections with encapsulated micro-organisms in this country (Musser et al 1990). A further role of antibodies is indicated by the high incidence of infections in persons with IgA deficiency (Hammarstrom and Smith 1999).

Genetically determined variability in the expression of pro-inflammatory (such as TNF-α) and anti-inflammatory cytokines (such as IL-10) likely determines the susceptibility to, or the severity of bacterial meningitis and sepsis. Cytokine production is clearly under genetic control. For example, approximately 60 % of the variation in TNF-α production upon *ex-vivo* stimulation of whole blood with endotoxin, and 75 % of the variation in IL-10 production is genetically determined (Westendorp et al 1997). In the TNF-α gene promoter two G to A transition polymorphisms at positions −308 and −238 have been described. Yet only few studies examined the influence of genetically determined variability in cytokine production on the outcome of severe sepsis and meningitis. Unfortunately, the results are not unequivocal and sometimes difficult to interpret (Nadel et al 1996, Stüber et al 1996, Westendorp et al 1997). This may be, because many factors are likely involved. Not only are absolute levels of cytokines important, but also the inducibility and stability of mRNA, as well as the capacity to down-regulate secretion. Interpretation is further complicated because polymorphisms in the TNF-α gene may not be functional, and because timing may influence the levels of measured cytokines. The latter point evidently favors the use of genetic tests.

As described above, defective fibrinolysis appears associated with a poor outcome of meningococcal shock. Concentrations of PAI-1, an inhibitor of fibrinolysis, were significantly higher in a group of non-survivors at a similar TNF-α concentration, which is one of the inducers of PAI-1 release (Kornelisse et al 1996). A single base pair insertion/deletion polymorphism in the promoter of the PAI-1 gene has been associated with differences in release of PAI-1. The promoter containing the 4G deletion allele produced six times more mRNA than the 5G insertion allele in response to IL-1β. The increase in PAI-1 transcription and in plasma PAI-1 activity associated with the 4G allele is related to a defective binding of a repressor of transcriptase to the polymorphic site, findings which together indicate that the polymorphism in the PAI-1 gene is indeed functional (Dawson et al 1993). A recent study indicated that patients with the 4G/4G genotype had significantly higher PAI-1 concentrations than those with the 4G/5G or 5G/5G genotype, and that they had an increased risk of death (OR = 2.0; 95 % CI = 1.0 − 3.8) (Hermans et al 1999). In this and another study, no difference was found in the frequencies of the three genotypes between patients with meningococcal disease and controls (Westendorp et al 1999). Thus the polymorphism is not associated with increased susceptibility to acquiring meningococcal disease, but high amounts of the inhibitor of fibrinolysis PAI-1 may be an important pathway promoting widespread intravascular thrombosis and shock. Therapeutic measures that reduce the concentration of PAI-1 or increase that of tissue plasminogen activator may therefore be beneficial in treating the disease.

A summary of some polymorphic genes associated with increased susceptibility to bacterial meningitis and sepsis is given in Table 8.

Table 8. Polymorphic genes associated with increased susceptibility to bacterial meningitis and sepsis

Gene or locus	Location	Polymorphism	Frequency	Population	Study	Phenotype	Ref.
FcγRIIa	aa 131	R/H		Caucasian	retrospective case control	R131 allele is associated with increased incidence meningococcal sepsis; poor IgG2 binding	Bredius et al 1994
complement deficiencies		several			several	increased incidence meningococcal disease; defective bacterial lysis	Figuerora and Densen 1991; Swart et al 1993
mannose-binding protein variants	aa 52, 54, and 57	several	0.12 – 0.25 (0.3)	Caucasian	case control (African)	increased susceptibility to meningococcal disease	Hibberd et al 1999
PAI-1	-675 bp	5G/4G	0.53/0.47	Caucasian	retrospective case control	4G allele is associated with high PAI-1 and poor outcome of meningococcal septic shock	Hermans et al 1999
TNF-α	-308bp	G/A			case control	A allele associated with increased risk on fatal meningococcal disease	Nadel et al 1996

References

Bax, W.A., Cluysemaer, O.J., Bartelink, A.K., Aerts, P.C., Ezekowitz, R.A., van Dijk, H. (1999). Association of familial deficiency of mannose-binding lectin and meningococcal disease. Lancet 354, 1094-1095.

Brandtzaeg, P., Joo, G.B., Brusletto, B., Kierulf, P. (1990). Plasminogen activator inhibitor 1 and 2, alpha-2-antiplasmin, plasminogen, and endotoxin levels in systemic menigococcal disease. Thromb. Res. 57, 271-278.

Brandtzaeg, P. (1996). Systemic meningococcal disease: clinical pictures and pathophysiological background. Rev. Med. Microbiol. 7, 63-72.

Bredius, R.G., Derkx, B.H., Fijen, C.A., de Wit T.P., de Haas, M., Weening, R.S., van de Winkel, J.G., Out, T.A. (1994). Fc gamma receptor IIa (CD32) polymorphism in fulminant menigococcal septic shock in children. J. Infect. Dis. 170, 848-853.

Cabellos, C., Macintyre, D.E., Forrest, M., Burroughs, M., Prasad, S., Tuomanen, E. (1992). Differing roles for platelet-activating factor during inflammation of the lung and subarachnoidal space. J. Clin. Invest. 90, 612-618.

Cartwright, K.A., Stuart, J.M., Jones, M.D., Noah, N.D. (1987). The Stonehouse survey: nasopharyngeal carriage of meningococci and *Neisseria lactamica*. Epidemiol. Infect. 99, 591-601.

Dawson, S.J., Wiman, B., Hamsten, A., Green, F., Humphries, S., Henney, A.M. (1993). The two allele sequences of a common polymorphism in the promoter of the plasminogen activator inhibitor-1 (PAI-1) gene respond differently to interleukin-1 in HepG2 cells. J. Biol. Chem. 268, 10739-10745.

Feldman, R.A., Koehler, R.E., Fraser, D.W. (1976). Race-specific differences in bacterial meningitis deaths in the United States, 1962-1968. Am. J. Publ. Health 66, 392-396.

Figueroa, J.E., Densen, P. Infectious diseases associated with complement deficiencies. Clin. Microbiol. Rev. 4, 359-395.

Floyd, R.F., Federspiel, C.F., Schaffner, W. (1974). Bacterial meningitis in urban and rural Tenessee. Am. J. Epidemiol. 99, 395-407.

Frank, M.M., Fries, L.F. (1991). The role of complement in inflammation and phagocytosis. Immunol. Today 12, 322-326.

Fraser, D.W., Geil, C.C., Feldman, R.A. (1974). Bacterial meningitis in Bernalillo County, New Mexico: a comparison with three other American populations. Am. J. Epidemiol. 100, 29-34.

Hammarstrom, L., Smith, C.I.E. (1999). Genetic approach to common variable immunodeficiency and IgA deficiency. In: Ochs, H.D., Smith, C.I.E., Puck, J.M. (eds.): Primary immunodeficiency diseases: A molecular and genetic approach. Oxford University Press, New York, pp. 250-262.

Hermans, P.W.M., Hibberd, M.L., Booy, R., Daramola, O., Hazelet, J.A., de Groot, R., Levin, M, and the Meningococcal Research Group. (1999). 4G/5G promoter polymorphism in the plasminogen-activator-inhibitor-1 gene and outcome of meningococcal disease. Lancet 354, 556-560.

Hibberd, M.L., Sumiya, M., Summerfield, J.A., Booy, R., Levin, M. (1999). Association of variants of the gene for mannose-binding lectin with susceptibility to meningococcal disease. Lancet 353, 1049-1053.

Kornelisse, R.F., de Groot, R., Neijens, H.J. (1995). Bacterial meningitis: Mechanisms of disease and therapy. Eur. J. Pediatr. 154, 85-96.

Kornelisse, R.F., Hazelet, J.A., Savelkoul, H.F.M., Hop, W.C.J., Suur, M.H., Borsboom, A.N.J., Risseeuw-Appel, I.M., van der Voort, E., de Groot, R. (1996). The relationship between plasminogen activator inhibitor-1, proinflammatory and counterinflammatory mediators in children with meningococcal septic shock. J. Infect. Dis. 173, 1148-1156.

Musser, J.M., Kroll, J.S., Granoff, D.M. (1990). Global genetic structure and molecular epidemiology of encapsulated *Haemophilus influenzae*. Rev. Infect. Dis. 12, 75-111.

Nadel, S., Newport, M.J., Booy, R. Levin, M. (1996). Variation in the tumour necrosis factor-α gene promoter region may be associated with death from meningococcal disease. J. Infect. Dis. 174, 878-880.

Stüber, F., Petersen, M., Bokelmann, F., Schader, U. (1996). A genomic polymorphism within the tumour necrosis factor locus influences plasma tumour necrosis factor-α concentrations and outcome of patients with severe sepsis. Crit. Care Med. 24, 381-384.

Swart, A.G., Fijen, C.A., te Bulte, M.T., Daha, M.R., Dankert, J. Kuijper, E.J. (1993). Complement deficiencies and meningococcal disease in The Netherlands. Ned. Tijdschr. Geneesk. 137, 1147-1152.

Wenger, J.D., Hightower, A.W., Facklam, R.R., Gaventa, S., Broome, C.V., and the Bacterial Meningitis Study group. (1990). Bacterial meningitis in the United States, 1986: report of a multistate surveillance study. J. Infect. Dis. 162, 1316-1323.

Westendorp, R.G.J., Langermans, J.A.M., Huizinga, T.W.J., Eloualis, A.H., Verweij, C.L., Boomsma, D.I., Vandenbrouke, J.P. (1997). Genetic influence on cytokine production and fatal meningococcal disease. Lancet 349, 170-173.

Westendorp, R.G.J., Hottenga, J.-J., Slagboom, P.E. (1999). Variation in plasminogen-activator-inhibitor-1 gene and risk of meningococcal septic shock. Lancet 354, 561-563.

Quagliarello, Scheld, W.M. (1992). Bacterial meningitis: pathogenesis, pathophysiology and progress. N. Engl. J. Med. 327, 864-872.

Chapter 18. Resistance to influenza virus

The Mx proteins

Influenza virus is primarily an infection of the upper respiratory tract and deeper airways. Annual epidemics occur seasonally and refinfection is common due to antigenic variation. Work on the genetic control of influenza virus infections is an example of how knowledge obtained in murine models may find his way to man. The work began more than 30 years ago with the observation that inbred mouse strains fell into distinct resistant and susceptible phenotypes upon experimental inoculation with several orthomyxoviruses. By analyzing F1 and backcross mice, Lindenmann (1964) subsequently demonstrated that a single autosomal locus, denoted as *Mx1*, controlled resistance. The gene symbol *Mx* is derived from myxovirus resistance. The gene segregated for dominant resistant (Mx^+) and recessive susceptible (Mx^-) alleles, and was later mapped to mouse chromosome 16 (Staeheli et al 1988). Subsequent work showed that several mouse strains originating from wild mice, in contrast to most standard laboratory mouse strains, carry the Mx^+ allele (Jin et al 1998).

A large number of studies in the 1970s and 1980s were directed towards elucidating the mechanisms of resistance and susceptibility. Macrophages, but not T- and B-cells, appeared to determine the response associated with the Mx^+ allele. This finding was later explained by the observation that the *Mx1* gene is regulated by interferon-α/β, but not by interferon-γ (Haller et al 1979, Haller et al 1980). Evidently these studies also stressed the significance of interferon-α/β in resistance to influenza virus. Indeed early studies had already indicated that the interferon system appears to limit viral spread before the development of a specific immune response.

The studies subsequently led to the identification of the *Mx1* gene product as a 72.5 kD protein present in interferon-treated macrophages from congenic BALB.A2G-Mx (Mx^+) mice, but not from interferon-treated macrophages from BALB/c (Mx^-) mice (Horisberger et al 1983). The gene was cloned and sequence analysis showed that non-functional *Mx* alleles have deletion (BALB/c) or point (CBA/J) mutations in the coding region (Haller et al 1986). The deletion genotype occurs in BALB/c and 33 other *Mx* strains. The deletion involves exons 9-11 and some flanking sequences, and results in synthesis of a shortened *Mx1* mRNA that encodes a truncated protein. The point mutation genotype occurs in CBA/J and 2 other Mx^- strains. Strains A2G (the strain in which the *Mx1* locus was originally described) and SL/NiA are Mx^+ (Staeheli et al 1988). The point mutation in strain CBA/J converts the lysine codon (AAA) in position 389 to a termination codon (TAA). The Mx^- phenotype of interferon-treated BALB/c and CBA/J cells is characterized by low levels of *Mx1* mRNA levels, which is probably due to decreased metabolic stabilities of the mutant mRNAs.

A second member of the *Mx* gene family was subsequently identified and denoted *Mx2* (Staeheli and Sutcliffe 1988). *Mx1* and *Mx2* have a high degree of sequence similarity and both map to chromosome 16. However, interferon-treated cells from virus-resistant strains do not express detectable amounts of *Mx2* mRNA. In many laboratory strains the *Mx2* mRNA carries an extra C residue at position 1366 that causes a translational frameshift (Staeheli and Sutcliffe 1988). Expression of a repaired *Mx2* cDNA in transfected cells renders them resistant to vesicular stomatitis virus, a rhabdovirus, but not to influenza virus (Zürcher et al 1992).

Subsequently, two human interferon-induced proteins (MxA and MxB) with homology to the murine *Mx1* protein were identified (Pavlovic et al 1992, Reeves et al 1988, Staeheli and Haller 1985). The genes were mapped within the region of human chromosome 21 (map position 21q22.3) that is homologous to mouse chromosome 16. Upon induction with interferon, MxA protects cultured cells against infections with influenza virus and vesicular stomatitis virus. The MxB protein is generally not detectable.

Function of the Mx protein

Influenza virus, in contrast to rhinovirus or other causes of common cold, is a potent inducer of interferon-α/β (Makela et al 1999). Binding of interferon-α/β to its receptor at the cell membrane triggers signaling pathways that include protein phosphorylation and activation of transcription factors, which in turn bind to specific DNA sequences known as interferon-stimulable response elements (Staeheli 1990). The Mx protein system is one of the interferon-inducible systems. It responds very fast. Other interferon-inducible proteins with antiviral effect include 2'-5' oligoadenylate synthetase and a protein kinase. Mx protein accumulates and is easy detectable within 4 h postinoculation. Both murine *Mx1* and human *MxA* bind guanosine triphosphate (GTP) and have intrinsic GTPase activity. The essential role of the *Mx1* protein in influenza resistance was demonstrated by transfection and transgenesis experiments. Transfection of the *Mx1* gene in Mx- cells was sufficient to transform them from a sensitive to a resistant phenotype (Staeheli et al 1986). Furthermore, microinjection of *Mx+* cells with an anti-Mx1 antibody blocks virus resistance (Arnheiter and Haller 1987). Transgenic mice that produce the Mx1 protein under the control of an interferon-responsive regulatory element produce mRNA levels of 60-70 % of normal *Mx*$^+$ mice and are completely protected against influenza infection (Arnheiter et al 1990). The Mx1 protein acts at an early stage of the viral infection cycle by inhibiting primary transcription of the virus. In contrast, the human *MxA* protein appears to inhibit a subsequent step in the cytoplasm of infected cells (Pavlovic et al 1992). The antiviral activity of *MxA* is not restricted to influenza virus. *MxA* also inhibits measles virus transcription in human brain cells (Schneider-Schaulies et al 1994).

While interferons are rapidly cleared from the circulation, the interferon-induced MxA protein can still be detected days after interferon production. The MxA protein may therefore be a useful marker for the biological activity of interferon-α/β and for the diagnosis of viral infections (Jakschies et al 1994).

Role of Mx proteins in antiviral resistance in humans

The role of the Mx proteins in antiviral resistance in humans is not yet clear. However, *in vitro* the multiplication of influenza viruses is suppressed in *MxA*-expressing cells, as indicated by a decrease in viral RNA synthesis, viral protein synthesis, virion production, and cytopathic effects (Marschall et al 2000). So far, Mx-defective mutations or polymorphisms have not been described in humans and in any other animal species except mice. It has been argued that patients with Down's syndrome may give an indication of the significance of the Mx system (Horisberger 1995). Patients with Down's syndrome are characterized by trisomy of chromosome 21, which encodes several genes of the Mx and interferon systems, and they might thus show a gene dosage effect. However, although in cultured cells of patients with Down's syndrome a gene dosage effect of interferon-inducible proteins can clearly be shown, these patients are even more susceptible to upper respiratory tract infections including influenza. This is probably explained by abnormalities of cellular and humoral immune responses in these patients (Horisberger 1995).

Evolutionary relationships between orthomyxoviruses and Mx proteins
Both the $Mx1^+$ and $Mx1^-$ alleles occur naturally and with comparable frequencies in wild mice (*Mus musculus domesticus*) in Europe and America, suggesting that the $Mx1^-$ allele is not necessary deleterious. The observations further suggest that selective forces, perhaps epidemics of orthomyxovirus infections or some heterozygous advantage, maintain the polymorphism. About 75 % of the mice are resistant to influenza viruses, and the resistant phenotype is indeed associated with synthesis of the *Mx1* protein (Haller et al 1987).
Mx1-related proteins have been identified in a large number of mammals, including rat, hamster, pig, horse, cow, and man, but also in chicken, fish, and yeast. Thus these *Mx* proteins are found in species that are naturally infected with influenza virus, and also in species that are not. Their significance in antiviral activity also differs among species. In mice, the *Mx1* protein is necessary and sufficient to protect against influenza virus and the resistance does not require a functioning immune system. In contrast, although *MxA* has antiviral activity in humans against influenza virus, induction of *MxA* (and interferons) is not sufficient to prevent influenza, and a functioning immune system is necessary for protection and recovery in this species. In fact, influenza virus appears to maintain itself in humans by continuously escaping the immune system by presenting varying epitopes to the human system. Of course these observations do not rule out the possibility that in humans the interferon and Mx systems are a first line of defense that induce an antiviral state, thus giving the specific immune system more time to respond. It has for example been documented that deficiencies in interferon production allow more extensive multiplication of influenza virus (Levin and Hahn 1985).
Some of the *Mx* proteins appear to have a broader antiviral profile than *Mx1* in mice, while others have no antiviral activity. Other GTP-binding proteins, such as Vps1p in yeast and dynamin in rat, are also related to *Mx1*. These proteins are synthesized constitutively and serve basic cellular functions, such as secretory and endocytotic processes (Arnheiter and Meier 1990, Horisberger 1995). Finally, *Mx* proteins have a carboxyterminal leucine zipper domain that functions as a common oligomerization element in regulatory and structural proteins (Melén et al 1992). Together these observations stimulate the speculation that the antiviral activity of the *Mx* proteins has evolved from other basic cellular functions in response to epidemic disease by (orthomyxo)virus infections.

Early defense by mannose-binding protein
Mannose-binding protein (MBP) selectively recognizes patterns of oligosaccharides around viral and bacterial pathogens. MBP also appears an important component of the innate preimmune host defense against certain strains of influenza A virus by neutralizing hemagglutinin and infectivity, and also by acting as an opsonin (Anders et al 1990, Hartshorn et al 1993). However, the influence of MBP gene variants on the course of influenza virus infection is unknown.

References

Anders, E.M., Hartley, C.A., Jackson, D.C. (1990). Bovine and mouse serum b inhibitors of influenza viruses are mannose-binding lectins. Proc. Natl. Acad. Sci. USA 87, 4485-4489.

Arnheiter, H., Haller, O. (1988). Antiviral state against influenza virus neutralized by microinjection of antibodies to interferon-induced Mx proteins. EMBO J. 7, 647-656.

Arnheiter, H., Meier, E. (1990). Mx proteins: antiviral proteins by chance or by necessity? New Biol. 2, 851-857.

Arnheiter, H., Skuntz, S., Noteborn, M., Meier, E. (1990). Transgenic mice with intracellular immunity to influenza virus. Cell 62, 51-61.

Haller, O., Arnheiter, H., Gresser, I., Lindenmann, J. (1979). Genetically determined, interferon-dependent resistance to influenza virus in mice. J. Exp. Med. 149, 601-662.

Haller, O., Arnheiter, H., Lindenmann, J., Gresser, I. (1980). Host gene influences sensitivity to interferon action selectively for influenza virus. Nature 283, 660-662.

Haller, O., Acklin, M., Staeheli, P. (1986). Genetic resistance to influenza virus in wild mice. Current Topics Microbiol. Immunol. 127, 331-337.

Haller, O., Acklin, M., Staeheli, P. (1987). Influenza virus resistance of wild mice: wild-type and mutant Mx alleles occur at comparable frequencies. J. Interferon Res. 7, 647-656.

Hartshorn, K.L., Sastry, K., White, M.R., Anders, E.M., Super, M., Ezekowitz, R.A., Tauber, A.I. (1993). Human mannose-binding protein functions as an opsonin for influenza A viruses. J. Clin. Invest. 91, 1414-1420.

Horisberger, M.A., Staeheli, P., Haller, O. (1983). Interferon induces a unique protein in mouse cells in cells bearing a gene for resistance to influenza virus. Proc. Nat. Acad. Sciences USA 80, 1910-1914.

Horisberger, M.A. (1995). Interferons, Mx genes, and resistance to influenza virus. Am. J. Respir. Crit. Care Med. 152, S67-S71.

Jakschies, D., Armburst, C., Clare, A., Nolte, K.U., Towbin, H. Deicher, H., von Wussow, P. (1994). Significant difference of the MxA-protein expression in human PBL of patients with viral and bacterial infections. J. Interferon Res. 14, S124.

Jin, H.K., Yamashita, T., Ochiai, K., Haller, O., Watanabe, T. (1998). Characterization and expression of the Mx1 gene in wild mouse species. Biochem. Genet. 36, 311-322.

Levin, S., Hahn, T. (1985). Interferon deficiency syndrome. Clin. Exp. Immunol. 60, 267-273.

Lindenmann, J. (1964). Inheritance of resistance to influenza in mice. Proc. Soc. Exp. Biol. Med. 116, 506-509.

Makela, M.J., Halminen, M., Ruuskanen, O., Puhakka, T., Prihonen, J., Julkunen, I., Ilonen, J., Lack of induction by rhinoviruses of systemic type I interferon production or enhanced MxA proteins expression during the common cold. Eur. J. Clin. Microbiol. Infect. Dis. 18, 665-668.

Marschall, M., Zach, A., Hechfischer, A., Foerst, G., Meier-Ewert, H., Haller, O. (2000). Inhibition of influenza C viruses by human MxA protein. Virus Res. 67, 179-188.

Melén, K., Ronni, T., Broni, B., Krug, R.M., von Bonsdorff, C.H., Julkunen, I. (1992). Interferon-induced Mx proteins form oligomers and contain a putative leucine zipper. J. Biol. Chem. 267, 24898-25907.

Pavlovic, J., Haller, O., Staeheli, P. (1992). Human and mouse Mx proteins inhibit different steps of the influenza virus multiplication cycle. J. Virol. 66, 2564-2569.

Reeves, R.H., O'Hara, B.F., Pavan, W.J., Gearhart, J.D., Haller, O. (1988). Genetic mapping of the Mx influenza virus resistance gene within the region of mouse chromosome 16 that is homolgous to human chromosome 21. J. Virol. 62, 4372-4375.

Schneider-Schaullies, S., Schneider-Schaullies, J., Schuster, A., Bayer, M., Pavlovic, J., ter Meulen, V. (1994). Cell type-specific MxA-mediated inhibition of measles virus transcription in human brain cells. J. Virol. 68, 6910-6917.

Staeheli, P., Haller, O. (1985). Interferon-induced human protein with homology to protein Mx of influenza-resistant mice. Mol. Cell. Biol. 5, 2150-2153.

Staeheli, P., Haller, O., Boll, W., Lindenmann, J., Weissmann, C. (1986). Mx protein: constitutive expression in 3T3 cells transformed with cloned Mx cDNA confers slesctive resitance to influenza virus. Cell 44, 147-158.

Staeheli, P., Grob, R., Meier, E., Sutcliffe, J.G., Haller, O. (1988). Influenza virus-susceptible mice carry Mx genes with a large deletion or a nonsense mutation. Moll. Cell. Biol. 8, 4518-4523.

Staeheli, P., Sutcliffe, J.G. (1988). Identification of a second interferon-regulated murine Mx gene. Mol. Cell. Biol. 8, 4524-4528.

Staeheli, P. (1990) Interferon-induced proteins and the antiviral state. Adv. Virus. Res. 38, 147-200.

Zürcher, T., Pavlovic, J., Staeheli, P. (1992). Mouse Mx2 protein inhibits vesicular stomatitis virus but not influenza virus. Virology 187, 796-800.

Chapter 19. Resistance to HIV/AIDS and other retroviruses

Human immunodeficiency virus-1 (HIV-1) was identified 20 years ago, and the epidemic now appears worse than ever predicted. An estimated 36 million people worldwide are currently living with HIV, and some 20 million people have already died. The worst of the epidemic is in sub-Saharan Africa, where, at the end of 2000, there were an estimated 25.3 million people living with HIV (Piot et al 2001).

Genetic variation of chemokine receptors and HIV infection
Obviously some individuals never seroconvert or become infected upon exposure to HIV-1. To these individuals belong members of the well-known risk groups, including gay men, intravenous drug users, prostitutes, neonates of infected mothers, and accidentally exposed health workers (Beretta 1999, Rowland-Jones et al 1995, Shearer et al 1996). The HIV-resistant phenotype has most often been ascribed to alterations in the CCR5 gene (a 32-bp deletion in both alleles of the CCR5 gene resulting in a total lack of CCR5 cell surface expression; discussed at length in chapter 7). Other genetic factors, HIV-specific immune responses, exposure to a defective virus, or combinations of these may further contribute to the HIV-resistant phenotype or to a slower progression to disease (Beretta 1999).

Evidently both virus and host strategies are intimately connected in influencing susceptibility to HIV-1 infection and the rate of disease progression. The main cellular receptor of HIV-1, CD4, is expressed on T helper lymphocytes, thymic precursors, macrophages, dendritic cells, and microglial cells in the brain. Because HIV-1 cannot replicate in non-human cells expressing human CD4, it became clear that additional entry factors were required to establish infection (Broder et al 1993, Lores et al 1992). Specific G-protein-coupled seven transmembrane spanning chemokine receptors have subsequently been found to function as coreceptors for HIV-1-entry. The CXC-chemokine receptor, CXCR4, was identified to be the coreceptor for the T-cell-line-tropic, fast replicating, syncytium-inducing (SI) variants of HIV-1 that emerge late during disease progression (Feng et al 1996).
Almost simultaneously, the receptor of the β-chemokines, CCR5, was identified as the coreceptor for primary, slow replicating, non-syncytium-inducing (NSI), macrophage-tropic variants (Alkhatib et al 1996, Deng et al 1996, Dragic et al 1996). These variants usually initiate HIV-1 infection and predominate in the asymptomatic phase of infection. A change in coreceptor use from CCR5 to multiple coreceptor usage, including CXCR4, often correlates with disease progression in HIV-1-infected individuals. In approximately half of the HIV-1-infected individuals, such SI T-cell-tropic variants emerge during the course of infection preceding progression to AIDS.
NSI variants are almost completely restricted to the use of CCR5. In contrast, primary SI variants may in addition to CXCR4 use CCR5, and sometimes CCR2b and CCR3, as coreceptors, but to a lesser extent (Connor et al 1997, Doranz et al 1996, He et al 1997, Ghorpade et al 1998). This multiple co-receptor usage makes that the ΔCCR5 deletion does not provide absolute protection against HIV infection. Homozygous ΔCCR5-carrying individuals may thus become infected through SI virus variants using the CXCR4 receptor.

HIV may infect microglial cells in the central nervous system by using CCR3 and CCR5, which receptors are expressed on these cells (He et al 1997, Ghorpade et al 1998). However, the *in vivo* relevance of the capability of HIV-1 to use other receptors than CCR5 and CXCR4 remains to be established.

Further illustrating the role of the CC chemokine receptors (CCRs) in the pathogenesis of HIV infection is the finding that acquired resistance to HIV may in part be based on a block of HIV infection via the production of CC chemokines. Evidence for such a mechanism was obtained by the observation that HIV-specific CD4+ T cells from exposed uninfected individuals carrying the wild-type CCR5 allele could produce high quantities of CC chemokines that were able to suppress the replication of the virus (Furci et al 1997). In addition, *in vitro* studies have shown that the CC-chemokines RANTES (and derivatives thereof), MIP-1α and MIP-1β can indeed block the entry of macrophage-tropic strains into susceptible cells (Margolis et al 1998, Zagury et al 1998).

Although some SI variants were able to use CCR2b as coreceptor, the *in vivo* function of CCR2b as coreceptor for HIV-1 is not clearly established. Nonetheless, individuals with a valine to isoleucine substitution at position 64 in CCR2b (CCR2-64I) at one or both alleles generally show a delayed disease progression (Kostrikis et al 1998, Rizzardi et al 1998). This mutation is however not associated with reduced transmission risk. The effect of the CCR2-64I mutation is less predominant and overshadowed by the protective effect of the CCR5 deletion (Hendel et al 1998). Moreover, because the disease-retarding effects of the CCR2-64I allele were found in African Americans, but not in Caucasians, there is clearly an effect of gene interactions (Mummidi et al 1998).

CCR2b does not appear to be a commonly used viral receptor and the 64I mutation does not alter CCR2b surface expression or its function as viral coreceptor (Lee et al 1998, Mariani et al 1999). Although individuals with the CCR2-64I mutation have a slightly reduced expression of CCR5 on CD4+ T cells, the CCR-64I effect on AIDS progression is not mediated by a negative effect on the CCR5 co-receptor function (Mariani et al 1999). Thus the mechanism through which the mutation in CCR2b exerts its effect on disease progression is not clear.

The CCR2-64I mutation is closely linked to a CCR5 promoter polymorphism (CCR5-59653T) on chromosome 3, indicating an evolutionary relationship between these genes. Indeed almost 97 % of individuals have the same genotype for both CCR2-64I and CCR5-59653T polymorphisms. The CCR2-64I and CCR5-59653T genetic variants are found in almost all populations studied: their allele frequencies are greatest (approximately 35 %) in Africa and Asia, but decrease in Northern Europe. The greater geographical distribution of this haplotype compared with that of CCR5-Δ32 suggests that it is a much older mutation whose origin predates the dispersal of modern humans (Martinson et al 2000).

Genetic variation at other loci and HIV infection
Approximately 20 – 30 % of the individuals known to have had high exposure to HIV and who are resistant to infection are homozygous for the ΔCCR5. Because the frequency of the homozygous mutant CCR5 allele carriers is 1 % or less, the attributable risk of this factor is indeed low. (The attributable risk estimates the contribution of a genetic factor to the burden of disease as if it were possible to remove the factor. For its calculation estimates of the various components of variance are required.) It is the-

refore clear that ΔCCR5 is not sufficient to explain the whole genetically determined variability to HIV infection, especially in Africa. In fact CCR5 polymorphisms can only account for a minority of cases of HIV resistance. Other genetic explanations for lower susceptibility or slower progression to disease have been found in the MHC system, SDF-1, haptoglobin variants, mannose-binding protein gene variant alleles, the c2 microsatellite close to the TNF-β gene, the CCR2 gene, and the vitamin D receptor genotype (Garred et al 1997, Hill 1998, Kaslow et al 1996, Khoo et al 1997, Smith et al 1997).

The MHC may explain a sixfold difference between groups with the shortest and the longest progression times to AIDS (Kaslow et al 1996). Maximum *HLA* heterozygosity of class I loci (A, B, and C) delayed AIDS onset, whereas individuals who were homozygous for one or more loci progressed rapidly to AIDS and death. The heterozygous advantage is probably explained by an increased repertoire of viral epitopes presented to CD8+ cytotoxic T cells. The HLA class I alleles *B*35* and *Cw*04* were associated with rapid development of AIDS in Caucasians. This association of alleles *B*35* and *Cw*04* with rapid AIDS development is less well explained, but may be due to altered killer cell function. Rapid disease progression has further been clearly associated with HLA-DR3, and in particular with the HLA-A1-B8-DR3 haplotype (Donald et al 1992). In contrast, HLA-B57 and HLA-B27 were associated with slow progression (Carrington et al 1999).

An interesting possibility is that alloantigenic immune responses may play a role in recognition and elimination of HIV-infected cells in some sexually exposed HIV-uninfected individuals, and in uninfected children born to HIV-infected mothers (Kiprov et al 1994, Mittleman and Shearer 1996). Such alloantigenic immune responses are directed towards cells displaying foreign histocompatibility antigens and lead to the destruction and elimination of these cells. Destruction of infected cells may clearly limit virus transmission. In a prospective cohort of highly exposed HIV-negative prostitutes in Nairobi, 43 individuals of 424 highly exposed individuals remained HIV-1 negative for 3 or more years. These HIV-negative sex workers had HLA types which were rare in the local population, and which may therefore have led to protection by strong alloantigenic responses. The reduced susceptibility to infection in this cohort was correlated with MHC class I alleles Aw28 (OR = 0.21; $P < 0.1$) and Bw70 (OR = 0.3; $P < 0.05$). (Fowke et al 1996).

Another polymorphism possibly influencing disease progression was found on the 3' untranslated region of the gene encoding the chemokine stromal cell-derived factor 1 (SDF-1) on chromosomal location 10q11.1. This gene encodes the natural ligand of CXCR4. However, the effects of allelic variants of SDF-1 on HIV progression are highly controversial. In different studies SDF1-3'A appeared associated with either an accelerated or a delayed clinical course of infection. SDF-1-3'A has a similar frequency in Asians and Caucasians (0.2), but is less frequent in African-Americans (0.06). The polymorphism is in a highly conserved region, suggesting that the polymorphism may have functional relevance (Hendel et al 1998, Van Rij et al 1998, Mummidi et al 1998, Schuitemaker 1999, Winkler et al 1998).

Several independent studies indicated that haptoglobin polymorphism plays a significant role in HIV replication and the prognosis of infection. Haptoglobin is a hemoglobin-binding protein that plays a role in iron-driven oxidative stress. The protein is expressed by a genetic polymorphism as three major phenotypes: 1-1, 2-1, and 2-2. The haptoglobin type distribution among Caucasians (HIV-infected or not) is approximately 18 % Hp 1-1, 50 % Hp 2-1, and 32 % Hp 2-2. Survival times were 4 years shorter, CD4 cell counts were lower, and plasma HIV-1 RNA levels (prior to antiviral therapy) were higher and increased more rapidly for the Hp 2-2 group than for the Hp 1-1 and Hp 2-1 groups. The Hp 2-2 type was associated with higher serum iron, transferrin saturation, and ferritin levels, and with low vitamin C concentrations, suggesting that a less efficient protection against hemoglobin/iron-driven oxidative stress directly stimulates HIV-1 replication (Delange et al 1998, Quaye et al 2000). As mentioned in chapter 16, the 2-2 phenotype is also associated with a higher risk of tuberculosis mortality. The high frequency of the 2-2 phenotype is therefore probably maintained as a balanced polymorphism through genetic pressures favoring the outcome of several inflammatory diseases, atherosclerosis, and autoimmune disorders in this phenotype (Langlois and Delanghe 1996).

The multigenic control of HIV infection clearly necessitates studies about gene interactions. So far, multigenic models of the interactions of genes in some Caucasian populations show independent effects of HLA and CCR genes, including the findings that risk-associated CCR genotypes predict about one sixth of rapid progressors, and that CCR and HLA genotypes together can predict more than two thirds of HIV-infected persons who are going to be rapid progressors (McNicholl et al 2000).

Animal models of retrovirus infections
Friend virus (FV) infection of adult immunocompetent mice is an interesting model for studying genetic resistance to infection by an immunosuppressive retrovirus (Hasenkrug and Chesebro 1997). In 1957 Charlotte Friend (Friend 1957) first described the virus that induced leukemia in adult mice. The erythroleukemias induced by the various strains of FV are multistage malignancies characterized by polyclonal proliferation of nonleukemic erythroid progenitor cells followed by a later stage in which there is clonal or oligoclonal expansion of malignant cells. The virus is pathogenic only in some inbred strains of mice, but not in rats or nonmurine species (Hoatlin and Kabat 1995). The virus induces leukemia in susceptible strains of mice within 3 weeks postinfection by site-specific proviral integration at the *Spi-1 (ets)* oncogene locus combined with inactivation or mutation of the p53 tumor suppressor gene. Both immunologic and non-immunologic mechanisms of resistance to FV infection have been elucidated.

Immunologic mechanisms in resistant strains of mice apparently induce a successful immune response quickly enough to prevent oncogenesis. At least four specific MHC class I and II alleles are necessary for recovery through immunological mechanisms, as well as one non-MHC gene, *Rfv-3*, which controls virus-specific neutralizing antibody responses through an unknown mechanism. These genetic requirements for recovery are in agreement with immunological requirements for cytotoxic T cell, helper T cell, and antibody responses. Each of these is essential for recovery from FV infection and thus reflects the complexity of the immune response required for recovery from this retroviral infection, findings that may be helpful in further understanding immunity to HIV. Experiments with MHC recombinant mice show that MHC regions H-2A,

E, D, and T are important for recovery from acute FV infection. Especially the H-2D region of the mouse MHC has a potent influence on recovery from FV infection, because it encodes the class I molecules that present viral antigens to CTLs. It further influences the kinetics of the virus-specific CD4+ helper T cell responsiveness and controls host susceptibility to FV-induced immunosuppression (Britt and Chesebro 1983, Morrison et al 1987).

Detailed linkage analysis using microsatellite markers and progeny tests of recombinant mice mapped the *Rfv-3* resistance gene to a 0.83 cM region of chromosome 15 (Super et al 1999). Mice require at least one resistance allele at this locus to make the antiviral neutralizing antibodies that are necessary to clear plasma viremia after FV infection. As mentioned above, this effect is necessary, but not sufficient, for recovery from leukemia. Of interest, *Rfv-3* appears to affect only the FV-specific antibody response and not responsiveness to other antigens (Morrison et al 1986). *Rfv-3* belongs to a cluster of genes that includes several cytokine receptor genes (amongst others *Hsf1, IL2rb, IL3rb,* and *Pdgfb*). This fine mapping now makes it feasible to attempt the physical cloning of *Rfv-3*.

In addition to the genes conferring immunological resistance, there are at least six polymorphic genes (including *Fv-1* to *Fv-6*) that confer resistance to infection through non-immunological mechanisms (Hasenkrug and Chesebro 1997, Hoatlin and Kabat 1995). Some of the mouse FV-control genes restrict replication and transmission of specific helperviruses by blocking cell-surface receptors (for example *Fv-4* or R^{mcf}) or by inhibiting steps of infection after penetration (for example *Fv-1*). *Fv-4* may do so because it encodes a truncated murine leukemia virus with *env* region. Thus a retrovirus that has integrated into the host's germ line, a provirus, appears to be beneficial for the host, because it protects against insertional mutagenesis and fatal retroviral disease caused by viruses of the same class. Indeed, much of the mammalian genome consists of endogenous retrovirus-like elements. Analogous genes have been found in avian species.

Fv-1, on mouse chromosome 4, has recently been identified. *Fv-1* prevents or delays spontaneous or experimentally induced viral tumors. Unlike receptor genes, where susceptibility to infection is dominant over resistance, or endogenous proviruses, *Fv-1* has multiple alleles and displays a pattern of codominance. This is most easily explained by postulating that the *Fv-1* gene product interacts directly with some virus component, presumably the viral capsid protein, to block the replication cycle. The two principal alleles are called *n* and *b*. These confer resistance to specific strains of FV. Mice carrying the *n* allele resist infection by B-tropic FV, and mice carrying the *b* allele are resistant to N-type virus. *n/b* heterozygotes are not infectable by either N- or B-type virus, but all three types of animals are sensitive to infection by the third virus type, called NB. *Fv-1* appears to act after virion entry into cells, i.e. before, during or after reverse transcription in some stage of entry, uncoating, or transport of the core structure, but before entry of viral DNA into the nucleus. *Fv-1* has been identified after positional cloning and subsequent selection of clones from yeast artificial chromosomes (YAC) libraries containing the region of interest. The *Fv-1^b*-positive YAC was then identified by its ability to confer partial resistance to N-tropic FV to cells in culture. Subsequently, a single open reading frame encoding a 459-amino acid protein was identified and shown to transfer resistance to N-tropic FV. A corresponding clone from *FV-1^{n/n}* mice conferred resistance to B-tropic FV, confirming the identification.

The alleles differ principally by a truncation and substitution of sequence at the carboxy terminus. Other species, including humans, have large numbers of sequences related to *Fv-1*, including the *gag* region of a human endogenous provirus-like element (HERV-L). Subsequent studies should now elucidate how *Fv-1* works (Best et al 1996, Coffin et al 1996).

Still other FV-control genes reduce the size (for example *W* [white-spotting] and *Sl* [Steel]) or affect the function of the target erythroblast cell population. *Fv-2* and *Fv-5* might conceivably control productive interaction of the viral oncogenic gp55 glycoprotein with the erythropoetin receptor or control the downstream pathway involved in the cellular response. The resistant allele of *Fv-2* is recessive and appears to block leukemia induction totally when present in its homozygous form (Chesebro et al 1990). *Fv-2* appears to encode a lineage-specific regulator of the mammalian cell cycle (Ben-David and Bernstein 1991). Finally, the *Fv-5^a* allele of CBA mice seems to act in a codominant manner in attenuating the activating effect of gp55 on erythroblasts (Hoatlin and Kabat 1995, Shibuya et al 1982).

References

Alkhatib, G., Combadiere, C., Broder, CC., Feng, Y., Kennedy, P.E., Murphy, P.M., Berger, E.A. (1996). CC CKR5: A RANTES, MIP-1α, MIP-1β receptor as a fusion cofactor for macrophage-tropic HIV-1. Science 272, 1955-1958.

Ben-David, Y., Bernstein, A. (1991). Friend virus-induced erythroleukemia and the multistage nature of cancer. Cell 66, 831-834.

Beretta, A. (1999). Natural resistance to HIV infection. Chapter XII in: Basic Science in HIV/AIDS: An update. Ed. by T.E. Mertens et al. WHO, pp. 96–98.

Best, S., LeTissier, P., Towers, G., Stoye, J.P. (1996). Positional cloning of the mouse retrovirus restriction gene *Fv1*. Nature 382, 826-829.

Britt, W.J., Chesebro, B. (1983). H-2D control of recovery from Friend virus leukemia: H-2D region influences the kinetics of the T lymphocyte response to Friend virus. J. Exp. Med. 157, 1736-1745.

Broder, C.C., Dimitrov, D.S., Blumenthal, R., Berger, E.A. (1993). The block to HIV-1 envelope glycoprotein-mediated membrane fusion in animal cells expressing human CD4 can be overcome by a human cell component(s). Virology 193, 483-491.

Carrington, M., Nelson, G.W., Martin, M.P., Kissner, T., Vlahov, D., Goedert, J.J., Kaslow, R., Buchbinder, S., Hoots, K., O'Brien, S.J. (1999). HLA and HIV-1: Heterozygote advantage and *B*35-Cw*04* disadvantage. Science 282, 1748-1752.

Coffin, J.M. (1996). Retrovirus restriction revealed. Nature 382, 762-763.

Connor, R.I., Sheridan, K.E., Ceradini, D., Choe, S., Landau, N.R. (1997). Change in coreceptor use correlates with disease progression in HIV-1-infected individuals. J. Exp. Med. 185, 621-628.

Delanghe, J.R., Langlois, M.R., Boelaert, J.R., Van Acker, J., Van Wanzeel, F., van der Groen, G., Hemmer, R., Verhofstede, C., De Buyzere, M., De Bacquer, D., Arendt, V., Plum, J. (1998). Haptoglobin polymorphism, iron metabolism and mortality in HIV infection. AIDS 12, 1027-1032.

Deng, H., Liu, R., Ellmeier, W., Choe, S., Unumatz, D., Burkhart, M., Di Marzio, P., Marmon, S., Sutton, R.E., Hill, C.M., Davis, C.B., Peiper, S.C., Schall, T.J., Littman, D.R., Landau, N.R. (1996). Identification of the major co-receptor for primary isolates of HIV-1. Nature 381, 661-666.

Donald, J.A., Rudman, K., Cooper, D.W. et al. 1992. Progression of HIV-related disease is associated with HLA DQ and DR alleles defined by restriction fragment length polymorphisms. Tissue Antigens 39, 241-248.

Doranz, B.J., Rucker, J., Yi, Y., Smyth, R.J., Samson, M., Peiper, S.C., Parmentier, M., Collman, R.G., Doms, R.W. (1996). A dual-tropic primary HIV-1 isolate that uses fusin and the β-chemokine receptors CKR-5, CKR-3, and CKR-2b as fusion cofactors. Cell 85, 1149-1158.

Dragic, T., Litwin, V., Allaway, G.P., Martin, S.R., Huang, Y., Nagashima, K.A., Cayanan, C., Maddon, P.J., Koup, R.A., Moore, J.P. Paxton, W.A. (1996). HIV-1 entry into CD4+ cells is mediated by the chemokine receptor CC-CKR-5. Nature 381, 667-673.

Feng, Y., Broder, C.C., Kennedy, P.E., Berger, E.A. (1996). HIV-1 entry cofactor: functional cDNA cloning of a seven-transmembrane, G-protein-coupled receptor. Science 272, 872-877.

Fowke, K.R., Nagelkerke, N.J.D., Kimani, J., et al. (1996). Resistance to HIV-1 infection among persistently seronegative prostitutes in Nairobi, Kenya. Lancet 348, 1347-1351.

Friend, C. (1957). J. Exp. Med. 105, 307-318.

Furci, L., Scarlatti, G., Burastero, S., Tambussi, G., Colognesi, C., Quillent, C., Longhi, R., Loverro, P., Borgonovo, B., Gaffi, D., Carrow, E., Malnati, M., Lusso, P., Siccardi, A.G., Lazzarin, A., Beretta, A. (1997). Antigen-diven C-C chemokine-mediated HIV-1 suppression by CD4(+) T cells from exposed uninfected individuals expressing the wild-type CCR-5 allele. J. Exp. Med. 186, 455-460.

Ghorpade, A., Qi Xia, M., Hyman, B.T. et al (1998). Role of the β-chemokine receptors CCR3 and CCR5 in nhuman immunodeficiency virus type 1 infection of monocytes and microglia. J. Virol. 72, 3351-3361.

Hasenbrug, K.J., Chesebro, B. (1997). Immunity to retroviral infection: the Friend virus model. Proc. Natl. Acad. Sci. USA 94, 7811-7816.

Hendel, H., Henon, N., Lebuanec, H., Lachgar, A., Poncelet, H., et al. (1998). Distinctive effects of CCR5, CCR2, and SDF1 genetic polymorphisms in AIDS progression. J. Acquir. Immune Defic. Syndr. Hum. Retrovirol. 19, 381-386.

Hoatlin, M.E., Kabat, D. (1995). Host-range control of a retrovial disease: Friend erythroleukemia. Trends Microbiol. 3, 51-57.

Garred, P., Madsen, H.O., Balsev, U., Hofmann, B., Pedersen, C., Gerstoft, J., Svejgaard, A. (1997). Susceptibility to HIV infection and progression of AIDS in relation to variant alleles of mannose-binding lectin. Lancet 349, 236-240.

He, J., Chen, Y., Farzan, M., Choe, H., Ohagen, A., Gartner, S., Busciglio, J., Yang, X., Hofmann, W., Newman, W., Mackay, C.R., Sodroski, J., Gabuzda, D. (1997). CCR3 and CCR5 are coreceptors for HIV-1 infection of microglia. Nature 385, 645-649.

Hill, A.V.S. (1998). The immunogenetics of human infectious diseases. Annu. Rev. Immunol. 16, 593-617.

Kaslow, R.A., Carrington, M., Apple, R., Park, L., Munoz, A., Saah, A.J., Goedert, J.J., Winkler, C., O'Brien, S.J., Rinaldo, C., Detels, R., Blattner, W., Phair, J., Erlich, H., Mann, D.L. (1996). Influence of combinations of human major histocompatibility complex genes on the course of HIV-1 infection. Nature Med. 2, 405-411.

Khoo, S.H., Pepper, L., Snowden, N., Hajeer, A.H., Valley, P., Wilkins, E.G., Mandal, B.K., Ollier, W.E. (1997). Tumour necrosis factor c2 microsatellite allele is associated with the rate of HIV disease progression. AIDS 11, 423-428.

Kiprov, D.D., Sheppard, H.W., Hanson, C.V. (1994). Alloimmunization to prevent AIDS? Science 263, 737-738.

Kostrikis, L.G., Huang, Y., Moore, J.P., Wolinsky, S.M., Zhang, L., Guo, Y., Deutsch, L., Phair, J., Neumann, A.U., Ho, D.D. (1998). A chemokine receptor CCR2 allele delays HIV-1 disease progression and is associated with a CCR5 promoter mutation. Nature Med. 4, 350-353.

Langlois, M.R., Delanghe, J.R. (1996). Biological and clinical significance of haptoglobin polymorphism in humans. Clin. Chem. 42, 1589-1600.

Lee, B., Doranz, B.J., Rana, S., Yi, Y., Mellado, M., Frade, J.M., Martinez-A.C., O'Brien, S.J., Dean, M., Collman, R.G., Doms, R.W. (1998). Influence of the CCR2-V64I polymorphism on human immunodeficiency virus type 1 coreceptor activity and on chemokine receptor function of CCR2b, CCR3, CCR5, and CXCR4. J. Virol. 72, 7450-7458.

Lores, P., Boucher, V., Mackay, C., Pla, M., Von Boehner, Jami, J., Barre-Sinoussi, F., Weill, J.C. (1992). Expression of human CD4 in transgenic mice does not confer sensitivity to human immunodeficiency virus infection. AIDS Res. Hum. Retroviruses 8, 2063-2071.

Margolis, L.B., Glushakova, S., Grivel, J.-C., Murphy, P.M. (1998). Blockade of CC chemokine receptor 5 (CCR5)-tropic human immunodeficiency virus-1 replication in human lymphoid tissue by CC chemokines. J. Clin. Invest. 101, 1876-1880.

Mariani, R., Wong, S., Mulder, L.C., Wilkinson, D.A., Reinhart, A.L., LaRosa, G., Nibbs, R., O'Brien, T.R., Michael, N.L., Connor, R.I., Macdonald, M., Busch, M., Koup, R.A., Landau, N.R. (1999). CCR2-64I polymorphism is not associated with altered CCR5 expression or coreceptor function. J. Virol. 72, 2450-2459.

Martinson, J.J., Hong, L., Karanicolas, R., Moore, J.P., Kostrikis, L.G. (2000). Global distribution of the CCR2-64I/CCR5-59653T HIV-1 disease-protective haplotype. AIDS 14, 483-489.

McNicholl, J.M., Downer, M.V., Udhayakumar, V., Alper, C.A., Swerdlow, D.L. (2000). Host-pathogen interactions in emerging and re-emerging diseases: A genomic perspective of tuberculosis, malaria, human immunodeficiency virus infection, hepatitis B, and cholera. Annu. Rev. Public Health 21, 15-46.

Mittleman, B.B., Shearer, G.M., Mother-to-infant transmission of HIV-type 1: Role of major histocompatibility antigen differences. AIDS Res. Hum. Retrovir. 12, 1397-1400.

Morrison, R.P., Earl, P.L., Nishio, J., Brooks, D.M., Chesebro, B. (1987). Different H-2 sub-regions influence immunization against retrovirus and immunosuppression. Nature 329, 729-732.

Morrison, R.P., Nishio, J., Chesebro, B. (1986). Influence of the murine MHC (H-2) on Friend leukemia virus-induced immunosuppression. J. Exp. Med. 163, 301-314.

Mummidi, S., Ahuja, S.S., Gonzalez, E., et al. (1998). Genealogy of the CCR5 locus and chemokine system gene variants associated with altered rates of HIV-1 disease progression. Nature Med. 4, 786-793.

Piot, P., Bartos, M., Ghys, P.D., Walker, N., Schwartländer, B. (2001). The global impact of HIV/AIDS. Nature 410, 968-973.

Quaye, I.K., Brandfl, J., Ekuban, F.A., Gyan, B., Ankrha, N.A. (2000). Haptoglobin polymorphism in human immunodeficiency virus infection: Hp0 phenotype limits depletion of CD4 cell counts in HIV-1 seropositive individuals. J. Infect. Dis. 181, 1483-1485.

Rizzardi, G.P., Morawetz, R.A., Vicenzi, E., Ghèzzi, S., Poli, G., Lazzarin, A., Pantaleo, G. (1998). CCR2 polymorphism and HIV disease. Swiss HIV cohort. Nature Med. 4, 252-253.

Rowland-Jones, S., McMichael A. (1995). Immune responses in HIV-exposed serone-gatives: have they repelled the virus? Curr. Opin. Immunol. 7, 448-455.

Schuitemaker, H. (1999). Scientific report CLB-Sanquin, Amsterdam.

Shearer, G.M., Clerici, M. (1996). Protective immunity against HIV infection: has nature done the experiment for us? Immunology Today 17, 21-24.

Shibuya, T., Niho, Y., Mak, T.W. (1982). Erythroleukemia induction by Friend leukemia virus. A host gene locus controlling early anemia or polycythemia and the rate of proliferation of late erythroid cells. J. Exp. Med. 156, 398-414.

Smith, M.W., Dean, M., Carrington, M., Winkler, C., Huttley, G.A., Lomb, D.A., Goedert, J.J., O'Brien, T.R., Jacobson, L.P., Kaslow, R., Buchbinder, S., Vittinghoff, E., Vlahov, D., Hoots, K., Hilgartner, M.W., O'Brien, S.J. (1997). Contrasting genetic influence of CCR2 and CCR5 variants on HIV-1 infection and disease progression. Science 277, 959-965.

Super, H.J., Hasenkrug, K.J., Simmons, S., Brooks, D.M., Konzek, R., Sarge, K.D., Morimoto, R.I., Jenkins, N.A., Gilbert, D.J., Copeland, N.G., Frankel, W., Chesebro, B. (1999). Fine mapping of the friend retrovirus resistance gene, Rfv3, on mouse chromosome 15. J. Virol. 73, 7848-7852.

Van Rij, R.P., Broersen, S., Goudsmit, J., Coutinho, R.A., Schuitemaker, H. (1998). The role of a stromal cell-derived factor-1 chemokine gene variant in the clinical course of HIV-1 infection. AIDS 12, F85-F90.

Winkler, C., Modi, W., Smith, M.W., Nelson, G.W., Wu, X., et al (1998). Genetic restriction of AIDS pathogenesis by an SDF-1 chemokine gene variant. Science 279, 389-393.

Zagury, D., Lachgar, A., Chams V., et al. (1998). C-C chemokines, pivotal in protection against HIV type 1 infection. Proc. Natl. Acad. Sci. USA 95, 3857-3861.

Chapter 20. Genetic factors influencing the outcome of viral hepatitis

Hepatitis B

The pathogenesis of hepatitis B virus (HBV) infection involves many steps, each of which may be under genetic control. Upon exposure to HBV, an unknown number of the exposed persons become infected. The majority of the infected persons clears the infection and develops protective antibodies (HbsAb). However, approximately 10 % of those infected become chronically infected. Chronic carriership may result in asymptomatic hepatitis B surface antigen (HbsAg) carriership or chronic active hepatitis, which in many cases – approximately 25 % - may lead to death from chronic liver failure or from hepatocellular carcinoma. HBV replicates in liver cells, but is not directly cytotoxic. The pathologic changes are therefore caused by the host's immune and inflammatory response.

Twin studies and segregation analysis indicated that the variable clinical outcome of HBV infection is significantly influenced by the host's genetic makeup (Lin et al 1989, Thursz 1997). For example, Chinese monozygotic twins were concordant for HBsAg carriage in 50 % of the cases, whereas dizygotic (and nontwinned siblings) were concordant in only 20 % of the families.

Several studies examined genetic factors controlling the immune response upon infection. Fathers of patients with chronic HBV infection have low levels of antibody to HbsAg after HBV infection, suggesting that genetic control of this antibody response may be an important factor in determining persistent hepatitis B surface antigen carriage (Hann et al 1982). Also the T helper cell responses are attenuated in patients with chronic HBV infection compared to those in patients with acute self-limiting infection (Ferrari et al 1990). MHC class II allele frequencies were therefore subsequently examined in several association studies. However, allele frequencies varied to a large extent in different populations, and the studies therefore failed to produce consistent results. Thus many HLA allele studies in a variety of populations have either reported no or weak or unconfirmed associations. As an example, allele HLA-DRB1*1302 and HLA-DR2 appeared associated with clearance of HBV infection in The Gambia and Quatar respectively, while HLA-DR7 appeared associated with persistent infection in Quatar. However, other studies found heterozygosity for any HLA class II genes (but not class I genes) associated with resistance to persistent infection, thus finding additional evidence for a mechanism that may have contributed to maintaining the polymorphism of the HLA system (Almarri and Batchelor 1994, Hohler et al 1997, Thursz et al 1995) (see also chapter 2). To explain the potential role of HLA-DR13 in protecting from chronic hepatitis B infection, its role in antigen presentation of the core antigen was examined *in vitro*. The HB core antigen induced a more vigorous CD4 T-cell-proliferative response *in vitro* when presented by HLA-DR13 than by other HLA-DR-DQ specificities (Diepolder et al 1998).

Because in acute hepatitis, but not in established chronic HBV carriership, HBV-specific MHC-restricted CD8+ CTL are readily detected, polymorphisms at the MHC class I loci may also be likely candidates to influence the outcome of HBV infection. However, similarly to MHC class II alleles, so far no clear association of MHC class I alleles with the development of chronic infection has been established (Thursz 1997). In HbsAg carriers, however, chronic active hepatitis has been associated with an incre- ased frequency of HLA-B35 and a decreased frequency of HLA-DQw1. Asymptomatic HbsAg carriage has been associated with HLA-B15 (Mota et al 1987, van Hattum et al 1987, Giani et al 1979).

As mentioned in chapter 8, mannose-binding protein (MBP) is an opsonizing protein that binds to mannose-terminated carbohydrate chain on bacterial, yeast, and virus- infected host cell walls. HBV also has a mannose-terminated carbohydrate chain on HbsAg (pre-S2 region), which could be a target for MBP (Thursz 1997). Thomas et al (1996) found that the MBP codon 52 mutant allele occurs more frequently in associa- tion with persistent HBV infection in Caucasians. This finding might explain that patients with chronic HBV infection have a relative defect in opsonization. Unfortu- nately, other reports failed to confirm the association between MBP alleles and susceptibility to chronic HBV (Hohler et al 1998).

Although the TNF2 allele is associated with increased TNF-α production (which has antiviral properties), the TNF2 allele has been associated with persistent HBV infec- tion. Because TNF-α mediates apoptosis, it is even possible that the higher levels of TNF-α production may actually induce tolerance. The finding may also partially explain the increased frequency of cerebral malaria in children with persistent HBV infection from sub-Saharan Africa, as TNF2 is also associated with cerebral malaria (Thursz 1997).

A number of chronic HBV carriers develops hepatocellular carcinoma. However, the rates at which this occurs vary widely in different areas. In addition to HBV, exposure to aflatoxin, and host genetic makeup appear cofactors in the development of hepa- tocellular carcinoma. Genetic variations in two liver enzymes (epoxide hydrolase and gluthatione S-transferase M1) that are responsible for the detoxification of aflatoxin B1 epoxides have been described (McGlynn et al 1995). The variants cause a slower rate of detoxification, and are associated with higher levels of aflatoxin B1-albumin adducts and with an increased risk of developing hepatocellular carcinoma. Thus a marked synergistic interaction between environmental and genetic risk factors in developing hepatocellular carcinoma was found.

Hepatitis C
As with HBV, infection with hepatitis C virus (HCV) may result in acute or chronic infection, respectively in approximately 20 and 80 % of the cases. Chronic HCV infec- tion may subsequently result in either progressive or non-progressive liver disease. This occurs again in a ratio of approximately 20 to 80 % of those experiencing chronic infection (Thursz 1997). Thus approximately 20 % of the chronically infected patients will progress to end-stage lever disease with cirrhosis. HCV is transmitted less easily from human-to-human than HBV. So far, no family clustering or twin studies of HCV infection have been reported, and studies examining an association of MHC class II alleles with the clearance of HCV infection gave inconsistent results (Congia et al 1996, Hohler et al 1997, Peano et al 1994, Tibbs et al 1996, Vitte et al 1995, Thursz et al 1999). However, in chronic HCV carriers HLA-DR5 appears to be associated with mil- der forms of liver disease, as measured by transaminase levels and histological exami-

nation of the severity of liver disease. Patients wild milder disease appear to have higher proliferative CD4+ T helper cell responses (Peano et al 1994, Tillmann et al 1996). Iron overload may adversely affect the outcome of liver disease. Therefore two mutations (Cys282Tyr and His63Asp) in the genetic hemochromatosis gene, designated HFE, were tested for association with hepatic iron accumulation and the activity and severity of hepatitis C infection. Equal numbers of hepatitis C patients and healthy controls carried the Cys282Tyr and the His63Asp mutations (approximately 4 and 24 % respectively). However serum iron levels and transferrin saturation (but not ferritin levels or liver iron content) were significantly higher in HCV-infected carriers than in HCV-infected non-carriers of HFE mutations. Scores for necroinflammatory activity and fibrosis in the liver were also significantly higher in patients with chronic HCV infection that carried HFE mutations (Martinelli et al 2000). Thus HFE gene mutations do not appear to influence the acquisition of HCV or the development of carriership, but they do appear to worsen the prognosis of liver disease in chronic HCV infection.

References

Almarri, A., Batchelor, J.R. (1994). HLA and hepatitis B infection. Lancet 344, 1194-1195.

Congia, M., Clemente, M.G., Dessi, C., Cucca, F., Mazzoleni, A.P., Frau, F., Lampis, R., Cao, A., Lai, M.E., De Virgiliis, S. (1996). HLA class II genes in chronic hepatitis C virus-infection and associated immunological disorders. Hepatology 24, 1338-1341.

Diepolder, H.,M., Jung, M.C., Keller, E., Schraut, W., Gerlach, J.T., et al. (1998). A vigorous virus-specific CD4+ T cell response may contribute to the association of HLA-DR13 with viral clearance in hepatitis B. Clin. Exp. Immunol. 113, 244-251.

Ferrari, C., Penna, A., Bertoletti, A. (1990). Cellular immune response to hepatitis B virus-encoded antigens in acute and chronic hepatitis B virus infection. J Immunol. 145, 3442-3449.

Giani, G., Chiaramonte, M., Venturi Pasini, C., Fagiolo, U., Naccarato, R. (1979). Hepatitis B surface antigenaemia and HLA antigens. N. Engl. J. Med. 296, 1056.

Hann, H.-W.L., Kim, C,Y., London, W.T., Whitford, P., Blumberg, B.S. (1982). Hepatitis B virus and primary hepatocellular carcinoma: family studies in Korea. Int. J. Cancer 1982, 30, 47-51.

Hohler, T., Gerken, G., Notghi, A., Lubjun, R., Taheri, H., Protzer, U., Loher, H.F., Schneider, P.M., Meyer zum Buschenfelde, K.H., Rittner, C. (1997). HLA-DRB1*1301 and 1302 protect against chronic hepatitis B. J. Hepatol. 26, 503-507.

Hohler, T., Gerken, G., Notghi, A., Knolle, P., Lubjun, R., Taheri, H., Schneider, P.M., Meyer zum Buschenfelde, K.H., Rittner, C. (1997). MHC class II genes influence the susceptibility to chronic hepatitis C. J. Hepatol. 27, 259-264

Hohler, T.M., Wunschel, M., Gerken, G., Schneider, P.M., Meyer Zum Buschenfelde, K.H., Rittner, C., (1998). No association between mannose-binding lectin alleles and susceptibility to chronic hepatitis B virus infection in German patients. Exp. Clin. Immunogenet. 15, 130-133.

Lin, T.M., Chen, C.J., Wu, M.M. et al (1989). Hepatitis B virus markers in Chinese twins. Anticancer Research 9, 737-742.

Martinelli, A.L, Franco, R.F., Villanova, M.G., Figueiredo, J.F., Secaf, M., Tavella, M.H., Ramalho, L.N., Zuculoto, S., Zago, M.A. (2000). Are haemachromatosis mutations related to the severity of liver disease in hepatitis C virus infection? Acta Haematol 102, 152-156.

McGlynn, K.A., Rosvold, E.A., Lustbader, E.D., Hu, Y., Clapper, M.L., Zhou, T., Wild, C.P., Xia, X.L., Baffoe-Bonnie, A., Ofori-Adjei, D., et al. (1995). Susceptibility to hepatocellular carcinoma is associated with genetic variation in the enzymatic detoxification of aflatoxin B. Proc. Natl. Acad. Sci USA 92, 2384-2387.

Mota, A.H., Fainboim, H., Terg, R., Fainboim, L. (1987). Association of chronic hepatitis and HLA B35 in patients with hepatitis B virus. Tissue Antigens 30, 238-240.

Peano, G., Menardi, G., Ponzzetto, A., Fenoglio, L.M. (1994). HLA-DR5 antigen. A genetic factor influencing the outcome of hepatitis virus infection? Arch. Intern. Med. 154, 2733-2736.

Thomas, H.C., Foster, G.R., Smiya, M., McIntosh, D., Turner, M.W., Summerfield, J.A. (1996). Mutation of gene of mannose-binding protein associated with chronic hepatitis B viral infection. Lancet 348, 1417-1419.

Thursz, M.R., Kwiatkowski, D., Allsopp, C.E.M., Greenwood, B.M. Thomas, H.C., Hill, A.V.S. (1995). Association between an MHC class II allele and clearance of hepatitis B virus in The Gambia. N. Engl. J. Med. 332, 1065-1069.

Thursz, M.R. (1997). Host genetic factors influencing the outcome of hepatitis. J. Viral Hepatitis 4, 215-220.

Thursz, M., Yallop, R., Goldin, R., Trepo, C., Thomas, H.C. (1999). Influence of MHC class II genotype on outcome of infection with hepatitis C virus. Lancet 354, 2119-2124.

Tibbs, C., Donaldson, P., Underhill, J., Thomson, L., Manabe, K., Williams, R. (1996). Evidence that the HLA DQA1*03 allele confers protection from chronic HCV infection in Northern European Caucasoids. Hepatology 24, 1342-1345.

Tillmann, H.L., Chen, D.F., Boker, K., et al (1996). Low frequency of HLA-DR5 in patients with HCV induced end-stage liver cirrhosis requiring transplantation. Human Immunol. 47, 122.

van Hattum, J., Schreuder, G.M.Th., Schalm, S.W. (1987). HLA antigens in patients with various courses after hepatitis B virus infection. Hepatology 7, 11-14.

Vitte, R.L., Fortier, C., Richardet, J.P., Grimbert, S., Trinchet, J.C., Beaugrand, M., Lepage, V., Raffoux, C. (1995). HLA antigens in patients with chronic hepatitis C. Tissue Antigens 45, 356-361.

Chapter 21. Resistance to malaria

Malaria is one of the most severe and widespread infectious diseases, which accounts for 300 to 500 million annual infections. The annual mortality is estimated between 1.5 to 2.7 million deaths, which mostly occurs in sub-Saharan Africa. Indeed it causes childhood morbidity and mortality well in excess of any other single infectious agent. Although most infections run a mild course, more than 3,000 children die of malaria each day. Malaria is most often transmitted by *Anopheles* mosquito vectors, but may also occur through blood transfusion. The variation in disease severity is remarkable. While the influence of parasite inoculum dose on disease severity is controversial, the influence of socio-economic variables is of limited relevance. Hence genetic variation in both host and parasite appear the key determinants of disease severity (Greenwood et al 1991, Korma et al 1995).

Malaria caused by *Plasmodium falciparum* and *P. vivax* is one of the best-studied examples of host genetics of resistance to a specific infectious disease. This probably reflects the easily identifiable geographical relationship between sickle cell anemia or the thalassemias and resistance to malaria, which is already known for more than 50 years (Haldane 1948) on the one hand, and the large number of polymorphic host molecules involved in the pathogenesis on the other hand. Additional evidence for a role for genetic factors in resistance to malaria comes from clear interethnic different infection rates in West Africa and from twin studies (Jepson et al 1997, Modiano et al 1996). The Fulani ethnic group, for example, showed consistently lower prevalences of malaria parasites, and fewer clinical episodes than two other ethnic groups living in the same area in Burkina Faso. The Fulani also showed stronger antibody responses. These genetic differences could not be attributed to differences in hemoglobin S (HbS) or HLA-B53 (see below), indicating that the genetic differences have another, still unknown, basis. Interestingly, the Fulani also have an increased prevalence of the hyperreactive malarious splenomegaly syndrome, which is a manifestation of hyper-responsiveness to malaria (Hill 1998, Modiano et al 1996).

Malaria is a severe disease, which may have exerted and probably still exerts a strong evolutionary force on human populations. This evolutionary force has resulted in phenotypic consequences for the red blood cell, the immune system, cytokines and several other systems. Indeed, more than 10 resistance genes have been identified to date (Hill 1996, Newport 1997). Most malaria resistance genes known so far have been identified through case-control studies (Table 9) (Hill et al 1991, Hill 1998, McGuire et al 1994, Rowe et al 1997, Williams et al 1996). Major protective loci include the sickle cell trait and the MHC class I, II, and III genes (Hill et al 1991, Jepson et al 1997, McGuire et al 1994). While carriage of HbS is associated with 90 % protection from severe malaria, the HLA-B53 antigen is associated with a 40 % protection (Hill et al 1991). Variation in the TNF-α promoter region was also found to be associated with resistance or susceptibility to malaria (McGuire et al 1994) (see chapter 10). Contrasting results have been reported for several polymorphisms, for example in the gene encoding mannose-binding protein (MBP) (Bellamy et al 1998, Luty et al 1998). Part of the genetic control of malaria occurs through specific immune responses. By comparing concordances of immune responses to malaria antigens in HLA-typed dizygous

Table 9. Genes associated with resistance or susceptibility to malaria

α-Globin (α-thalassemia)
β-Globin
Glucose-6-phosphate dehydrogenase
Spectrin
Erythrocyte band 3 protein
Glycophorin A
Glycophorin B
HLA-B
HLA-DRB1
HLA-DQB1
TNF-α
ABO blood group
Duffy chemokine receptor
ICAM-1
CD36
iNOS

Adapted from Hill (1998) and Aitman et al (2000).

and monozygous twins, it was possible to estimate the magnitude of the effects of MHC and non-HMC genes on these responses. Although the non-MHC effects were more striking, significant effects of the MHC on responses to many malarial antigens were also identified (Hill 1996).

Hemoglobinopathies and red cell polymorphisms
Hemoglobin is located in red blood cells where it is the transporter of O_2 and CO_2. In adults normal hemoglobin consists of two α- and two β-chains, each containing approximately 140 amino acids. The four subunits each contain an iron-containing heme group and function together as tetramere. Both the α and β globin gene clusters, located respectively on chromosomes 16 and 11, contain a number of well-defined, non-random patterns of restriction length polymorphisms (RFLPs). Hemoglobinopathies are the commonest single gene disorders in man, affecting up to 80 % of some populations, and consist of the structural hemoglobin variants and the α- and β-thalassemias (Weatherall 1996).
The three structural hemoglobin variants, S, C and E, reach very high frequencies. These hemoglobinopathies have probably been generated rather recently in our evolutionary history (Flint et al 1993). A SNP in the gene encoding the β chain of human hemoglobin, resulting in the substitution of valine for glutamic acid at the 6th position of the β chain, is responsible for the sickle cell trait. Interestingly, there is evidence that suggests that hemoglobin S may have arisen twice. The high frequency of these mutations probably results from heterozygous advantage. As mentioned above, heterozygosity for hemoglobin S provides 90 % protection against severe anemia and cerebral involvement due to *P. falciparum* infection. Another variant of hemoglobin, hemoglobin E, occurs at high frequencies in the eastern parts of the Indian subcontinent, the Union of Myanmar, and South East Asia. This structural hemoglobin variant displays the phenotype of mild β-thalassemia and also confers protection against *P. falciparum* malaria (Flint et al 1993). However, it is still uncertain whether hemoglobin C, which is common in West Africa, is protective.

The thalassemias are complex diseases that are associated with hemolytic anemia. Evidence has accumulated that malaria is the major selective force that has maintained the α- and β-thalassemias at very high frequencies in many parts of the world. Areas where thalassemia occurs almost completely overlap with areas where malaria is, or has been. Exceptions to this rule could be explained by gene flow, the action of drift on small populations, and other mechanisms that are still compatible with the selective force of thalassemia (Weatherall 1996, Flint et al 1993). In addition, in every part of the world with malaria there are completely different patterns of α and β globin gene mutations. Thus these mutations have arisen locally and have reached a high frequency by selection. As mentioned, there are two α globin genes per haploid genome. The commonest forms of α-thalassemia result from deletion of either one or both of these genes, respectively called α^+-thalassemia and α^0-thalassemia. These may be present in homozygous or heterozygous form. At least six different forms of α^+-thalassemia and two forms of α^0-thalassemia have been identified.

α^+-Thalassemia is a mild disease and very common in many tropical countries and appears to give only partial protection to malaria. Yates et al (1995) suggest that only the homozygotes are resistant to *P. falciparum* malaria to a level of about 10 to 25%. While sickle cell anemia and the β-thalassemias are most probably balanced polymorphisms based on heterozygous advantage, α-thalassemia may represent a transient polymorphism, because deleterious effects of the mild form of α-thalassemia could not be demonstrated. This form α-thalassemia would therefore have gone to fixation, if the selective pressure of malaria had continued. Indeed in the Tharu population of Nepal it has reached near fixation with a gene frequency of 0.8 (Modiano et al 1991). In contrast to α^+-thalassemia, total lack of α globin expression (homozygous α^0-thalassemia) results in severe disease, which is usually fatal during gestation or early in life.

In addition to the α^+-thalassemia mutations, there are at least 120 different β- thalassemia mutations. β-Thalassemia major (Cooley's anemia) results from a homozygous defect resulting in a complete lack of β-globin expression. It is characterized by severe hemolytic anemia. β-Thalassemia occurs in a number of Mediterranean at-risk populations (Greek and Turkish Cypriots, Greeks, Continental Italians and Sardinians), the Middle East, the Indian subcontinent, and South East Asia. Homozygous carriers develop severe anemia, splenomegaly, and skeletal modifications. They fail to thrive and, if untreated, usually die within the first decade of life. A successful genetic preventive program, based on heterozygote screening and prenatal diagnosis, has therefore been initiated in several of these populations (Cao et al 1996). A case-control study from Liberia showed that β-thalassemia provides partial protection against clinical malaria, including severe forms, caused by *P. falciparum* (Willcox et al 1983).

The mechanisms by which the mutations afford protection to *P. falciparum* malaria have only partially been elucidated. Infected red blood cells from individuals with the HbS trait may sickle more readily than non-parasitized cells and are therewith more prone to phagocytotic destruction (Luzatto et al 1970). Furthermore, invasion and growth of the parasites are inhibited when the cells are under reduced oxygen conditions (Pasvol and Weatherall 1978).

Thalassemic red blood cells infected with *P. falciparum* bind significantly more antibody from the sera of patients with acute malaria than normal cells (Luzzi et al 1991), either because they more efficiently expose malarial antigens, or because they express red-cell neoantigens related to senescence more effectively. Also interaction between susceptibility to the non-lethal *P. vivax*, acting as a natural cross-reacting vaccine in communities where both parasites are prevalent, and resistance to *P. falciparum* has been proposed as an immunologic basis of the protection afforded by α^+-thalassemia (Williams et al 1996). Interestingly, the α^+-thalassemia trait also appears to protect (both homozygotes and heterozygotes) against hospital admission with infections other than malaria (Allen et al 1997).

In addition to hemoglobin mutations, other defects in red cell structure and metabolism afford protection against malaria. Glucose-6-phosphate dehydrogenase (G6PD) deficiency is a frequent X-linked disorder that affords protection to heterozygous females and hemizygous males (Ruwende et al 1995). G6PD participates in the hexose monophosphate shunt of glycolysis. G6PD deficiency reduces severe malaria in heterozygous females and hemizygous males for 46 – 58 %. This disorder also affects anti-malaria treatment (see chapter 27).

The mutation in the gene encoding the band 3 protein (chromosomal location 17q21-q22) of the red cell membrane that is the cause of Melanesian ovalocytosis is lethal for homozygotes, but is protective against malaria for heterozygotes. The mutation, a 27 bp deletion in the gene encoding band 3 protein, is very common in Melanesia where sickle cell anemia is absent (Jarolim et al 1991, Schofield et al 1992). In contrast, the band 3 deletion is absent in Africans. This is a clear illustration that resistance genes may differ strikingly in different populations, thus prohibiting the pooling of data from different populations in genetic studies. The mutation is associated with a lower incidence of both *P. falciparum* and *P. vivax* parasitemia and lack of malaria-associated mortality.

Several blood group antigens and other antigens on the surface of red blood cells have been associated with protection or susceptibility to malaria. As mentioned (see chapter 7), people in sub-Saharan Africa who lack the Duffy blood group antigen appear to resist malaria caused by *P. vivax*. The absence of the Duffy antigen is the major resistance gene for *P. vivax* malaria in Africa, but not in Asia. Furthermore, the Lewis (a) phenotype is associated with protection, while the Kidd Jk (b) phenotype is associated with susceptibility to severe disease from *P. falciparum* malaria (Miller 1994, Weatherall 1996).

Rowe et al (1997) raised the possibility that the complement receptor 1 (CR1) on erythrocytes play a role in the formation of rosettes, the binding of infected to uninfected erythrocytes. This phenomenon is associated with severe malaria. Erythrocytes with a common African CR1 polymorphism (S1(a-)) have a reduced adhesion to malarial antigen and may therefore protect against severe malaria. Subsequent studies, however, found no association with polymorphisms in the gene coding for CR1 (as well as in genes coding for the interleukin receptor IL-1RA and the vitamin D receptor) (Bellamy et al 1998, Luty et al 1998).

Another critical step in the infection by *P. falciparum* is the adhesion of parasitized red blood cells to capillary endothelium. The human protein CD36 is a major endothelial receptor for *P. falciparum*-infected red blood cells. CD36 is further expressed on monocytes and platelets. CD36 may contribute to disease by sequestering infected red blood cells and by inhibiting the immune response to the parasite. Several mutations that result in CD36 deficiency have been described, of which the T1264G stop mutation in exon 10 is the most frequent. This mutation accounts for 80 % of mutant alleles. The mutation may have arisen on a single founder chromosome, because it is in strong linkage disequilibrium with a single allele of the CD36 intron-3 microsatellite in four distinct African populations. Many people of African origin, but few whites, are deficient in CD36. Unexpectedly, mutations that cause CD36 deficiency appeared associated with severe malaria, particularly cerebral malaria. The frequency of detected mutations, in heterozygous and homozygous state, is higher in patients with severe malaria than in controls (OR = 1.54; 95 % CI 1.09 – 2.15, P = 0.01). CD36 deficiency might contribute to disease because of reduced sequestering of red blood cells in organs outside the brain. The mechanism by which the CD36 mutations have been maintained at high frequencies in Africans (7.5 - 18.5 % in heterozygous or homozygous state) and in other races has not been elucidated. However, other prevalent infections might be an explanation (Aitman et al 2000). CD36 has a wide variety of functions and so partial or complete CD36 deficiency might alter the host response in many processes (Craig et al 2001). Evidently a mutation with a beneficial effect on one function might have a deleterious effect on another.

Other red cell variants that have been associated with partial protection of the red cell against invasion with *P. falciparum in vitro* are variants of glycophorin A, B, or C (Hill et al 1992, Miller 1994). However, there are likely other protective genetic mechanisms involved in malaria that so far have not been studied in detail, and that are associated with red cell function or not. For example, there is small but significant association between blood group O and resistance to severe disease (Hill et al 1992). Furthermore Miller (1994) suggested that the common genetic form of hemochromatosis in sub-Saharan Africa may give heterozygous advantage in resisting malaria, because the parasite is able to detoxify intracellular heme-derived iron and needs extracellular iron for its development.

Immunogenetic mechanisms

In a large case control study, the HLA class I antigen HLA-B53 afforded 40 % clinical protection from both cerebral malaria and severe malarial anemia, but intriguingly without reducing mild disease or infection rates. This antigen occurs frequently in Africa, but is seldom elsewhere. Interestingly, the global distribution of HLA-B53 remarkably resembles that of the protective HbS variant. Investigations of the molecular mechanism underlying the protective role of HLA-B53, an approach denoted as reverse genetics, has subsequently led to the identification of a conserved malarial epitope from a liver stage specific antigen that induces cytotoxic T lymphocytes (Aidoo et al 1995, Hill et al 1992).

In addition to the HLA class I system, the class II system has been implicated in protection from malaria. A single HLA class II DR-DQ haplotype, HLA-DRB1*1302-DQB1*0501 is also independently associated with protection against severe malarial anemia. These HLA types are both common in The Gambia where these studies have been carried out, suggesting that they have an evolutionary advantage in malaria-endemic areas. Further supporting the significance of the association was the finding that Gabonese children that carry HLA-DQB1*0501 had stronger Th1 responses to *P.*

falciparum liver stage antigen type 1 compared with non-carriers (May et al 2001). Thus in contrast to the severe forms of malaria, no clear-cut associations were found between HLA and mild malaria or parasite densities in several studies. This is not well understood, but heterogeneity in host resistance or diversity in parasite virulence might explain the difference in the rate of reduction between severe and mild disease (Bennet et al 1993, Gupta and Hill 1995, Hill 1996). Several studies also gave indications that different HLA associations may occur in different populations, probably reflecting local differences in parasite antigens, transmission, or other factors (Hill 1996).

The single base change in the promoter at position −308 from the transcriptional start site (see chapter 10) of the TNF-α gene is associated with a markedly increased risk of cerebral malaria and a fatal outcome of the disease, which is due to increased plasma levels of this proinflammatory cytokine. The influence of the TNF-α polymorphism is confined to homozygotes and is independent of HLA-B53 status (Kwiatkowski 1995, McGuire et al 1994). The frequency of this gene variant, 0.16 in The Gambia, indicates that the disadvantage in susceptibility to malaria for homozygotes is balanced by some, not yet identified, heterozygote advantage. Homozygosity for other TNF promoter polymorphisms, respectively at positions −238 and −376, have also been found to be associated with cerebral malaria (but not anemia) in other populations (McNicholl et al 2000, Weatherall 1996). Such findings may pave the way to new therapies directed at this pro-inflammatory cytokine. Unfortunately, however, trials of anti-TNF-α therapy have so far been unsuccessful.

In addition to the role of TNF-α variants in the inflammatory response itself, TNF-α-genotypic variants might also influence malaria severity by upregulating the *P. faliciparum* adhesion receptor, intercellular adhesion molecule-1 (ICAM-1), on endothelial cells of blood vessels. Upregulating of ICAM-1 could subsequently lead to increased accumulation of parasites in the brain microvascular endothelial cells. This has been suggested because certain genetic variants of ICAM-1 were associated with high risks of developing cerebral malaria. A particular variant of ICAM-1 (ICAM-1[Kilifi]) with a point mutation in the N-terminal domain of the protein was found in high frequencies in Africans in coastal Kenya and was unexpectedly found more frequently in homozygous form in severe cases of malaria than in controls (Fernandez-Ryes et al 1997). However, in The Gambia no effect of this variant on malaria susceptibility was detected, and in Gabon the variant was even associated with protection (Bellamy et al 1998, Craig et al 2001). The ICAM-1[Kilifi] mutation has a range of effects, including reduced avidity for LFA-1 and fibrinogen, and reduced rhinovirus and *P. falciparium* adhesion (Craig et al 2000). Mutation of ICAM-1 might thus have very varying effects on inflammatory responses and infection outcome. In combination with complex interactions with parasite-determined virulence factors and local conditions affecting parasite transmission, the outcome of selection pressure on ICAM-1 might therefore have been very different. Both temporal and regional variations in selective forces might therefore be likely. Thus ICAM-1 (and other host receptors such as CD36) might have been subject to both balancing and frequency-dependent selection.

Nitric oxide (NO) plays an important role in the resistance to malaria and a variety of other organisms. NO is induced by TNF-α, IFN-α, and IFN-γ, while IL-10 suppresses NO synthesis. NO activity has anti-plasmodial activity *in vitro*, but excess NO production has been suggested to contribute to the pathogenesis of cerebral malaria. In the promoter region of the gene encoding inducible NO synthase (iNOS, NOS2), a G-954C SNP and a (CCTTT)n pentanucleotide repeat polymorphism have been identified. In Gabon and in the Gambia (but not in Tanzania), the C allele at position −954 and long allele pentanucleotide microsatellite polymorphisms (>11) were associated with protection against mild *P. falciparum* malaria. In five specific ethnic population groups from Africa, Europe, Asia, and the Caribbean there were highly significant differences in allele frequencies of the pentanucleotide repeat, suggesting a strong selective pressure from malaria or other infectious diseases operating on this locus (Levesque et al 1999, Luty 2001, Migot-Nabias et al 2000, Xu et al 2000).

Conclusion

Malaria has exerted strong selective forces on human populations. Very different mutations associated with a degree of protection occur at very different frequencies in different ethnic populations, and, intriguingly, certain host genotypes reduce severe malaria without reducing mild disease. The genetics of malaria susceptibility therefore clearly illustrates many important, but still partly understood principles at work in the co-evolution of man and microbes, such as balanced and transient polymorphisms. In addition, the work has pointed to important functional pathways that are involved in the pathogenesis of disease or in recovery from disease. Important functional pathways include the adhesion of the parasite to the Duffy antigen (*P. vivax*), the increased fragility of infected mutant red cells, and the increased presentation of antigens on infected mutant red cells to the immune system, which facilitates their binding by antibody. The protective role of HLA-B53 (*P. falciparum*) indicates the significance of CD8+ cytotoxic T lymphocytes in reducing severe disease. Moreover, the identification of protective HLA antigens and the determinants selected by them points to an approach by which protective T cell epitopes may be identified and used in vaccine development. TNF-α promoter polymorphisms underline the detrimental role of a severe inflammatory response in malaria.

References

Aidoo, M., Lalvani, A., Allsopp, C.E.M., Plebanski, M., Meisner, S.J., Krausa, P., Browning, M., Morris-Jones, S., Gotch, F.M., Fidock, D.A. et al. (1995). Identification of conserved antigenic components for a cytotoxic T lymphocyte-inducing vaccine against malaria. Lancet 345, 1003-1007.

Aitman, T.J., Cooper, L.D., Norsworthy, P.J., Wahid, F.N., Gray, J.K., Curtis, B.R., McKeigue, P.M., Kwiatkowski, D., Greenwood, B.M., Snow, R.W., Hill, A.V., Scott, J. (2000). Malaria susceptibility and CD36 mutation. Nature 405, 1015-1016.

Allen, S.J., O'Donnell, A., Alexander, N.D., Alpers, M.P., Peto, T.E.A., Clegg, J.B., Weatherall, D.J. (1997). α+-Thalassemia protects children against disease caused by other infections as well as malaria. Proc. Natl. Acad. Sci. USA 94, 14736-14741.

Bellamy, R., Kwiatkowski, D., Hill, A. (1998). Absence of an association between intercellular adhesion molecule 1, complement receptor 1, and interleukin 1 receptor antagonist gene polymorphisms in a West African population. Trans. R. Soc. Trop. Med. Hyg. 92, 312-316.

Bennet, S., Allen, S.J., Olerup, P., Jackson, D.J., Wheeler, J.G., Rowe, P.A., Riley, E.M., Greenwood, B.M. (1993). Human leucocyte antigen (HLA) and malaria morbidity in a Gambian community. Trans. Royal Soc. Trop. Med. Hyg. 87, 286-287.

Cao, A., Rosatelli, M.C., Galanello, R. (1996). In: Variation in the human genome. Wiley, Chichester (Ciba Foundation Symposium 197), 137-155.

Craig, A., Fernandez-Reyes, D., Mesri, M., McDowall, A., Altieri, D.C., Hogg, N, Newbold, C. (2000). A functional analysis of a natural variant of intercellular adhesion molecule-1 (ICAM-1 Kilifi). Hum. Mol. Genet. 9, 525-530.

Craig, A., Hastings, I., Pain, A., Roberts, D.J (2001). Genetics and malaria-more questions than answers. Trends in Parasitol. 17, 55-56.

Fernandez-Reyes, D., Craig, A.G., Kyes, S.A., Peshu, N., Snow, R.W., Berendt, A.R., Marsh, K., Newbold, C.I. (1997). A high frequency African coding polymorphism in the N-terminal domain of ICAM-1 predisposes to cerebral malaria in Kenya. Hum. Mol. Genet. 112, 1357-1360.

Flint, J., Harding, R.M., Boyce, A.J., Clegg, J.B. (1993). The population genetics of the haemoglobinopathies. Clinic. Haematology 6, 215-262

Haldane, J.B.S. (1948). The rate of mutation of human genes. Proceedings of the Eigth International Congress of Genetics and Heredity. Hereditas 35 (Suppl.), 267-273.

Greenwood, B.M., Marsh, K., Snow, R. (1991). Why do some children develop severe malaria? Parasit. Today 8, 239-242.

Gupta,. S., Hill, A.V.S. (1995). Dynamic interactions in malaria: host heterogeneity meets parasite polymorphism. Proc. Roy. Soc. Lond. B. 261, 271-277.

Hill, A.V.S., Allsopp, C.E.M., Kwiatkowski, D., Anstey, N.M., Twumasi, P., Rowe, P.A., Bennet, S., Brewster, D., McMichael, A.J., Greenwood, B.M. (1991). Common West African HLA antigens are associated with protection from severe malaria. Nature 352, 595-600.

Hill, A.V.S., Elvin, J., Willis, A.C., Aidoo, M., Allsopp, C.E.M., Gotch, F.M., Gao, X.M., Takiguchi, M., Greenwood, B.M., Townsend, A.R. et al. (1992). Molecular analysis of the association of HLA-B53 and resistance to severe malaria. Nature 360, 434-439.

Hill, A.V.S. (1996). Host genetics and infectious disease. Parasitology 112 (Suppl.), S23-S26.

Hill, A.V.S. (1996). Genetic susceptibility to malaria and other infectious diseases: from the MHC to the whole genome. Parasitology 112 (Suppl.), S75-S84.

Hill, A.V.S. (1998). The immunogenetics of human infectious diseases. Annu Rev. Immunol. 16, 593-617.

Jarolim, P., Palek, J., Amato, D., Hassan, K., Sapak, P., Nurse, G.T., Rubin, H.L., Zhai, S., Sahr, K.E., Liu, S.C. (1991). Deletion in erythrocyte band 3 gene in malaria-resistant Southeast Asian ovalocytosis. Proc. Natl. Acadam. Sci. USA 88, 11022-11026.

Jepson, A., Banya, Sisay-Joof, F., Hassan-King, M., Nunes, C., Bennett, S., Whittle, H. (1997). Quantification of the relative contribution of major histocompatibility complex (MHC) and non-MHC genes to human immune responses to foreign antigens. Infect. Immun. 65, 872-876.

Jepson, A., Sisay-Joof, F., Banya, W., Hassan-King, Frodsham, A., Bennet, S., Hill, A.V.S., Whittle, H. 1997. Genetic linkage of mild malaria to the MHC in Gambian children. Br. Med. J. 315, 96-97.

Koram, K.A., Bennett, S., Adiamah, J.H., Greenwood, B.M. (1995). Socio-economic determinants are not major risk factors for severe malaria in Gambian children. Trans. Roy. Soc. Trop. Med. Hyg. 89, 151-154.

Kwiatkowski, D. (1995). Malarial toxins and the regulation of parasite density. Parasitol. Today 11, 206-212.

Levesque, M.C., Hobbs, M.R., Anstey, N.M., Vaughn, T.N., Chancellor, J.A. Pole, A., Perkins, D.J., Misukonis, M.A., Chanock, S.J., Granger, D.L., Weinberg, J.B. (1999). Nitric oxide synthase type 2 promoter polymorphisms, nitric oxide production, and disease severity in Tanzanian children with malaria. J. Infect. Dis. 180, 1994-2002.

Luty, A.J. (2001) Personal communication.

Luty, A.J., Kun, J.F., Kremsner, P.G. (1998). Mannose-binding lectin plasma levels and gene polymorphisms in *Plasmodium falciparum* malaria. J. Infect. Dis. 178, 1221-1224.

Luzzato, L., Nwachuku, P.S., Reddy, S. 91970). Increased sickling of parasitised erythrocytes is the mechanism of resistance against malaria in the sickle cell trait. Lancet 1, 319-321.

Luzzi, G.A., Merry, A.H., Newbold, C.I., Marsh, K., Pasvol, G., Weatherall, D.J. (1991). Surface antigen expression on *Plasmodium falciparum*-infected erythrocytes is modified in α and β thalassaemia. J. Exp. Med. 173, 785.

May, J., Lell, B., Luty, A.J., Meyer, C.G., Kremsner, P.G. (2001). HLA-DQB1*0501-restricted Th1 type immune responses to *Plasmodium falciparum* liver stage antigen 1 protect against malaria anemia and reinfections. J. Infect.Dis. 183, 168-172.

McGuire, W., Hill, A.V., Allsopp, C.E., Greenwood, B.M., Kwiatkowski, D. (1994). Variation in the TNF-alpha promoter region associated with susceptibility to cerebral malaria. Nature 371, 508-510.

McNicholl, J.M., Downer, M.V., Udhayakumar, V., Alper, C.A., Swerdlow, D.L. (2000). Host-pathogen interactions in emerging and re-emerging diseases: A genomic perspective of tuberculosis, malaria, human immunodeficiency virus infection, hepatitis B, and cholera. Annu. Rev. Public Health 21, 15-46.

Migot-Nabias, F., Mombo, L.E., Luty, A.J., Dubois, B., Naias, R., Bisseye, C., Millet, P., Lu, C.Y., Deloron, P. (2000). Human genetic fators related to susceptibility to mild malaria in Gabon. Genes Immun. 1, 435-441.

Miller, L.H., (1994). Impact of malaria on genetic polymorphism and genetic diseases in Africans and African Americans. Proc. Natl. Acad. Sci. USA 91, 2415-2419.

Modiano, G., Morpurgo, G., Terrenato, L. et al (1991). Protection against malaria morbidity: near-fixation of the α-thalassemia gene in a Nepalese population. Am. J. Hum. Genet. 48, 390-397.

Modiano, D., Petrarca, V., Sirima, B.S., Nebié, I., Diallo, D., Esposito, F., Coluzzi, M. (1996). Different responses to *Plasmodium falciparum* malaria in West African sympatric ethnic groups. Proc. Natl. Acad. Sci. USA 93, 13206-13211.

Newport, M.J. (1997). Unraveling the genetic basis of susceptibility to infectious diseases. Curr. Opinion Inf. Dis. 10, 233-238.

Pasvol, G., Weatherall, D.J., (1978). A mechanism for the protective effect of haemoglobin S against *P. falciparum* malaria. Nature 274, 701-703.

Rowe, J.A., Moulds, J.M., Newbold, C.I., Miller, L.M. (1997). *P. falciparium* rosetting mediated by a parasite variant erythrocyte membrane protein and complement receptor 1. Nature 388, 292-295.

Ruwende, C., Khoo, S.C., Snow, R.W., Yates, S.N.R., Kwiatkowski, D., Gupta, S., Warn, P., Allsopp, C.E.M., Gilbert, S.C., Peschu, N., Newbold, C.I., Greenwood, B.M., Marsh, K., Hill, A.V.S. (1995). Natural selection of hemi- and heterozygotes for G6PD deficiency in Africa by resistance to severe malaria. Nature 376, 246-249.

Schoffield, A.E., Reardon, D.M., Tanner, M.J. (1992). Defective anion transport activity of the abnormal band 3 in hereditary ovalocytic red blood cells. Nature 355, 836-838.

Weatherall, D.J. (1996). Host genetics and infectious disaese. Parasitology 112, S23-S29.

Willcox, M., Bjorkman, A., Brohult, J., Pehrson, P.O., Rombo, L., Bengtsson, E. (1983). A case-control study in northern Liberia of *Plasmodium falciparum* malaria in haemoglobin S and β-thalassemia traits. Ann. Trop. Med. Parasitol. 77, 239-246.

Williams, T.N., Maitland, K., Bennett, S., Ganczakowski, M., Peto, T.E., Newbold, C.I., Bowden, D.K., Weatherall, D.J., Clegg, J.B. (1996). High incidence of malaria in α-thalassaemic children. Nature 383, 522-525.

Xu, W., Humphries, S., Tomita, M., Okuyama, T., Matsuki, M., Burgner, D., Kwiatkowski, D., Liu, L., Charles, I.G. (2000). Survey of the allelic frequency of a NOS2A promoter microsatellite in human populations: assessment of the NOS2A gene and predisposition to infectious disease. Nitric Oxide 4, 379-383.

Yates, S.N.R., Snow, R.W., Allsopp, C.E.M., Newton, C.R.J.C., Kwiatkowski, D., Palmer, D., Peschu, N., Greenwood, B.M., Marsh, K., Newbold, C.I., Hill, A.V.S. (1995). Resistance of homozygotes but not heterozygotes for α+ thalassemia to severe malaria: implications for the time depth of malarial selection. Proc. British Soc. Parasitol. 5th Malaria Meeting, 13-14.

Chapter 22. Trypanotolerance

Species of the flagellated protozoan *Trypanosoma* cause disease in humans, cattle, and small ruminants. The disease is called sleeping sickness in people and *nagana* in cattle. In man *T. brucei gambiense* and *T. brucei rhodesiense* cause African trypanosomiasis or sleeping sickness, which disease is restricted to equatorial Africa by its vector, the tsetse fly. *T. cruzi* causes South American trypanosomiasis or Chagas' disease, which is transmitted by a non-flying vector, the "kissing" bug. In cattle three trypanosome species are important: *T. vivax*, *T. congolense*, and *T. brucei*. An important feature of trypanosomes is their remarkable degree of antigenic variation, which is based on their capacity to switch between approximately 1,000 genes coding for variable surface glycoproteins, thus enabling the parasite to survive in their infected host.

People
African trypanosomiasis may be lethal to people, but little is known about possible human genetic susceptibility to trypanosomiasis. Familial clustering has been reported, but appears to be a consequence of shared exposure rather than of genetic susceptibility (Khonde et al 1997).

Cattle
Trypanosomiasis is arguably the most important debilitating disease of domestic livestock in Africa. Livestock production in tsetse-affected areas therefore relies heavily on vector control, the use of trypanocidal drugs, and exploitation of the genetic resistance exhibited by indigenous breeds. One to two weeks following infection with trypanosomes, susceptible cattle develop intermittent fever and anemia. They are often bitten repeatedly by tsetse flies and thus become infected with different kinds of trypanosomes. Infection in domestic livestock impairs the animal's immune system and causes severe anemia, weight loss, and reproductive disorders. Infected animals may die if not treated. Before dying they may deteriorate for months. Trypanosomiasis occurs in more than a third of the continent. About 45 million cattle (30 % of the continent's cattle) and similar number of goats and sheep are at risk of infection. Losses of meat and milk production and tractive power, in addition to the costs of programs to control the disease, are estimated to total US $1,340 million each year (Koudandé 2000).

In Africa little overlap exists between cattle-producing areas and tsetse-infested areas, except in regions of West and Central Africa where a few indigenous cattle breeds exist that are genetically resistant to trypanosomiasis. Resistance to the effects of trypanosomiasis (trypanotolerance) is for example well documented among several indigenous taurine breeds of West African cattle (*Bos taurus*), notably the N'Dama breed. This trait was recognized by farmers in the N'Dama as early as in the 1920s. Other resistant breeds are the Savannah Shorthorn, the Dwarf West African Shorthorn, the Lagune, the Muturu, and the Baoulé. These breeds have been in Africa for 5,000 to 7,000 years (Murray et al 1984). In the 1980s and 1990s the trypanotolerant cattle populations increased, resulting in N'Dama herds in nearly all western and central African countries, which could be the source of further dissemination of genetic material (d'Ieteren 1993). The reasons why N'Dama cattle have not spread further to the east of Zaire are not clear and not documented. Yet most disease-resistant breeds occur in small numbers (they constitute approximately 6 % of the cattle population of Africa) and they have a relatively low productivity. N'Dama cattle, for example, produce only one-quarter to one-tenth of the milk and meat of breeds kept in industrialized countries. However, the more numerous and productive exotic livestock breeds, such as the Zebu and Boran (*Bos indicus*), and European cattle breeds, are highly susceptible to trypanosomiasis ("trypanosusceptible") and other tropical diseases (streptothricosis, ticks, heartwater, anaplasmosis, babesiosis for example) (Mwangi et al 1998). Also East African cattle are very susceptible to trypanosomiasis.

Since long it has been suggested that the trypanotolerant trait has a genetic base. In comparative studies with N'Dama, Zebu, and their F1 crosses, the crossbred expressed an intermediate level of resistance compared with the two parent breeds (Chandler 1952). Unfortunately no markers linked to trypanotolerance are known in cattle at present to assist in breeding programs. However, there is evidence that trypanotolerance is not only a breed characteristic, particularly of the N'Dama, but that it is also a heritable trait within the breed, thus opening new opportunities for improved productivity through selection for trypanotolerance and production traits (d'Ieteren and Trail 1993). At present, with no opportunity for marker-assisted introgression, this would be a feasible option, next to using F1 crosses between local trypanotolerant breeds and European dairy cattle, to improve production in heavily tsetse-challenged areas.

In trypanotolerant N'Dama cattle parasitemia lasts shorter, and after about 20 weeks they are parasitemic only periodically. The ability to resist the development of anemia during infection, as assessed by packed red cell volume percent (PCV), has been shown to correlate with the capacity to be productive. PCV during trypanosome affection appears affected by the humoral immune response. The trypanotolerant N'Dama cattle indeed produce higher level of antibodies against both variant and invariant parasitic antigens (Murray et al 1984). Thus regulation of the PCV appears a key trait of trypanotolerance and has opened the possibility of using PCV values as selection criterion for trypanotolerance (Murray et al 1990). T cells may further play a critical role in mounting an effective immune response against infection. T cell responses in infected N'Dama cattle started earlier and were maintained longer than in infected Boran cattle. Moreover, T cells from infected N'Dama cattle recognized a broader spectrum of trypanosome antigens (Paling and Dwinger 1993). It has further been found that the sialic acid content of erythrocytes decreases during trypanosome infection, causing abnormal aging of the cells and their clearance by phagocytotic cells. This mechanism likely contributes to the development of anemia. Trypanotolerant cattle were found to have more sialic acid in their red blood cells and were also able to maintain higher complement levels during trypanosome infection in comparison with susceptible cattle (Koudandé 2000).

Trypanotolerant cattle are also indicated to be well adapted to other unfavorable environmental factors affecting production, such as heat stress and water restriction (d'Ieteren 1993). There is thus a large need to ally the disease resistance of tropical livestock with the productivity of exotic breeds. The characteristics of the, sometimes rare, local trypanotolerant breeds and populations point to their potential useful role as a source for genes of great usefulness to future animal breeders. This underscores the need to conserve livestock biodiversity.

Small ruminants
Several breeds of small ruminants indigenous to West and Central Africa are also called trypanotolerant, because they survive and remain healthy and productive in tsetse-infested areas. Despite their small size, they may thus significantly contribute to the production of animal protein in tsetse-infested areas in Africa. These breeds include Djallonke sheep and West African Dwarf goats. They constitute 6 % and 11 % of the sheep and goat populations in these areas, respectively. Following infection with *T. congolense* these breeds show a very low mortality rate and they maintain their ability to gain weight. However, their control of parasitemia and anemia appears less than in trypanotolerant cattle, and the trait is therefore often referred to as "resilience", rather than real "resistance". Resilience is defined as the ability to maintain relatively undepressed production while subjected to parasitic challenge. This resilience appears based on innate resistance, leading to quick hematological changes and antibody responses upon infection. However, despite this trait, *T. congolense* infection in these animals affects their reproductive performance through hormonal imbalance, abortions, stillbirths, and reduced sperm quality. Djalonke sheep appear to be more tolerant to infection than the West African Dwarf goats. Sahelian sheep, in contrast to the Djallonke sheep, are trypanosusceptible, and crosses between Sahelian and Djallonke sheep confirmed that the trypanotolerant trait in Djallonke sheep is genetically linked. Such crosses may be important for meat production thanks to their larger size and good growth. In addition to genetic differences between breeds, also large individual differences in reproductive response related to the infection were found among Djallonke sheep.

In addition to genetic factors, several other factors contribute to *T. congolense* infection rates and disease severity, including feeding success of tsetse flies and different host-vector interactions. Apart from resilience to the effects of trypanosomiasis, there are strong indications that Djalonke sheep and West African Dwarf goats also possess resistance or resilience to other diseases, such as helminthosis, tick infections, and tick-borne diseases (Osaer and Goossens 1999).

Mice
Also laboratory strains of mice show different susceptibilities to trypanosomiasis, which were shown to be under multigenic control (Kemp et al 1995, Kemp et al 1997). A linkage analysis of survival following trypanosome challenge was performed by selective genotyping in large F2 populations, produced by crossing the resistant C57BL/6 with susceptible A/J and BALB/c inbred mouse lines. Regions on mouse chromosomes 5 and 17 were found to be important in determining resistance in both crosses, while an additional region on chromosome 1 showed evidence of involvement in only one cross. The three loci, designated *Tir1*, *Tir2*, and *Tir3*, have a large effect on resistance to trypanosomiasis and account for most of the variation in both F2 populations (Kemp et al 1997, Iraqi et al 2000). Furthermore, evidence of a powerful genetic imprinting effect was shown in backcross mice (Kemp et al 1995). (The term imprinting is used to describe the phenomenon by which certain genes function differently, depending on whether they are maternally or paternally derived. Thus it confers a functional change in particular alleles at the time of gametogenesis deter-

mined by the sex of the parent.) Introgression of the three loci in a recipient mouse strain by a backcross strategy, as a model to study marker-assisted introgression in cattle, appeared feasible and effective in relatively short time (Koudandé 2000).

The physiologic mechanisms by which these loci confer resistance have not been elucidated. However, cytokines likely play an important role. Differential host responses to *T. cruzi* in genetically resistant and susceptible mice provided evidence for a critical role of IL-10 and TGF-β in determining susceptibility. High levels of IL-10 appeared to increase susceptibility to *T. cruzi* infection (Morrissey 1994, Silva et al 1991). IL-10 may do so by downregulating cell-mediated and Th1 cytokine responses.

References

Chandler, R.I. (1952). Comparative tolerance of West African N'Dama cattle to trypanosomiasis. Ann. Trop. Med. Parasit. 46, 127-134.

d'Ieteren, G.D.M. Trypanotolerant livestock, a sustainable option for increasing livestock production in tsetse-affected areas. In: Towards increased use of trypanotolerance: current research and future directions, Ed: G.J. Rowlands, A.J. Teale, Nairobi, Addis Ababa. 1993, pp. 3-11.

d'Ieteren, G.D.M., Trail, J.C.M. Field research on measurement and use of trypanotolerance criteria to enhance trypanotolerant cattle.1. ILCA's achievements and future plans. In: Towards increased use of trypanotolerance: current research and future directions, Ed: G.J. Rowlands, A.J. Teale, Nairobi, Addis Ababa. 1993, pp. 23-28.

Iraqi, F., Clapcott, S.J., Kumari, P., Haley, C.S., Kemp, S.J., Teale, A.J. (2000). Fine mapping of trypanosomiasis resistance loci in murine advanced intercross lines. Mamm. Genome 11, 645-648.

Kemp, S.J., Darvasi, A., Soller, M., Teale, A.J. (1995). Genetic control of resistance to trypanosomiasis. Fourth International Veterinary Immunology Symposium, University of California, p. 59.

Kemp, S.J., Iraqi, F., Darvasi, A., Soller, M., Teale, A.J. (1997). Localization of genes controlling resistance to trypanosomiasis in mice. Nat. Genet. 16, 194-196.

Khonde, N., Pepin, J., Niyonsenga, T., De Wals, P. (1997). Familial aggregation of *Trypanosoma brucei gambiense* trypanosomiasis in a very high incidence community in Zaire. Trans. R. Soc. Trop. Med. Hyg. 91, 521-524.

Koudandé, O.D. (2000). Introgression of trypanotolerance genes. Thesis Wageningen University.

Morrissey, P.J. (1994). IL-10 mediates susceptibility to *Trypanosoma cruzi* infection. J. Immunol. 153, 3135-3140.

Murray, M., Trail, J.C., Davis, C.E., Black, S.J. (1984). Genetic resistance to African trypanosomiasis. J. Infect. Dis. 149, 311-319.

Murray, M., Trail, J.C.M., D'Ieteren, G.D.M. (1990). Trypanotolerance in cattle and prospects for the control of trypanosomiasis by selective breeding. Rev. Sci Tech. Off. Int. Epiz. 9, 369-386.

Mwangi, E.K., Stevenson, P., Ndung,'U, J.M., Stear, M.J., Reid, S.W., Gettinby, G., Murray, M. (1998). Studies on host resistance to tick infestations among trypanotolerant *Bos indicus* cattle breeds in east Africa. Ann. N. Y. Acad. Sci., 849, 195-208.

Osaer, S., Goossens, B. (1999). Trypanotolerance in Djallonke sheep and West African dwarf goats in the Gambia. Thesis, University of Utrecht.

Paling, R.W., Dwinger, R.H. (1993). Potential of trypanotolerance as a contribution to sustainable livestock production in tsetse affected Africa. Veterinary Quarterly 15, 60-67.

Silva, J.S., Twardzik, D.R., Reed, S.G. (1991). Regulation of *Trypanosoma cruzi* infections *in vitro* and *in vivo* by transforming growth factor β (TGF-β). J. Exp. Med. 174, 539-545.

Chapter 23. Post infection complications: the examples of reactive arthritis and Guillain-Barré syndrome

After recovery of an acute infectious disease, some people may develop chronic complications, even when the causative micro-organism is no longer detectable. These chronic sequelae are not always accompanied with clinical disease. Chronic carriership may for example occur, as in viral hepatitis and latent herpes virus infections, and after years this may or may not result in chronic liver disease or reactivation disease, respectively. Other chronic consequences of infections include gastric ulcer and cancer after *Helicobacter pylori* infection, cervical cancer after infection with certain types of human papilloma virus (see chapter 24), Guillain-Barré's syndrome after many infections, notably measles virus and *Campylobacter* infections, and reactive arthritis triggered by infections of the gastrointestinal tract due to *Yersinia, Salmonella or Campylobacter* spp., and of the genitourinal tract due to *Chlamydia*. There is much controversy and research on the question whether *Chlamydia pneumoniae* infection, a common respiratory tract infection, might promote the development of atherosclerosis. A genetic component in determining the occurrence of these, sometimes rare, complications is obvious, especially when autoimmune mechanisms are involved.

Reactive arthritis
Reactive arthritis requires a triggering infection in a susceptible host. The incidence of reactive arthritis after food poisoning by *Salmonella* or *Yersinia* spp. varies considerably (1-10% for *Salmonella*), for reasons which are unclear. Susceptibility is mainly related to HLA-B27 (with very high odds ratios [ORs] of 14 – 30), though approximately one third of the reactive arthritis patients are HLA-B27 negative. In addition, not all HLA-B27 positive patients (even in one family) develop arthritis after infection. This partial influence of HLA-B27 may point to the influence of the other HLA haplotype, or to other modifying genetic or environmental factors (Møller 1990).
The pathogenesis of the disorder is for the most part not defined. However, several groups detected cell-mediated immune responses directed against reactive arthritis-associated organisms in the synovial fluid of affected joints. The microbial antigens recognized by microbe-specific MHC-class II-restricted CD4+ T cells appear mainly cationic, leading to the suggestion that cationic bacterial proteins disseminate from the site of infection, are trapped in cartilage matrix, and subsequently provoke joint inflammation. In addition to cationic proteins, heat shock proteins, notably from the hsp60 family, have been identified as targets of the immune response to *Salmonella, Yersinia,* and *Chlamydia*. Bacterial hsp60-specific T cells may cross-react with the homologous human hsp60 and constitute an autoimmune response in the joint (Life et al 1991). In *Chlamydia*-induced reactive arthritis there is also evidence of dissemination of small numbers of organisms to the joint (Ford et al 1985, Van den Berg et al 1984). New findings suggested that also other viable bacteria might be present inside the joints, causing the disease and triggering the inflammatory response (Beutler and Schumacher 1997). In the joints a Th2 cytokine pattern is found, which probably reflects a failure to mount an appropriate defensive Th1 response, allowing the bacteria to persist in the joint (Müller et al 1998).
So far the role of HLA-B27 in the pathogenesis of reactive arthritis has not been explained (Ikeda and Yu 1998). HLA-B27-restricted, organism-specific CD8+ T cells have

been isolated from affected joints, but the bacterial antigens they recognize are unknown. HLA-B27 appears further related to the intensity of the humoral immune response following infection (Bohemen et al 1986). Because HLA-B27 is not unequivocally related to reactive arthritis, cross reactivity between bacterial antigenic epitopes and HLA-B27 can only be part of a multifactorial process leading to this disease (Bohemen et al 1985). Additional factors, genetic or not, are therefore likely also required for the development of reactive arthritis.

Guillain-Barré syndrome

A variety of viral and bacterial infections, as well as immunization with non-infectious material, has been associated with the Guillain-Barré syndrome, an inflammatory demyelinating condition of the peripheral nervous system. Especially *Campylobacter jejuni* infections are associated with the syndrome. The syndrome is considered to be immune-mediated. Characteristics of the disease are nerve infiltration of leukocytes and autoantibodies of the IgG isotype directed against nerve constituents (Van der Pol et al 2000). Intravenous immune globulin has been proven to be effective as therapy. Infections appear to trigger the development of anti-ganglioside antibodies, especially antibodies directed against GM1, GM1b, GM2, and GQ1b, and these may provoke an autoimmune response (Jacobs et al 1997a). Bacterial infections may trigger such responses by molecular mimicry, because their LPS displays ganglioside-like epitopes (Jacobs et al 1997b). Differences in the specificities of these anti-ganglioside antibodies appear to account for differences in clinical and pathologic presentation of the disease.

As in many immune-mediated diseases, genetic factors may contribute to susceptibility and severity of the disease. A genetic predisposition to the disease is also suggested by a high familial incidence, notably in children born to consanguineous parents (Barr-Joseph et al 1991). However, conflicting results have been reported regarding the role of the HLA system in Guillain-Barré syndrome. One genetic study demonstrated significantly increased frequencies of the HLA-A3 and HLA-B8 antigens in patients with Guillain-Barré syndrome, with ORs of 9.6 and 4.6 respectively (Hafez et al 1985). However, other studies failed to identify differences in HLA and T-cell receptor gene polymorphisms between patients and controls (Ma et al 1998). These conflicting results may in part be explained by linkage of TNF-α polymorphisms to some of the HLA polymorphisms. Indeed, a role for TNF-α in the pathogenesis of the disease was suggested, because Japanese patients with Guillain-Barré syndrome following *Campylobacter jejuni* infection had a significantly higher frequency of the 100-base pair (TNFa2) allele of the TNFα microsatellite marker (24.4 % in patients vs. 9.4 % in controls; p = 0.001). This allele is associated with high TNF-α production (Ma et al 1998). Earlier studies have also shown that serum concentrations of TNF-α increased in patients with Guillain-Barré syndrome and that serum levels correlated with disease severity. Alternatively, the higher frequency of the TNFa2 allele may suggest the involvement of another gene, in close disequilibrium to the TNF-α gene, in the susceptibility to Guillain-Barré syndrome following *Campylobacter jejuni* infection. This gene may be encoded either in the HLA region or in the telomeric end of the class III region, where a number of genes involved in inflammation have been identified (Gruen and Weissman 1997, Ma et al 1998).

An association of IgG receptor IIa alleles with Guillain-Barré syndrome suggests a role for leukocytes as effector cells in the pathogenesis. FcγRIIa-H131 homozygosity (see chapter 12) was significantly increased in patients as compared to controls (OR 2.45; 95% CI 1.12 to 5.36). This allotype is capable of initiating efficient cellular effector functions. Furthermore, FcγRIIa-H131 homozygous patients had a higher risk for severe disease than did patients with other genotypes (OR = 18.57; 95 % CI 1.95 to 176.49) (Van der Pol et al 2000).

References

Barr-Joseph, G., Etzioni, A., Hemli, J., Gershoni-Baruch, R. (1991). Guillain-Barré syndrome in three siblings less than 2 years old. Arch. Dis. Child 66, 1078-1079.

Beutler, A.M., Schumacher, H.R. Jr. (1997). Reactive arthritis: is it a useful concept? Br. J. Clin. Pract. 51, 169-172.

Bohemen, C.G. van, Lionarons, R.J., Bodegom, P. van, Dinant, H.J., Landheer, J.E., Nabbe, A.J., Grumet, F.C., Zanen, H.C. (1985). Susceptibility and HLA-B27 in post-dysenteric arthropathies. Immunology 56, 377-379.

Bohemen, C.G. van, Nabbe, A.J., Landheer, J.E., Grumet, F.C., Mazurkiewicz, E.S., Dinant, H.J., Lionarons, R.J., Bodegom, P.C. van, Zanen, H.C. (1986). HLA-B27M1M2 and high immune responsiveness to *Shigella flexneri* in post-dysenteric arthritis. Immunol. Lett. 13, 71-74.

Ford, D.K., daRoza, D., Schulzer, M. (1985). Lymphocytes from the site of disease but not blood lymphocytes indicate the cause of arthtitis. Ann. Rheum. Dis. 44, 701-710.

Gruen, J.R., Weissman, S.M. (1997). Evolving views of the major histocompatibility complex. Blood 90, 4252-4265.

Hafez, M., Nagaty, M., Al-Tonbary, Y., el-Shennawy, F.A., el-Mongui, A., el-Sallab, S., Attia, S. (1985). HLA-antigens in Guillain-Barré syndrome. J. Neurogenet. , 285-290.

Ikeda, M., Yu, D.T. (1998). The pathogenesis of HLA-B27 arthritis: role of HLA-B27 in bacterial defense. Am. J. Med. Sci. 316, 257-263.

Jacobs, B.C., van Doorn, P.A., Groeneveld, J.H., Tio-Gillen, A.P., Van der Merche, F.G. (1997a). Cytomegalovirus infections and anti-GM2 antibodies in Guillain-Barré syndrome. J. Neurol. Neurosurg. Psychiatry 62, 614-643.

Jacobs, B.C., Endtz, H.P., van der Meche, F.G., Hazenberg, M.P., de Klerk, M.A., van doorn, P.A. (1997b). Humoral immune response against *Campylobacter jejuni* lipopolysaccharides in Guillain-Barré and Miller Fisher syndrome. J. Neuroimmunol. 79, 62-768.

Life, P.F., Bassey, E.O.E., Gaston, J.S.H. (1991). T-cell recognition of bacterial heat-shock proteins in inflammatory arthritis. Immunol. Rev. 121, 113-135.

Ma, J.J., Nishiura, M., Mine, H., Kuroki, S., Nukina, M., Ohta, M., Saji, H., Obayashi, H., Kawakami, H., Saida, T., Uchiyama, T. (1998). Genetic contribution of the tumor necrosis factor region in Guillain-Barré syndrome. Ann. Neurol. 44, 815-818.

Øller, P. (1990). HLA-B27 is necessary but not sufficient. Scand. J. Rheumatol. Suppl. 87, 131-133.

Müller, B., Gimsa, U., Mitchison, N.A., Radbruch, A., Sieper, J., Yin, Z. (1998). Modulating the Th1/Th2 balance in inflammatory arthritis. Springer Semin. Immunopathl. 20, 181-196.

Van den Berg, W.B., Van de Putte, L.D.A., Zwarts, W.A., Joosten, L.A.B. (1984). Electrical charge of the antigen determines intra-articular antigen handling and chronicity of arthritis in mice. J. Clin. Investig. 74, 1850-1859.

Van der Pol, W.-L., Van den Berg, L.H., Scheepers, R.H., Van der Bom, J.G., Van Doorn, P.A., Van Koningsveld, R., Van de Broek, M.C., Wokke, J.H., Van de Winkel, J.G. (2000). IgG receptor IIa alleles determine susceptibility and severity of Guillain-Barré syndrome. Neurology 25, 54, 1661-1665.

Chapter 24. Host genes, infection, and cancer: the example of cervical cancer

One of the most threatening sequelae of infection is cancer. Several examples of infections causing cancer are now well known. One of the most frequently occurring cancers, cervical carcinoma, is strongly associated with infection by oncogenic or high-risk types, notably types 16 and 18, of human papilloma virus (HPV) (Stoler 2000). In developing countries this cancer is one of the leading causes of death. However, only a small fraction of infected women develop cancer, suggesting that genetic, immunological, viral and other factors may contribute to the progression to cervical cancer. Notably cellular immune surveillance appears important in the control of HPV infection and the development of neoplasia. Other types of HPV have been associated with other types of cancers, especially at laryngeal, tonsilar, nasal, and other head and neck sites. Vaccine trials for oncogenic HPVs have been started.

Data from the Swedish national registers allowed Magnusson et al (1999) to compare the incidence of cervical cancer in relatives of cases of cervical cancer and randomly selected age-matched controls. A significant familial clustering occurred among biologic, but not adoptive, relatives. The relative risk for biologic mothers to cases was 1.83, whereas for adopted mothers the relative risk was not significantly different from 1. The relative risk for biological full sisters to cases was 1.93, and for nonbiological sisters to cases was 1. Half-sisters to cases had a relative risk of 1.45, which is a clear indication that the familial aggregation seen in the first-degree relatives is mainly caused by genetic factors. Because the relative risk for half-sisters to cases having a father or a mother in common was similar, the difference in risk for half-sisters is not caused by vertical transmission of HPV. These findings indicate a familial risk for cervical cancer that is similar in magnitude as in other forms of cancer, such as prostate cancer, for which a genetic contribution has long been known (Magnusson et al 1999).

In precursor lesions, HPV is usually found in an episomal (=extra-chromosomal) location. In malignancy, however, the HPV genome is usually integrated. Integration of HPV DNA is probably an important step in tumor progression by inducing chromosomal abnormalities or by altering gene expression near chromosomal breakpoints and fragile sites. Several authors localized integration sites in regions containing proto-oncogenes, such as *SRC*, *RAF*, and *MYC*, indicating that at least in some genital tumors, cis-activation of cellular oncogenes by HPV may be involved in the malignant transformation of the cervical cells (Durst et al 1987, Popescu et al 1987). In addition, activating mutations in the gene designated *FGFR3* (for fibroblast growth factor receptor 3) were found in human bladder, as well as in cervix carcinomas. The most frequent mutation in *FGFR3* was a ser249 to cys mutation (Cappellen et al 1999).
HPV may not only affect proto-oncogenes, but tumor suppressor genes as well. Hampton et al (1994) searched for tumor suppressor genes on chromosome 11 after *in vitro* studies showed that this chromosome contains a gene or genes capable of suppressing tumorigenicity in cell lines derived from cervical carcinomas. These authors used 16 highly polymorphic markers to compare matched DNA samples from noninvolved tissues and tumors. Fourteen of the 32 patients examined demonstrated clonal gene-

tic alterations resulting in loss of heterozygosity for 1 or more of the markers. Because 7 of the clonal genetic alterations on chromosome 11 were specific to the long arm, and because these overlapped with other allelic deletions, it was possible to conclude that 11q22-q24 contains at least 1 tumor-suppressor gene relevant to cervical carcinoma (Hampton et al 1994). The interaction of HPV with the tumor suppressor gene p53 is described below.

Although conflicting data have been reported, it appears that polymorphisms of the MHC and p53 genes may influence the outcome of HPV infection and progression to cervical cancer. The p53 gene is a tumor suppressor gene that regulates cell proliferation. The HPV E6 protein from oncogenic strains, which is required for viral replication in host cells, binds p53 and mediates its degradation by the ubiquitin pathway (Maciag and Villa 1999). The E6 protein of HPV, as well as the E7 protein, has cell-transforming properties *in vitro.* It has been suggested, but not yet unequivocally proven, that a polymorphism at codon 72, Pro or Arg, on exon 4 of the p53 gene influences the susceptibility of p53 to degradation, and therewith the risk of developing cervical cancer (Brady et al 1999, Giannoudis et al 1999, Storey et al 1998, Tachezy et al 1999, Yamashita et al 1999, Zehbe et al 1999).
The final outcome of the E6/p53 interaction may be complex, because also HPV E6 variants may differ in their ability to degrade p53. Interaction of E6 variants with different p53 variants might therefore explain some of the conflicting results reported. For example, Van Duin et al (2000) reported that neither specific HPV 16 E6 variants, nor specific p53 genotypes were associated with disease progression. However, HPV-16 E6 350T variants were significantly over-represented in p53 Arg homozygous women with cervical cancer. Thus infection with this specific viral variant in p53 Arg/Arg women appears to confer a higher risk of cervical cancer.

Because presentation of viral peptides to T cells occurs in the context of HLA molecules, several studies attempted to analyze genetic polymorphisms of both HPV and HLA and the clinical outcome of infection (Bontkes et al 1998, Helland et al 1994). In one study HLA-B44 frequency was increased in patients with a progressive lesion (OR = 9 [4.6-17.5]), while HLA-DRB1*07 frequency was increased among HPV 16-positive patients compared with HPV-negative patients. (OR = 5.9 [3.0-11.3]). Thus immunogenetic factors associated with susceptibility to HPV 16 infection appear different from those associated with disease progression. While multiple HPV 16 variants can induce disease progression, the role of HPV 16 variants in disease susceptibility and progression in relation to HLA polymorphism is unclear (Bontkes et al 1998).
In addition, but not contra imagination, Apple et al (1994) found that HLA associations with cervical carcinoma are HPV-type specific. In their study, certain HLA class II haplotypes, such as HLA-DRB1*1501-DQB1*0602 were significantly associated with cervical carcinoma, while HLA-DR13 haplotypes were negatively associated. These associations were not observed among cases with cervical carcinoma containing HPV types other than HPV 16, arguing that the associations were HPV 16-type specific. The combined OR for HLA-DRB1*1501-DQB1*0602 and HPV 16 was actually 75 ($p < 10^{-7}$). Because the HLA-DRB1 and HLA-DQB 1 alleles in the HLA-DRB1*1501-DQB1*0602 haplotype are in strong linkage disequilibrium, it is difficult to determine which locus is responsible for HPV 16-induced cancer susceptibility. However, different disease associations with other haplotypes suggested that susceptibility and resistance to carcino-

ma may be more influenced by the HLA-DRB1 locus, and the studies further postulated a role for amino acid position 57. Although larger studies are needed to exactly define the significance of HPV type–HLA associations, DR-DQ analysis of patients with low-grade cervical dysplasia coupled with HPV typing may in the near future predict regression or progression of dysplasia to carcinoma.

In conclusion, while adoption, family, and population studies clearly point to genetic factors contributing to susceptibility and disease progression of HPV-associated cervical cancer, the genetic basis still appears elusive. While host genes involved in tumor suppression and antigen presentation are likely candidate genes, the interaction between variant host and viral genes determining these events appears to be complex and not yet unraveled. The final outcome of this interaction therefore remains to be exactly defined.

References

Apple, R.J., Erlich, H.A., Klitz, W., Manos, M.M., Becker, T.M., Wheeler, C.M. (1994). HLA DR-DQ associations with cervical carcinoma show papillomavirus-type specificity. Nature genet. 6, 157-162.

Bontkes, H.J., van Duin, M., de Gruijl, T.D., Duggan-Keen, M.F., Walboomers, J.M., Stukart, M.J., Verheijen, R.H., Helmerhorst, T.J., Meijer, C.J., Scheper, R.J., Stevens, F.R., Dyer, P.A., Sinnott, P., Stern, P.L. (1998). HPV 16 infection and progression of cervical intra-epithelial neoplasia: analysis of HLA polymorphism and HPV 16 E6 sequence variants. Int. J. Cancer 78, 166-171.

Brady, C.S., Duggan-Keen, M.F., Davidson J.A., Varley, J.M., Stern, P.L. (1999). Human papillomavirus type 16 E6 variants in cervical carcinoma: relationship to host genetic factors and clinical parameters. J. Gen. Virol. 80, 3233-3240.

Cappellen, D., De Oliveira, C., Ricol, D., Gil Diez de Medina, S., Bourdin, J., Sastre-Garau, X., Chopin, D., Thiery, J.P., Radvanyi, F. (1999). Frequent activating mutations of FGFR3 in human bladder and cervix carcinomas. Nature Genet. 23, 18-20.

Duin, M. van, Snijders, P.J., Vossen, M.T., Klaassen, E., Voorhorst, F. Verheijen, R.H., Helmerhorst, T.J., Meijer, C.J., Walboomers, J.M. (2000). Analysis of human papillomavirus type 16 E6 variants in relation to p53 codon 72 polymorphism genotypes in cervical carcinogenesis. J. Gen. Virol. 81, 317-325.

Durst, M., Croce, C.M., Gissmann, L., Schwarz, E., Huebner, K. (1987). Papilloma virus sequences integrate near cellular oncogenes in some cervical carcinomas. Proc. Natl. Acad. Sci. USA 84, 1070-1074.

Hampton, G.M., Penny, L.A., Baergen, R.N., Larson, A., Brewer, C., Liao, S., Busby-Earle, R.M.C., Williams, A.W.R., Steel, C.M., Bird, C.C., Stanbridge, E.J., Evans, G.A. (1994). Loss of heterozygosity in cervical carcinoma: subchromosomal localization of a putative tumor-suppressor gene to chromosome 11q22-q24. Proc. Natl. Acad. Sci. 91, 6953-6957.

Helland, A., Borresen, A.L., Kristensen, G., Ronningen, K.S. (1994). DQA1 and DQB1 genes in patients with squamous cell carcinoma of the cervix: relationship to human papillomavirus infection and prognosis. Cancer Epidemiol. Biomarkers Prev. 3, 479-486.

Giannoudis, A., Graham, D.A., Southern, S.A., Herrington, C.S. (1999). p53 codon 72 ARG/PRO polymorphism is not related to HPV type or lesion grade in low- and high-grade squamous intra-epithelial lesions and invasive squamous carcinoma of the cervix. Int J. Cancer 24, 66-69.

Maciag, P.C., Villa, L.L. (1999). Genetic susceptibility to HPV infection and cervical cancer. Braz. J. Med. Res. 32, 915-922.

Magnusson, P.K.E., Sparen, P., Gyllensten, U.B. (1999). Genetic link to cervical tumours. Nature 400, 29-30.

Popescu, N.C., Amsbaugh, S.C., DiPaolo, J.A. (1987). Human papillomavirus type 18 DNA is integrated at a single chromosome site in cervical carcinoma cell line SW756. J. Virol. 51, 1682-1685.

Stoler, M.H. (2000). Human papillomaviruses and cervical neoplasia: a model for carcinogenesis. Int. J. Gynecol. Pathol. 19, 16-28.

Storey, A., Thomas, M., Kalita, A., Harwood, C., Gardiol, D., Mantovani, F., Bruer, J., Leigh, I.M., Matlashewski, G., Banks, L. (1998). Role of a p53 polymorphism in the development of human papillomavirus-associated cancer. Nature 393, 229-234.

Tachezy, R., Mikyskova, I., Salakova, M., Van Ranst, M. (1999). Correlation between human papillomavirus-associated cervical cancer and p53 codon 72 arginine/proline polymorphism. Hum. Genet. 105, 564-566.

Yamashita, T., Yaginuma, Y., Saithoh, Y., Kawai, K., Kurakane, T., Hayashi, H., Ishika-wa, M. (1999). Codon 72 polymorphism of p53 as a risk factor for patients with human papillomavirus-associated squamous intrae-pithelial lesions and invasive cancer of the uterine cervix. Carcinogenesis 20, 1733-1736.

Zehbe, I., Voglino, G., Wilander, E., Genta, F., Tommasino, M. (1999). Codon 72 polymorp-hism of p53 and its association with cervical cancer. Lancet 354, 218-219.

Chapter 25. Genetic predisposition to transmissible spongiform encephalopathies

Transmissible spongiform encephalopathies (TSEs) or prion diseases comprise Creutzfeldt-Jakob disease (CJD) of humans, scrapie of sheep, and bovine spongiform encephalopathy (BSE) in cattle. These are rare neurodegenerative disorders, which are characterized by an accumulation in the brain of an abnormal, protease-resistant form of the host-encoded prion protein (PrP). This pathological accumulation results from a conformational change of the normal protein PrP^c into PrP^{sc}. The normal function of PrP is unknown. The abnormal form induces conversion of PrP^c into PrP^{sc} and neurodegeneration when experimentally or accidentally introduced in recipient hosts. This feature explains the transmissible character of these diseases, as illustrated by animal-to animal transmission in cattle and sheep, the recent introduction of BSE in cattle, probably by meat offal from scrapie-affected sheep, and by iatrogenic contamination in humans. BSE is even believed to have led to a new variant (nv) of spongiform encepalopathy in humans. Presently it is completely speculative to predict how many people will dye of this CJDnv. In humans CJD may occur spontaneously, but has also been transmitted by the administration of growth hormone and gonadotrophin extracted from human cadaveric pituitary glands, and by organ transplants. However, spontaneous or sporadic CJD may also have an environmental origin (Deslys et al 1994). Data on maternal transmission and transmissibility across species are still deficient (Ridl and Baker 1999).

Considerable attention has been given to the host PrP gene and susceptibility to disease. Several studies indicate a very substantial genetic component in the TSEs, and the PrP gene appears to confer the main component of this susceptibility. Sixteen % of humans with spontaneous CJD harbor mutations in the PrP gene. In addition, a coding polymorphism (129 Met/Val) defines a predisposing factor to sporadic and iatrogenic CJD (Collinge et al 1991, Laplanche et al 1995). In France people with iatrogenic or sporadic CJD all harbor the homozygous methionine codon 129 genotype, compared with only 50 % in the healthy control population (p < 0.00002) (Deslys et al 1994). People homozygous for methionine at codon 129 of PrP also appear particularly susceptible to CJDnv (Gordon 1999).

Also in sheep natural scrapie is associated with polymorphisms in the PrP gene. In sheep, at least 10 different mutually exclusive polymorphisms are present in PrP (Bossers et al 2000). The incidence of natural scrapie is particularly associated with polymorphisms at codons 136, 154, and 171 of the PrP gene. In Suffolk sheep, codon 171 is the codon at which most variation is found; RR171 is thought to be associated with resistance to developing the clinical signs of disease, while QQ171 is associated with susceptibility to the disease (Hunter et al 1997, O'Doherty et al 2000). However, the apparent linkage between Q (glutamine) at codon 171 and scrapie is not completely recessive, because some scrapie cases are heterozygous Q/R. Another polymorphism predisposing to great risk of disease is VV136. However, in heterozygous VA136 Cheviot sheep scrapie occurred only in a specific subgroup, depending on animals being heterozygous for other polymorphisms at codons 154 or 171 (Hunter et al 1996). V136 is usually a rare allele in healthy control sheep, but is not an absolute requirement for scrapie to develop (Hunter et al 1996, Hunter et al 1997). Subsequent studies in other breeds confirmed that the PrP^{VRQ} allelic variant (VRQ at codon positions 136, 154, and 171) is associated with increased susceptibility for scrapie in these breeds as well, thus irrespective of genetic background (Belt et al 1996). Homozygous PrP^{VRQ}/ PrP^{VRQ} are at greatest risk and also have the shortest survival time of about 2 years. Sheep carrying the PrP^{ARR} allele were found to be resistant.

Susceptibility of sheep for scrapie may be predicted by *in vitro* conversion of their PrP protein into its protease-resistant form (Bossers et al 2000). In addition, specific strains of the natural "scrapie-agent" might differ in binding affinity for PrP or in their ability to convert PrP^c into PrP^{sc}. These and other results indicate that genetic susceptibility to TSEs and differences in survival time are indeed directly related to the structure of the prion protein by determining its ability to convert into its abnormal structure. Studies using transgenic mice expressing targeted alterations in PrP and challenged with a mouse-adapted strain of BSE confirm that PrP is the major gene controlling susceptibility as reflected by incubation time (Moore et al 1998).

Selective breeding strategies have been initiated in order to trying to eliminate the disease from sheep flocks (O'Doherty et al 2000). However, epidemiological studies have indicated that progress from selective breeding may be slow, even in closed flocks. This may be due to heterozygous rams continuously supplying susceptible alleles to their offspring. The best strategy may therefore be a combination of control measures, including slaughter of lambs born to infected ewes in combination with selectively breeding from resistant rams (Dawson et al 1998, O'Doherty et al 2000, Stringer et al 1998, Woolhouse et al 1998).

While scrapie has been known for over 200 years in sheep across the world, BSE is a new "emerging" disease. This difference may be responsible for the observation that the cattle PrP gene has, so far, failed to show any association between PrP alleles and susceptibility to BSE (Hunter 1997). This discrepancy therewith illustrates our ignorance of the mechanisms responsible for maintaining PrP polymorphisms and even suggests that TSEs may not be a major factor in maintaining them. In addition, sheep from different breeds (Cheviot, Suffolk, Merino, Poll Dorset) from Australia and New Zealand, both scrapie-free countries, were found to harbor scrapie-susceptible genotypes (Hunter and Cairns 1998). This finding illustrates that scrapie in sheep (apparently in contrast to inherited or familial prion diseases in humans) is not simply solely a genetic disease and that environmental or accidental transmission of the "scrapie-agent" is a prerequisite for developing the disease.

Inherited or familial prion disease
In addition to sporadic or iatrogenic CJD in humans, families with inherited prion diseases have been identified. Approximately 10 % of CJD cases are familial and genetic linkage studies have clearly demonstrated a causal relationship between different PrP mutations and degenerative neurological disease. These disorders have been roughly divided in those resembling Creutzfeldt-Jakob disease (CJD) and those resembling Gerstmann-Sträusler-Scheinker disease (GSS).

CJDs are recognized by subacute progression of dementia and motor signs, myoclonus, periodic triphasic discharges on EEG, and widespread spongiform degeneration in the cerebral gray matter. GSSs are characterized by slowly progressive cerebellar signs, variable spongiform degeneration of the gray matter, and widespread deposition of PrP immunopositive amyloid plaques. However, although afflicted family members with the same mutation usually have a considerable overlap in clinical symptoms, they may also show vastly different clinical presentations.

Familial CJD has been found in patients with an insertion of variable numbers of octapeptide repeats at codon 67, 75, or 83 in the N-terminal domain of PrP, and with mutations at codons 178, 180, 200, 210, and 232 in the region of the third and fourth helical domains (DeArmond and Prusiner1996, Prusiner 1996). The E200K substitution has been found among Israeli Jews of Libyan origin, in Slovaks originating from Orava in Czechoslovakia, in a cluster of familial cases in Chile, in a large German family living in the United States, as well as in British and Japanese families. Although researchers have argued that this mutation originated in a Sephardic Jew whose descendants migrated from Spain and Portugal at the time of the inquisition, it is more likely that the E200K mutation has arisen independently multiple times by the deamidation of a methylated CpG (a methylated deoxycytosine [C] coupled to deoxyguanosine [G] through a phosphodiester bond) (Prusiner 1996).

Mutations at codons 102, 105, and 117 in the first α-helical domain of PrP are associated with GSS. Another mutation, at codon 145, results in a premature stop codon. Two other mutations, also associated with GSS, occur in the third and fourth helical domains at codons 198 and 217. The mutation at codon 102 (a P to L mutation) has been found in ten different families in nine different countries, including the original GSS family (Prusiner 1996). The LOD score exceeds 3.

References

Belt, P.B.G.M., Bossers, A., Schreuder, B.E.C., Smits, M.A. (1996). PrP allelic variants associated with natural scrapie. In: Bovine Spongiform Encephalopthy; The BSE dilemma, ed. Gibbs, C.J. (Springer, New York), pp. 294-305.

Bossers, A., de Vries, R., Smits, M.A. (2000). Susceptibility of sheep for scrapie as assessed by *in vitro* conversion of nine naturally occurring variants of PrP. J. Virol. 74, 1407-1414.

Collinge, J., Palmer, M.S., Dryden, A.J. (1991). Genetic predisposition to iatrogenic Creutzfeldt-Jakob disease. Lancet 337, 1441-1442.

Dawson, M., Hoinville, L.J., Hosie, B.D., Hunter, N. (1998). Guidance on the use of PrP genotyping as an aid to the control of clinical scrapie. Scrapie Information Group. Vet. Rec. 142, 623-625.

DeArmond, S.J., Prusiner, S.B. (1996). Transgenetics and neuropathology of prion diseases. In: Prions Prions Prions, Current Topics in Microbiology and Immunology 207, Ed. by Capron et al, Springer Verlag, Berlin, Heidelberg, pp. 125-146.

Deslys, J.P., Marce, D., Dormont, D. (1994). Similar genetic susceptibility in iatrogenic and sporadic Creutzfeldt-Jakob disease. J. Gen. Virol. 75, 23-27.

Gordon, N. (1999). New variant Creutzfeldt-Jacob disease. Int. J. Clin. Pract. 53, 456-459.

Hunter, N., Foster, J.D., Goldmann, W., Stear, M.J., Hope, J., Bostock, C. (1996). Natural scrapie in a closed flock of Cheviot sheep occurs only in specific PrP genotypes. Arch. Virol. 141, 809-824.

Hunter, N. (1997). PrP genetics in sheep and the applications for scrapie and BSE. Trends Microbiol. 5, 331-334.

Hunter, N., Moore, L., Hosie, B.D., Dingwall, W.S., Greig, A. (1997). Association between natural scrapie and PrP genotype in a flock of Suffolk sheep in Scotland. Vet. Rec. 140, 59-63.

Hunter, N., Goldmann, W., Foster, J.D., Cairns, D., Smith, G. (1997). Natural scrapie and PrP genotype: case-control studies in British sheep. Vet. Rec. 141, 137-140.

Hunter, N., Cairns, D. (1998). Scrapie-free Merino and Poll Dorset sheep from Australia and New Zealnd have normal frequencies of scrapie-susceptible PrP genotypes. J. Gen. Virol. 79, 2079-2082.

Laplanche, J.L., Beaydry, P., Ripoll, L., Launay, J.M. (1995). Prion protein: structure, functions and polymorphisms associated with human spongiform encephalopathies. Pathol. Biol (Paris), 43, 104-113.

Moore, R.C., Hope, J., McBride, P.A., McConnell, I., Selfridge, J., Melton, D.W., Manson, J.C. (1998). Mice with gene targetted prion protein alterations show that *Prnp*, *Sinc* and *Prni* are congruent. Nature Genetics 18, 118-125.

O'Doherty, E., Aherne, M., Ennis, S., Weavers, E., Hunter, N., Roche, J.F., Sweeney, T. (2000). Detection of polymorphisms in the prion protein gene in a population of Irish Suffolk sheep. Vet. Rec. 146, 335-338.

Prusiner, S.B. (1996). Human prion diseases and neruodegenration. In: Prions Prions Prions, Current Topics in Microbiology and Immunology 207, Ed. by Capron et al, Springer Verlag, Berlin, Heidelberg, pp.1-17.

Ridley, R.M., Baker, H.F. (1999). Big decisions based on small numbers: lessons from BSE. Veterinary Quarterly 21, 86-92.

Stringer, S.M., Hunter, N., Woolhouse, M.E.J. (1998). A mathematical model of the dynamics of scrapie in a sheep flock. Mathematical Biosciences 153, 79-98.

Woolhouse, M.W.J., Stringer, S.M., Matthews, L., Hunter, N., Anderson, R.M. (1998). Epidemiology and control of scrapie within a sheep flock. Proceedings of the Royal Society, London, B265, 1205-1210.

Part IV   Application of infectious disease genetics in human and veterinary medicine

Chapter 26. Use of genetic information in preventive medicine, public health, and patient treatment

The future of public health

Public health involves the activities that are directed on the assessment of the health and health needs of the population, the development of health policy, and the assurance and evaluation of health programs (Institute of Medicine 1988, Khoury 1997). Public health therefore deals with diseases that affect a large proportion of the population and results in collective prevention directed at the general population. It deals with disease, first, when there are large numbers of the population suffering disability and death from disease against which private medical practice is making little headway. And second, when there is any reason to hope that the incidence of these diseases may be altered by measures applicable to the general population, even if these measures are not yet known (Keys 1953). For that purpose it is not only necessary to know the occurrence of disease, but also to know the occurrence of risk factors. Surveillance programs are thus a corner stone of public health programs. Knowledge on particular genotypes in racially or geographically defined populations that affect susceptibility to infectious disease may therefore modify surveillance and control programs. Indeed by influencing risk factors (genetic or not), it is possible to diminish the disease burden of the population.

As more knowledge will become available on the impact of susceptibility genotypes and their interaction with other environmental factors, the quality of public health forecast programs would likely improve. Data on frequencies of variably susceptible genotypes in populations, their spatial structuring, and eventually their evolutionary stability can thus help in forecasting disease spread and disease load. Quantitative data on allele frequencies in different populations, the role of interacting cofactors, and their impact on disease, disability and death are therefore needed to direct and improve prevention efforts and disease interventions. Evidently the value of such information increases when an underlying model of the genetic susceptibility, a linked genetic marker or a responsible genetic variant, or both, are known. Note however that for population-based prevention strategies direct identification of the allele(s) responsible for susceptibility is required, while for family-based prevention indirect approaches such as linkage and segregation analysis may suffice. While alleles with a strong effect on disease may be rare, the effects of alleles with a modest effect may be more important from a public health perspective. Quantitative estimates of risk can be developed for susceptibility genes with incomplete penetrance (Khoury et al 1993). For example, disease risks can be stratified according to the number of susceptibility alleles carried by individuals. For quantitative traits researchers routinely emphasized means and standard errors of the means, disregarding potentially informative individuals with extreme responses (Omenn 2000). However, the increasing attention to genetics in epidemiological studies will likely focus the attention more to heterogeneity of subpopulations. Some examples follow.

As noted in chapter 14, the presence of blood group O is associated with more severe symptoms among persons infected with *Vibrio cholerae* O1. A study from Peru tried to estimate the significance of this finding in determining resource needs during cholera epidemics. During a cholera epidemic in this country, a country with a high blood group O prevalence, many people needed rehydration therapy. Among people with cholera, those with blood group O sought care, were admitted to the hospital, and required rehydration therapy more often than those of other blood groups. The researchers concluded that health officials should therefore consider the prevalence of blood group O when estimating the number of severely ill persons and resources needed during cholera epidemics (Swerdlow et al 1994).

Su et al (2000) estimated the effects of three mutations in genes encoding chemokine receptors or their ligands (SDF1-3'A, CCR-64I, and CCRΔ32), each with presumed (SDF1-3'A) or proven protective effects against HIV-1 infection, on AIDS onset in 70 different populations. The large allele frequency differences of each of the mutations translate into an extensive variation in risk of AIDS onset (relative hazard), with the highest relative hazard in South-East Asia and Africa, evidently also the areas where AIDS is known to be more prevalent.

Prevention and control

One may argue that genetic factors predisposing to disease appear less prone to therapeutic or preventive measures. However, genetic predispositions to disease, including infectious disease, should not be seen as immutable causes of disease. Instead researchers should seek to modify gene expression or the many other factors that interact and influence the risk and severity of disease. Genetically based differences in susceptibility to infection and in response to vaccination are of obvious relevance for the design and evaluation of disease control programs. A classification of individuals and populations based on frequency of susceptibility genes may thus for example facilitate risk management. The basic idea (and the challenge!) is that prevention programs can be made more effective by targeting them toward individuals, families, and population subgroups with different genetic susceptibilities.

Also the evaluation of prevention programs may make use of genetic information. For example, in many countries the effects of mass vaccination campaigns are evaluated in large-scale serological surveys, such as for example the PIENTER project (De Melker and Conyn-van Spaendonck 1998). Because several studies have indicated that antibody responses are genetically determined, for example after vaccination against measles and rubella, such surveys should take full account of these differences. The fraction of non-responders due to genetic factors is for example important for the estimation of the proportion of the community that must be immunized to eliminate infection (Anderson 1988). In such estimates, knowledge of the proportion non-responders has been neglected. Thus, stratification and adjustment for genetic risk factors may refine the design and analysis of vaccination trials. Evidently a further challenge is to optimize prevention strategies for non-responders to vaccination.

A further application of genetic information would be the design of control programs specifically tailored to those most susceptible to severe forms of disease or most amenable to preventive or therapeutic measures. Hepatitis B virus infection is characterized by persistent infection of only a proportion of those primarily infected. These persistently infected carriers have an inadequate immune response that allows continued viral replication and persistent viremia. They are important in maintaining virus transmission. Because carriership partly depends on genetic factors, it would be feasible and advantageous to use this information to design and implement strategies aimed at preventing persistent infection.

As discussed earlier, helminth parasitic infections tend to be highly aggregated so that a few individuals harbor most of the worm burden in a population. The difference between "wormy" and "non-wormy" people is partly genetically determined. The "wormy people" are most important in transmission of the infection. Control strategies may be very effective in reducing morbidity and transmission if they are specifically designed on the reduction of the worm burden in these persons (Anderson 1988). An additional advantage of selective treatment may that the emergence of drug resistant strains of the parasites is delayed or diminished.

Implications of genetic information for patient treatment
Genetic studies may identify key host proteins that are involved in the pathogenesis of infectious disease. Further studies may establish their mode of action, their influence on the course of disease, and whether they are a potential target for therapeutics, either in all affected patients or in a genetically identified subset. For example, defective fibrinolysis appears a poor prognostic factor for the course of meningococcal sepsis. Patients with the 4G/4G genotype of the plasminogen activator inhibitor-1 (PAI-1) promoter have higher PAI-1 concentrations, a defective fibrinolysis, and an increased risk of death (Hermans et al 1999). PAI-1 is not only an inhibitor of fibrinolysis, but also of the protein C pathway (Gladson et al 1989). Acquired protein C deficiency also appears an important pathway in meningococcal sepsis (Smith et al 1997). Protein C is a natural anticoagulant and also has important anti-inflammatory activity. Future studies should now elucidate whether therapy aimed at reducing the concentration of PAI-1 or enhancing the concentrations of tissue plasminogen activator (t-PA) or protein C are beneficial in all patients with meningococcal sepsis or only in those with one or two copies of the 4G allele.

The identification of TNF-α associations with several infectious diseases, including malaria, suggests using anti-TNF-α agents in the treatment of these diseases, particularly in those individuals with high TNF-α expression levels. Studies using monoclonal antibody therapy and new synthetic macromolecules directed against TNF-α are underway. Also polymorphisms in other cytokine genes, such as IL-2, IL-4, and IL-10, affect infectious disease susceptibility, knowledge that may translate in therapies directed at these cytokines or their receptors.

Examining patients for genetic variants predisposing for infection may especially be warranted when effective treatment or prevention can be offered. For example, homozygosity or heterozygosity for mutations in the gene encoding for mannose-binding protein (MBP) yields low plasma concentrations of MBP and defects in opsonization and phagocytosis, which may lead to immunodeficiencies and recurrent infections (Garred et al 1994). MBP deficiency is a frequent cause of immunodeficiency reaching frequencies up to 10 % (chapter 8). Because MBP mutations represent a lifelong risk factor, and because the deficiency may be treated by purified or recombinant MBP, screening for MBP variants may be warranted in patients with recurrent infections (see chapter 8). However, the definition of MBP deficiency is not yet clear, and it also unclear whether MBP gene variants always result in low serum concentrations or loss-of-function of MBP (Kuijper et al 2000). Genetic screening of MBP variants therefore appears superior to quantification of serum MBP concentrations. Similarly, patients with deficiency for immunoglobulin subclasses may be treated successfully.

Complement deficiencies appear most important in protection against encapsulated bacteria, such as *Neisseria meningitidis, Streptococcus pneumoniae,* and *Haemophilus influenzae* (see chapter 8). Patients with these deficiencies may benefit from vaccination with effective polysaccharide or conjugate vaccines against these pathogens. Screening for complement deficiencies may especially be warranted in patients with meningococcal infection, because relapse of infection is ten times more common in these patients. Recurrences are seen in 45 % of affected complement-deficient individuals and usually involve uncommon meningococcal serogroups (Ross and Densen 1984). In addition, it may be indicated to screen family members of patients with rare complement and other immunodeficiency defects for these disturbances. Affected family members may be cautioned and offered additional vaccinations against these diseases. For that purpose vaccines directed at meningococci A, C, W135, and Y, as well as against 23 serotypes pneumococci are available.

Patients may further be tested for polymorphisms that are associated with rate of disease progression. Well-known examples of variants associated with disease progression are the CCR5Δ32 heterozygosity and CCR5 promoter region polymorphisms associated with delayed progression of HIV-1 infected people to AIDS (Beretta 1999, McDermott et al 1998).

Screening for polymorphisms and genetic disorders
Apart from hypothesis-driven research, the question may arise, as much debated in other fields of medicine, whether individuals should be examined for the presence of polymorphisms or genetic disorders associated with, or responsible for a high susceptibility for one or more infections. The question arises many scientific, ethical, legal, psychological, and economical (in terms of cost/benefit analyses) concerns. Broadly stated, genetic screening for susceptibility loci will be indicated and accepted more when the chance on infection is higher, when the clinical presentation of the genetic defect is more severe, and when better preventive or interventive measures are available. To contribute to the discussion, I consider it useful to make a division of potential screening programs in relation to the moment of infection.

1. Screening for genetic susceptibility before the onset of infection.

Because most infections, at least in well developed countries, do not have a very high impact in terms of incidence and severity, screening for susceptibility loci in the general population (for predictive purposes) will often not be indicated, accepted by the public, or be cost-effective. Moreover, exposure to a pathogen or clinical severity may be low thanks to other measures such as vaccination. This may be different in specific subpopulations or in families with already known mutations. For example, it may be indicated to screen families with a high incidence of infections or families with known immunodeficiencies. Important parameters in this respect are the frequency of the genetic trait in the (sub)population, the lifetime risk of disease, and the relative risk associated with the genetic trait (Khoury et al 1993). Furthermore, for genetic testing to be useful effective preventive measures or treatment should be available. Genetic testing is also important for those who do not carry the relevant mutation. These individuals are spared from unnecessary medical procedures and anxiety. An example to illustrate that genetic testing may add in preventing or minimizing the effects of disease is that of sickle cell anemia. Certain subpopulations have a high frequency of this disorder. The patients may suffer life-threatening bacterial (mostly pneumococal) infections, which may be prevented by antibiotic prophylaxis and vaccination. Homozygous sickle cell anemia patients may suffer from such severe infections already during their first half-year of life. (Evidently the HbS mutation illustrates the complexity of pathogen-host interactions, in that HbS heterozygosity reduces the risk for malaria, while HbS homozygosity results in sickle cell disease and enhanced susceptibility for bacterial infections.)

2. Screening for genetic susceptibility when infection is suspected, imminent, or proven.

In certain cases knowledge of the presence of susceptibility genes (including immune deficiencies) may add value in the treatment or prevention of disease, for example in meningococcal disease. To be useful the assays should be very quick, sensitive and specific, and effective therapeutics should preferably be available. For example, bone marrow transplants are sometimes given to patients with immune disorders. As mentioned in chapter 8, leukocyte adhesion defects may result in a very variable clinical picture depending on the degree of CD18 deficiency. A mild clinical syndrome is associated with CD11-CD18 concentrations of approximately 10 % of the normal concentrations, while total lack of CD11-CD18 results in a very severe clinical picture. Bone marrow transplantation has been successful in some of these patients. Other patients with phagocytic dysfunctions have benefited from leukocyte transfusions.
Screening for mutations affecting drug metabolism may be indicated to optimize therapy with antibiotics. Such a screening could lead to a therapeutic regime tailored to the individual, and could thus reduce adverse events, reduce the chance on treatment failure, and the risk of developing drug-resistant variants. If screening for NAT2 genotypes (chapter 27) were cost-effective in the prevention or control of tuberculosis, persons could be genotyped before starting therapy with isoniazid.

3. Screening for genetic susceptibility after recovery from infection.

Persons presenting with or recovered from a specific disease may have a clear genetic predisposition, knowledge of which may be advantageous in preventing the same or related infection in the same individual or its relatives.
For illustration, persons who experienced infection with meningococci, especially the less common serotypes, have a high probability of complement deficiencies as well as a high chance on reinfection. Screening of such persons and their relatives for these genetic defects may thus lead to early antibiotic treatment or vaccination.
Patients with hypo- or agammaglobulinemia may be given repeated injections of immunoglobulins to decrease the incidence and mortality associated with recurrent infections with encapsulated organisms, such as *Streptococcus pneumoniae* or *Haemophilus influenzae*. These organisms require opsonization for adequate phagocytosis. Infections in these patients commonly result in sinusitis, otitis media, bacterial pneumonia, and bacteriemia. Pneumococcal vaccine (Pneumovax) may be given to those

patients capable of mounting some immune responses to provide a degree of protection against overwhelming infections with this organism (see also the Recommendations of the Advisory Committee on Immunization Practices (ACIP): Use of vaccines and immune globulins in persons with altered immunocompetence. MMWR Morb. Mortal. Wkly Rep. 1993, 42(RR-4), 1-18).

Defects in cytochrome b or NADPH oxidase lead to deficient oxygen metabolism in phagocytes. Phagocytosed bacteria are not killed due to a lack of oxygen radicals, and chronic granulomatous disease may follow (see chapter 8). Most patients are detected before they are 5 years old when they suffer from infections caused by staphylococci, Gram-negative bacteria, and fungi. Causal therapy of this disorder is impossible, but early and prophylactic treatment with antibiotics and antimycotics clearly benefit the prognosis of these patients (Roos et al 1993). These genetic disorders may even be detected before birth, and carriers of autosomally recessive chronic granulomatous disease may be traced.

Family members of persons who experienced severe or fatal BCG disease (see chapter 10) may be examined for mutations in the genes encoding IL-12, the IL-12 receptor, or the IFN-γ receptor before giving them BCG vaccine.

References

Anderson, R.M. (1988). The population biology and genetics of resistance to infection. In: Genetics of resistance to bacterial and parasitic infection. Ed. by D. Wakelin and J.M. Blackwell, Taylor and Francis, London, Philadelphia, New York.

Beretta, A. (1999). Natural resistance to HIV infection. Chapter XII in: Basic Science in HIV/AIDS: An update. Ed. by T.E. Mertens et al. WHO, pp. 96–98.

De Melker, H.E., Conyn-van Spaendonck, M.A. (1998). Immunosurveillance and the evaluation of national immunization programmes: a population-based approach. Epidemiol. Infect. 121, 637-643.

Garred, P., Madsen, H.O., Hofmann, B., Svejgaad, A. (1995). Increased frequency of homozygosity of abnormal mannan-binding-protein alleles in patients with suspected immunodeficiency. Lancet 346, 941-943.

Gladson, C.L., Schleef, R.R., Binder, B.R. Loskutoff, D.J., Griffin, J.H. (1989). A comparison between activated protein C and des-1-41-light chain-activated protein C in reactions with type 1 plasminogen activator inhibitor. Blood 74, 173-181.

Hermans, P.W.M., Hibberd, M.L., Booy, R., Daramola, O., Hazelet, J.A., de Groot, R., Levin, M, and the Meningococcal Research Group. (1999). 4G/5G promoter polymorphism in the plasminogen-activator-inhibitor-1 gene and outcome of meningococcal disease. Lancet 354, 556-560.

Institute of Medicine, Committee for the Study of the Future of Public Health. (1988). National Academy Press, Washington, DC.

Keys, A. (1953). Atherosclerosis: A problem in newer public health. J. Mount Sinai Hosp. 20, 118-139.

Khoury, M.J., Beaty, T.H., Cohen, B.H. Fundamentals of genetic epidemiology. Monographs in epidemiology and biostatistics, volume 22, Oxford University Press, New York, Oxford, 1993.

Khoury, M.J. (1997). Genetic epidemioloy and the future of disease prevention and public health. Epidemiol. Rev. 19, 175-180.

Kuijper, E.J., Drogari-Apiranthitou, M., Kuijpers, T.W., Fijen, C.A. (2000). Immunologie in de medische praktijk. XXVIII. Gevoeligheid voor meningokokkenziekte door een familiair tekort aan mannosebindend lectine. Ned. Tijdschr. Geneesk. 144, 2079-2080.

McDermott, D.H., Zimmerman, P.A., Guignard, F., Kleeberger, C.A., Leitman, S.F., the Multicenter AIDS Cohort Study (MACS), and Murphy, P.M. (1998). CCR5 promoter polymorphism and HIV-1 disease progression. The Lancet 352, 866-870.

Omenn, G.S. (2000). The genomic era: A crucial role for the public health sciences. Environmental health perspectives. 108, A204-A205.

Roos, D., De Boer, M., De Klein, A., Bolscher, B.G., Weening, R.S. (1993). Chronic granulomatous disease: mutations in cytochrome b558. Immunodeficiency 4, 289-301.

Ross, S.C., Densen, P. (1984). Complement deficiency states and infection: epidemiology, pathogenesis and consequences of neisserial and other infections in an immune deficiency. Medicine (Baltimore) 63, 243-273.

Smith, O.P., White, B., Vaugan, D., Rafferty, M., Claffey, L., Lyons, B., Casey, W. (1997). Use of protein-C concentrate, heparin, and haemofiltration in meningococcus-induced purpura fulminans. The Lancet 350, 1590-1593.

Su, B., Sun, G., Lu, D., Xiao, J., Hu, F., Chakraborty, R., Deka, R., Jin, L. (2000). Distribution of three HIV-1 resistance-conferring polymorphisms (SDF1-3'A, CCR-64I, and CCRΔ32) in global populations. Eur. J. Hum. Genet. 8, 975-979.

Swerdlow, D.L., Mintz, E.D., Rodriguez, M., Tejada, E., Ocampo, C., et al. (1994). Severe life-threatening cholera associated with blood group O in Peru: implications for the Latin American epidemic. J. Infect. Dis. 170, 468-472.

Chapter 27. Pharmacogenetics in the treatment of infectious disease

Individuals vary in their response to drugs and this variation is partly genetically determined. Variation may for example be caused by differences in absorption, distribution, receptor binding, and elimination of drugs. Most drugs are metabolized in the liver, where they may be hydroxylated, demethylated, or conjugated. Genetic polymorphisms in these reactions have been described. The study of genetically determined variations in drug response is known as pharmacogenetics. Although there are some examples of classic Mendelian inheritance of drug response, most often drug responses appear to behave as complex traits that are influenced by interactions of different gene products with environmental factors. Most genetic variation of drug effects is caused by polymorphism in the genes coding for drug-metabolizing enzymes, and to a lesser extent in genes coding for receptors. The frequency of polymorphic alleles coding for metabolizing enzymes varies considerably between populations of different origin. Such differences are important in drug development and may lead to prescriptions and dosage recommendations tailored to each individual (Meyer 1999).

One of the best known clinically relevant polymorphisms concerns those in the cytosolic enzyme *N*-acetyltransferase-2, NAT2, which metabolizes a variety of arylamine and hydrazine drugs. Several clinically relevant polymorphisms in NAT2 have been described leading to differences in the rate of acetylation. The frequency of slow and rapid acetylator phenotypes varies considerably among ethnic groups (Evans 1993). Approximately 30 to 40 % of Caucasians and Africans are fast acetylators, in contrast to 80 – 90 % of the Asians and Eskimos. Slow acetylators are homozygous, while intermediate and fast acetylators may harbor one or two "fast" alleles. Missense mutations of the NAT2 gene are responsible for instability and inefficiency of NAT2, and three variants of NAT2 account for about 95 % of slow acetylator status. Slow NAT2 acetylators are at risk of developing adverse reactions to sulfonamides and isoniazid.
Isoniazid has a bactericidal action on *M. tuberculosis* and remains one of the major antituberculosis agents. The drug affects many aspects of the bacteria's metabolism, including the enzyme mycolase synthetase (Jacobs 1994). Based on genetic differences in acetylation, the speed of elimination of isoniazid appears trimodal in distribution in a South African population (Parkin et al 1997). The NAT2*12 allele, which codes for fast acetylation, has a high frequency in the population studied. Homozygous fast acetylators of isoniazid have a shorter exposure to therapeutic concentrations of the drug than slow or intermediate acetylators (Parkin et al 1997). These individuals are thus at risk for developing less favorable results when isoniazid is used to treat tuberculosis. They may also develop drug-resistant mycobacteria during therapy when given typical once-weekly isoniazid dosage regimens and may thus require higher therapeutic doses. Slow acetylators, on the other hand, are at risk for developing peripheral neuropathy during treatment. Although this genetic difference appears not a major factor in determining the success of antituberculosis therapy, there may be an influence under suboptimal therapeutic conditions and on the development of resistant bacteria. In clinical practice when the drug is administered daily and usually in combination with other antibiotics there is little need to determine the patient's

inactivator status. However, if isoniazid is included in once-weekly intermittent regimens for the prevention of tuberculosis, it is important to determine the patient's inactivator status to ensure its effectiveness. Several simple laboratory methods have been described for that purpose. Other drugs, such as pyridoxine, have also been added to isoniazid treatment to circumvent problems due to acetylator phenotype. Hepatic disorders may further impair isoniazid metabolism. In addition, there are indications that during severe hepatic disorders the activity of NAT2 may be disturbed more in fast acetylators than in slow acetylators.

Humans have approximately 60 enzymes of the cytochrome P450 type, which are involved in detoxification of many toxic substances. Polymorphisms in these enzymes may cause different rates of detoxification. One of these enzymes, CYP2C19, is involved in the metabolism of the anti-malaria drug proguanil, which has to be metabolized into its active form cycloguanil. However, 13 to 23 % of Asians and 2 to 5 % of Caucasians are slow metabolizers of this drug due to non-active alleles of CYP2C19 (de Morais et al 1994). Seven alleles of CYP2C19 without activity have been described (www.ncbi.nlm.nih.gov/Omim*124020 Cytochrome P450, subfamily IIC, polypeptide19; CYP2C19). It is very likely that anti-malaria treatment using proguanil is ineffective in such slow metabolizers.
In contrast, treatment with omeprazol, a gastric acid inhibitor, against infection due to *Helicobacter pylori* appears to be more effective in homozygous slow metabolizers than in heterozygous intermediate or homozygous fast metabolizers. Combination therapy with omeprazol and amoxicilline appeared effective respectively in 100, 60, and 29 % of these persons (Furata et al 1998).

Host gene polymorphisms may further influence anti-malaria treatment in another way. Glucose-6-phosphate dehydrogenase (G6PD) deficiency is an inherited, sex-linked trait that predisposes the host to hemolytic anemia after exposure to certain anti-malaria drugs such as chloroquine and primaquine. Prior to use of these drugs, it is therefore often recommended to test males for G6PD deficiency.

Patients with chronic hepatitis B virus (HBV) carriership respond differently to therapy with interferon-α. HBV DNA is cleared in approximately 40 % of them (Thursz et al 1997). In patients responding to therapy, CD4+ T helper cell responses and CD8+ CTL responses increase substantially in those who seroconvert from HBV "e" antigen positive (HbeAg) to antibody to HbeAg (HbeAb) positive. The MHC alleles HLA-DR6 and HLA-DR3 appeared associated with response to therapy. Because DR6 is the serological "supertype" of the allele HLA-DRB1*1302, an allele associated with clearance of HBV after infection, the finding suggest that this allele may also influence the response to interferon-α therapy (Scully et al 1990, Thursz et al 1997).
Likewise, very few patients treated for hepatitis C virus (HCV) infection, approximately 25 %, achieve a sustained remission. Similarly as with treatment of HBV, patients who do respond show significantly increased CD4+ T helper cell responses. In a small study, the MHC class II serotype HLA-DR7 appeared associated with the sustained response to interferon-α therapy (Brandhagen et al 1996).

Dideoxynucleosides, including azidothymidine (AZT), are potent inhibitors of HIV reverse transcriptase and are a common component of highly active anti-retroviral therapy (HAART). Resistance to chemotherapy with these drugs is a serious impediment to antiviral efficacy and negatively affects the outcomes of HIV-infected patients. Although drug-resistant HIV strains resulting from genetic mutation have emerged, some patients don't respond to therapy even though they do not harbor drug-resistant strains of the virus. Schuetz et al (1999) found that overexpression and amplification of the multiple drug resistance protein 4 (MRP4) gene correlated with ATP-dependent efflux of these drugs from T cells and, thus, with resistance of the cells to these drugs. MRP4 is a transporter that is directly linked to the efflux of nucleoside monophosphate analogs from mammalian cells. Peripheral blood mononuclear cells from several HIV-infected and HIV-uninfected persons showed considerable variation in the expression of MRP4. Future studies are needed to determine whether people with high MRP4 expression indeed respond poorly to therapy.

In conclusion, while pharmacogenetics at the moment appears in its infancy, it is likely that pharmaceutical companies will develop drugs that are either effective in the large majority of the population, or that will be offered together with a diagnostic test to that part of the population that may respond favorably to the drug. Together these developments may render pharmaceutical treatment more tailored to the individual, and therewith safer, more efficacious, and more cost-effective.

References

Brandhagen, D.J., Poterucha, J.J., Gross Jr., J.B., Gossard, A.A., Therneau, T.M., Persing, D.H. (1996). The effects of HLA haplotype on interferon response in patients with chronic hepatitis C. Dig. Dis. Wkly. 282, A-71.

de Morais, S.M.F., Wilkinson, G.R., Blaisdell, J., et al (1994). The major genetic defect responsible for the polymorphism of S-mephenytoin in humans. J. Biol. Chem. 269, 15419-15422.

Evans, D.A.P. (1993). Genetic factors in drug therapy. Clinical and molecular pharmacogenetics. Cambridge University Press.

Furata, T., Ohashi, K., Kamata, T., et al (1998). Effect of genetic differences in omeprazol metabolism on cure rates for *Helicobacter pylori* infection and peptic ulcer. Ann. Intern Med. 129, 1027-1030.

Jacobs, R.F. (1994). Recommendations for infection control in outpatient care for the prevention of tuberculosis. Pediatr. Infect. Dis. J. 13, 939-940.

Meyer, U.A. (1999). Medically relevant genetic variation of drug effects. In: Evolution in health and disease. Ed. by S.C. Stearns, Oxford University Press, pp. 41-49.

Parkin, D.P., Vandenplas, S., Botha, F.J.H., Vandenplas, M.L., Seifart, H.I., van Helden, P.D., van der Walt, B.J., Donald, P.R., van Jaarsveld, P.P. (1997). Trimodality of isoniazid elimination. Phenotype and genotype in patients with tuberculosis. Am. J. Resp. Crit. Care Med. 155, 1717-1722.

Schuetz, J.D., Connelly, M.C., Sun, D., Paibir, S.G., Flynn, P.M., Srinivas, R.V., Kumar, A., Fridland, A. (1999). MRP4: A previously unidentified factor in resistance to nucleoside-based antiviral drugs. Nat. Med. 5, 1048-1051.

Scully, L., Brown, D., Lloyd, C., Shein, K., Thomas, H. (1990). Immunological studies before and during interferon therpay: Identification of factors predicting response. Hepatology 12, 1111-1117.

Thursz, M.R. (1997). Host genetic factors influencing the outcome of hepatitis. J. Viral Hepatitis 4, 215-220.

Chapter 28. Genetics and the response to vaccination

Vaccines have a variable efficacy, and this may in part be determined by genetic differences in the vaccine recipients. Because antibody production after vaccination is normally supported by MHC class II-restricted T helper cell responses, MHC class II phenotype is a likely candidate influencing the antibody response upon vaccination. More specifically, if among an individual's MHC class II molecules (two each of DP, DQ, and DR), none binds a particular antigenic peptide, that individual will be a non-responder and will subsequently not mount an CD4+ T helper cell response. Evidently nonresponse to vaccination may not only be a threat to the vaccinated individual, but it may also be a threat for control programs aimed at reducing or preventing the transmission of a pathogen in a population, such as measles virus, poliovirus and hepatitis B virus. Such vaccination campaigns must obtain a threshold level of protected individuals, below which the pathogen will be unable to spread.

Genetic backgrounds of non-response to vaccination have been most extensively examined for measles and hepatitis B virus. However, the mechanisms leading to non-response may be at work in other vaccinations as well. Study of non-response therefore may provide important considerations for vaccine research and development, and clinical practice. Such studies may stress the need to develop vaccines that are universally immunogenic in heterogeneous populations. When significant associations between polymorphisms and non-response to vaccination are known, the frequency of the polymorphism may predict the number of nonresponders and hence the level of herd immunity in a population. Preferably, however, the study of non-response should lead to novel ways of circumventing it. In the example of hepatitis B vaccination this has, among others, been attempted by extending the amount of epitopes presented to the immune system (see below).

When non-response may be due to the lack of particular MHC alleles, researchers may overcome this by assuring that a (novel) vaccine is presented by at least one of the most common alleles in each population. If the number of epitopes required to afford protection is small, such an approach may indeed lead to a "universal vaccine". This approach is being used in the search for an effective vaccine against HIV, to be based on the recognition of infected cells by CD8+ cytotoxic T lymphocytes. Therefore, researchers attempted to find the 10 most common HLA class I alleles, at least one of which would be found in every person or at least in a large percentage of the world's population. The 10 HLA alleles that are present in the majority of people in different populations include HLA-A1, -A2, -A24, -A30, -A31, -A33, -A68, and HLA-B35, -B44, and -B70. Nonetheless, in the different populations 4 to 18 % of the people did not carry any of these alleles and so would probably not be protected by the "universal vaccine". However, further optimizations may be possible by extending and refining this set of HLA alleles (Bodmer 1996).

Measles

Measles is one of the most transmissible viruses known. Currently, despite vaccination programs in many countries, the virus causes 1 to 1.5 million of deaths per year woldwide, most of which are in third-world countries. Although reports of measles outbreaks in vaccinated populations have highlighted the role of vaccine failure, the rate of complete vaccine failure appears to be low (Anders et al 1996). However, vaccinated individuals clearly differ in the magnitude of their antibody response. Primary vaccine failure occurs with the currently used vaccine strains between 2 and 10 % of the time. Seronegativity after measles vaccination appears to cluster among related family members. Slight interracial differences in antibody responses have also been reported, which however are not necessarily explained by genetic differences (Poland and Jacobson 1998, Poland et al 1999).

The antibody response to measles vaccine is clearly influenced by the presence of maternal antibodies, as well as by class I and II HLA genes. One of the most consistent findings is that non-responders in general show higher homozygosity rates at different loci.

The human HLA class II locus DQA1 is only moderately polymorphic compared to other HLA class II loci. In a study in which 81 of 881 vaccinated individuals failed to develop an antibody response following vaccination, the non-responders had an excess of HLA-DQA1*05 alleles, while strong responders had an excess of HLA-DQA1*01 alleles (Hayney et al 1998). The non-responders were also more often homozygous at this locus and thus suggest a role of heterozygous advantage in the antibody response to this vaccine. Non-responders were also less likely to carry HLA-DRB1*13 alleles than were the strong responders (Hayney et al 1996). Interestingly, the absence of this allele has also been associated with increased susceptibility to other infectious diseases, i.e. hepatitis B infection and severe malaria in the Gambia. The presence of this allele has also been negatively associated with the development of cervical cancer in Hispanic women infected with human papilloma virus type 16. Whether HLA-DRB1*13 has a specific restricted influence on the processing and presentation of certain antigens, or functions more globally in a pathway common to the processing and presentation of many antigens remains to be established (Hayney et al 1996).

Several class I alleles (HLA-B13, -B44, and -C5) were also associated with non-response to measles vaccination. As for the class II alleles, non-responders were also more likely to be HLA-B homozygous than normal and strong responders were (OR = 2.1 and 3.7 respectively). There was further evidence for an allele-dose response phenomenon for HLA-B7, which allele was associated with a strong antibody response (Poland et al 1998).

Finally, non-responders were also more homozygous at position 665 of the gene encoding the transporter associated with antigen processing 2 (TAP2) (OR = 5.0) (Hayney et al 1997). There were no significant interactions between TAP2 and DR homozygosity, suggesting an independent, additive effect of TAP2 homozygosity at position 665 on the antibody response following measles vaccination. Although the molecular mechanisms between these associations remain to be defined, the studies clearly indicate a genetic influence on the antibody response to vaccination. The most consistent finding appears the negative influence of homozygosity at three different loci.

Hepatitis B

The lack of response to hepatitis B vaccination is a problem for persons at risk of acquiring hepatitis B infection, particularly those who work in the health care. Up to 20 % of vaccinated persons may fail to respond or respond poorly. There is no evidence for silent infection with HBV in nonresponders, indicating that immunotolerance or immunosuppression induced by latent HBV infection is not responsible for nonresponsiveness (Zuckerman 1996). Importantly, nonresponders to hepatitis B vaccines remain fully susceptible to infection. Especially in the Far East the non-response rate is high (Thursz 1997).

The response to hepatitis B vaccination appears similarly influenced by MHC class I, II, and III (or serum complement) alleles, and by the homozygosity or heterozygosity of certain alleles (Alper et al 1989, Stachowski et al 1995). The influence of the major histocompatibility complex to the vaccination response was also observed in mice (Waters 1998), which may thus be an interesting animal model to examine the molecular base of non-response. Non-responders carry more HLA-A1, -B8, -DR3, -DR7, and -DQ2 alleles, and homozygotes for these alleles are almost exclusively found among non-responders. Differences are even more marked when extended (interaction) haplotypes are taken into account. For example, the extended MHC haplotypes HLA-A1-B8-Bfs-C4AQO-C4B1-DR3-DQ2 and HLA-A1-B8-BfF-C4A6-C4B2-DR3-DQ2 are almost exclusively found in non-responders (Alper et al 1989, Dondi et al 1996, Langö-Warensjö et al 1998, Martinetti et al 1995, Stachowski et al 1995, Vingerhoets et al 1995). In the Far East, nonimmune responsiveness upon immunization with plasma-derived HbsAg vaccine is associated with the extended MHC haplotype HLA-Bw54CREG-C4 RFLP (6.5 kb + 12.0 kb)-DR4-DRw53-DQw4 (DQA1*0301-DQB1*0401) (Hatae et al 1992). This extended HLA haplotype controls nonresponsiveness as a dominant genetic trait. The association of the polymorphisms in Bf, C4A and C4B complement serum components with the hepatitis B vaccination response stresses the role of the complement system in response to foreign peptides.

A number of the HLA haplotypes associated with non-response to hepatitis B vaccination are involved in autoimmune disease, which may point to a mechanistic explanation of their role in non-responsiveness (Martinetti et al 1995). Some mechanisms of nonresponse have been examined in more detail. The defect in nonresponders appears to be in their helper T cells rather than in antigen-presenting cells, because antigen processing and presentation, and MHC class II binding of HbsAg peptides were normal (Desombere et al 1995, McNicholl et al 2000, Salazar et al 1995).

In conclusion, the presence or absence of certain MHC alleles, even in heterozygous form, and their interactions influence the quality and quantity of the response to hepatitis B vaccination.

The response to hepatitis B vaccination is further influenced by haptoglobin type. Haptoglobin is an antioxidant and hemoglobin-binding protein, which is expressed by a genetic polymorphism as three major phenotypes: 1-1, 2-1, and 2-2. Functional differences between these phenotypes are explained by modulation of oxidative stress, prostaglandin synthesis, and immune response (Langlois and Delanghe 1996). Persons with a 2-2 haptoglobin phenotype produced significantly lower hepatitis B antibody titers than those having a 1-1 or 2-1 haptoglobin phenotype (Louagie et al 1993).

As indicated, the identification of protective HLA antigens and the determinants selected by them may point to an approach by which protective T cell epitopes may be identified and used in vaccine development, while the study of non-response may provide the solution to circumvent it. Examining non-response following recombinant hepatitis B vaccination has been fruitful, because a number of persons that fail to respond to the conventional vaccine may respond to vaccines in which additional viral sequences (pre-S1 and pre-S2) have been included (McDermott et al 1998). The pre-S1 and pre-S2 domains appear to have an important immunogenic role in augmenting anti-HBs responses and circumventing genetic nonresponsiveness to the S antigen, thus eliciting antibodies that prevent the attachment of the virus to hepatocytes and that are effective in viral clearance. Indeed, Milich et al (1986) demonstrated in mice that the immune responses to pre-S1, pre-S2, and S regions appear independent of MHC-linked regulation, assuring fewer genetic nonresponders to a vaccine containing all three antigenic regions. Nonetheless, in humans a small proportion of the vaccinees still remained non-responders upon additional vaccination with vaccine in which the pre-S1 and pre-S2 proteins were included. Homozygosity and heterozygosity of the HLA-DRB1*0701-DQB1*0202 genotype were associated with this poor response (McDermott et al 1999ab). Other researchers have succeeded in overcoming non-response to hepatitis B vaccination in mice by using nucleic acid vaccination or novel adjuvants (Brunel et al 1999, Schirmbeek et al 1995).

BCG tuberculosis vaccine

As indicated in chapter 10, persons with rare mutations in the genes encoding IL-12, the IL-12 receptor, and the IFN-γ receptor may be unable to clear the live attenuated BCG tuberculosis vaccine. Disseminated, sometimes fatal BCG disease may follow. Hence BCG and similar live-vector vaccines are contraindicated in persons with these disorders (McNicholl et al 2000). It may therefore be indicated to examine family members of these persons before administering them such vaccines.

References

Alper. C.A., Kruskall, M.S., Marcus-Bagley, D., Craven, D.E., Katz, A.J., Brink, S.J., Dienstag, J.L., Awdeh, Z, Yunis, E.J. (1989). Genetic prediction of nonresponse to hepatitis B vaccine. N. Engl. J. Med. 321, 708-712.

Anders, J.F., Jacobson, R.M., Poland, G.A., Jacobson, S.J., Wollan, P.C. (1996). Secondary failure rates of measles vaccines: a metaanalysis of published studies. Pediatr. Infect. Dis. 15, 62-66.

Bodmer, J. (1996). World distribution of HLA alleles and implications for disease. In: Variation in the human genome. Wiley, Chichester (Ciba Foundation Symposium 197), 233-258.

Brunle, F., Darbouret, A., Ronco, J. (1999). Cationic lipid DC-Chol induces an improved and balanced immunity able to overcome the unresponsiveness to the hepatitis B vaccine. Vaccine 17, 2192-2203.

Desombere, I., Hauser, P., Rossau, R., Paradijs, J., Leroux-Roels, G. (1995). Nonresponders to hepatitis B vaccine can present envelope particles to T lymphocytes. J. Immunol. 154, 520-529.

Dondi, E., Finco, O., Mantovani, V., Mele, Ruberto, G., Cuccia, M. (1996). Invovlement of HLA and C4 in the non responsiveness to hepatitis B vaccine. Fund. Clin. Immunol. 4, 73-78.

Hatae, K., Kimura, A., Okuba, R., et al. (1992). Genetic control of nonresponsiveness to hepatitis B virus vaccine by an extended HLA haplotype. Eur. J. Immunol. 22, 1899-1905.

Hayney, M.S., Poland, G.A., Jacobson, R.M., Schaid, D.J., Lipskey, J.J. (1996). The influence of the HLA-DRB1*13 on measles vaccine response. J. Investig. Med. 44, 261-263.

Hayney, M.S., Poland, G.A., Dimanlig, P., Schaid, D.J., Jacobson, R.M., Lipsky, J.J. (1997). Polymorphism of the TAP2 gene may influence antibody response to live measles vaccine virus. Vaccine 1997 15, 3-6.

Hayney, M.S., Poland, G.A., Jacobson, R.M., Rabe, D., Schaid, D.J., Jacobson, S.J. Lipsky, J.J. (1998). Relationships of HLA-DQA1 alleles and humoral antibody following measles vaccination. Int. J. Infect. Dis. 2, 143-146.

Langlois, M.R., Delanghe, J.R. (1996). Biological and clinical significance of haptoglobin polymorphism in humans. Clin. Chem. 42, 1589-1600.

Langö-Warensjö, A. Cardell, A., Lindblom, B. (1998). Haplotypes comprising subtypes of the DQB1*06 allele direct the antibody response after immunisation with hepatitis B surface antigen. Tissue Antigens 52, 374-380.

Louagie, H., Delanghe, J., Desombere, I., De Buyzere, M., Hauser, P., Leroux-Roels, G. (1993). Haptoglobin polymorphism and the immune response after hepatitis B vaccination. Vaccine 11, 1188-1190.

Martinetti, M., Cuccia, M., Daielli, C., Ambroselli, F., Gatti, C., Pizzochero, C., Belloni, C., Orsolini, P., Salvaeschi, L. (1995). Anti-HBV neonatal immunization with recombinant vaccine. Part II. Molecular basis of the impaired alloreactivity. Vaccine 13, 555-560.

McDermott, A.B., Cohen, S.B., Zuckerman, J.N., Madrigal, J.A. (1998). Hepatitis B third-generation vaccines: improved response and conventional vaccine non-response—evidence for genetic bases in humans. J. Viral Hepat. Suppl. 2, 9-11.

McDermott, A.B., Cohen, S.B., Zuckermann, J.N., Madrigal, JA (1999a). Human leukocyte antigens influence the immune response to pre-S/S hepatitis B vaccine. Vaccine 17, 330-339.

McDermott, A.B., Madrigal, JA, Sabin, C.A., Zuckermann, J.N., Cohen, S.B. (1999b). The influence of host factors and immunogenetics on lymphocyte responses to Hepagene vaccination. Vaccine 17, 1329-1337.

McNicholl, J M., Downer, M.V., Udhayakumai, V., Alper, C.A., Swerdlow, D.L. (2000). Host-pathogen interactions in emerging and re-emerging diseases: A genomic perspective of tuberculosis, malaria, human immunodeficiency virus infection, hepatitis B, and cholera. Annu. Rev. Public Health 21, 15-46.

Milich, D.R., McLachlan, A., Chisari, F.V., Kent, S.B., Thornton, G.B. (1986). Immune response to the pre-S(1) region of hepatitis B surface antigen (HbsAg): A pre-S(1)-specific T cell response can bypass nonresponsiveness to the pre-S(2) and S regions of the HbsAg. J. Immunol. 137, 315-322.

Poland, G.A., Jacobson, R.M. (1998). The genetic basis for variation in antibody response to vaccines. Curr. Opin. Pediatr. 10, 208-215.

Poland, G.A., Jacobson, R.M., Schaid, D., Moore, S.B., Jacobson, S.J. (1998). The association between HLA class I alleles and measles vaccine-induced antibody response: evidence of a significant association. Vaccine 16, 1869-1871.

Poland, G.A., Jacobson, R.M., Colbournem S.A., Thampy, A.M., Lipsky, J.J., Wollan, P.C., Roberts, P., Jacobsen, S.J. (1999). Measles antibody seroprevalence rates among immunized Inuit, Innu and Caucasian subjects. Vaccine 17, 1525-1531.

Salazar, M., Deulofeut, H., Granja, C., Deulofeut, R., Yunis, D.E., Marcus-Bagley, D, Awdeh, Z., Alper, C.A., Yunis, E.J. (1995). Normal HbsAg presentation and T-cell defect in the immune response of nonresponders. Immunogentics 41, 366-374.

Schirmbeek, R., Bohm, W., Ando, K., Chisari, F.V., Reimann, J. (1995). Nucleic acid vaccination primes hepatitis B virus surface antigen-specific cytotoxic T lymphocytes in nonresponder mice. J. Virol. 69, 5929-5934.

Stachowski, J., Kramer, J., Fust, G., Maciejewski, J., Baldamus, C.A., Petranyi, G.G. (1995). Relationship between the reactivity to hepatitis B virus vaccination and the frequency of MHC class I, II and III alleles in hemodialyis patients. Scand. J. Immunol. 42, 60-65.

Thursz, M.R. (1997). Host genetic factors influencing the outcome of hepatitis. J. Viral Hepatitis 4, 215-220.

Vingerhoets, J., Goilav, C., Vanham, G., Kerstens, L., Muylle, L., Kegels, E., Van Hoof, J., Plot, P., Gigase, P. (1995). Non-response to a recombinant pre-S2-containing hepatitis B vaccine: association with the HLA-system. Ann. Soc. Belg. Med. Trop. 75, 125-129.

Waters, J.A. (1998). Evidence for the genetic basis of non-response in mice. J. Viral Hepat. Suppl. 2, 1-4.

Zuckerman, J.N. (1996). Nonresponse to hepatitis B vaccines and the kinetics of anti-HBs production. J. Med. Virol. 50, 283-288.

Chapter 29. Improving genetic disease resistance in farm animals

Infectious diseases in farm animals are controlled by vaccination, hygienic measures, optimization of management, preventive medication, and, evidently unique to veterinary medicine, the culling of infected animals (or even animals at-risk for infection) as a means to prevent pathogen transmission. In specific circumstances these may all be very powerful measures. Using genetic resistance may also be advantageous, because it may limit the number of treatments and vaccinations. By limiting the use of antibiotics in animal husbandry, the development of antimicrobial resistance may be minimized. In addition, genetically resistant animals may excrete no or small amounts of the pathogen, therewith reducing the transmission of the pathogen. This may occur not only among resistant animals, but from resistant animals to, and among susceptible animals as well. This latter effect may even exceed that of individual genetic improvement. Evidently a major advantage is that genetic improvements will be transferred to the next generations. But even if genetic improvement does not result in disease resistance, selection may be directed on affording resilience, the capacity to remain productive despite infection. Genetic improvement of disease resistance is further important, as several of the disease control measures and breeding programs for production traits likely reduce the natural selection for disease resistance. For example, turkeys selected for rapid growth during 28 generations were more susceptible to *Pasteurella multocida*, the cause of fowl cholera, than non-selected controls (Nestor et al 1996).

Genetic variation is a prerequisite for successful selection and has been found in animals and poultry exposed to a wide range of viral, bacterial and parasitic pathogens. Constraints of genetic improvement for disease resistance are the required long-term investments, the relatively slow progress, competition between genetic improvement for production traits and disease resistance, and lack of knowledge about inheritance of disease resistance (Van der Zijpp and Sybesma 1989). Successful genetic improvement of farm animals therefore apparently requires major gene effects or large heritabilities.

Major questions regarding improving genetic disease resistance are whether the balance between costs and benefits will be positive, and whether a practical procedure for measuring resistance in candidate breeding animals is available. Because the benefits of a breeding program are expressed in economic terms, the benefits of a breeding program should be expressed as reduced production loss, rather than increase in resistance.

Exploiting genetic variation between and within breeds

Genetic improvement of farm animals may be based on using the variation between breeds, within breeds, or combinations thereof. A relatively rapid, simple, and sustainable means of genetic improvement is to exploit breed differences in resistance to infection. If one breed is markedly superior to all competing breeds in the selection objective, then this breed should be used. This strategy is called breed substitution. For example, the Red Maasai sheep in East Africa are not only more resistant to infection with nematodes, such as *Haemonchus contortus,* but they are also more productive under moderate to severe parasite challenge. The breed is therefore the population of choice for farmers in East Africa (Stear and Wakelin 1998). Other methods of exploiting genetic variation between breeds are systematic crossbreeding and the creation of a composite population, for example by using marker assisted introgression and selection. Marker-assisted selection is especially useful for moving alleles from one line or breed to another by selecting alleles from one breed and the remainder of the marker alleles from the other breed. A beneficial allele from a donor line may thus be introduced in a recipient line in three phases. First, crossing donor and receptor lines produces F1 progeny. Second, backcrossing of the crossbred progeny to the recipient line is done during several generations to recover as much as possible alleles at other loci from the recipient line, while maintaining the beneficial allele. Third, individuals from the last backcross are intercrossed to create a new line that is homozygous for the beneficial allele from the donor line.

Several approaches to exploit genetic resistance within breeds are amenable:
- Direct selection for disease resistance against single or more infectious diseases. This may especially be advantageous for diseases for which development of protective vaccines has met with little success, such as parasitic diseases (Frisch et al 2000).
- Selection for "robustness", which may be very variable under different environmental conditions and therefore needs further specification, for example by defining an index that combines several parameters. Because selection for specific resistance to "all" diseases is impossible, this approach may have its long-term benefits. Traditionally farmers who selected healthy animals for reproduction have followed this approach without knowing its effectiveness.
- Selection for markers flanking a gene variant associated with or responsible for resistance or selection for the gene variant itself. Marker-assisted selection can be used to select for increased frequency of the allele of value. Marker-assisted selection is especially useful when disease resistance is difficult to measure, for example because natural exposure is infrequent or because experimental inoculation is expensive. Markers thus provide a means to estimate genetic potential even when the animals do not express the trait. A major constraint on marker-assisted selection is the high costs of DNA testing.
- Selection for immune markers. Although immunological mechanisms are clearly amenable to selection (Biozzi et al 1979, Qureshi and Miller 1991), it is often unknown how they relate to resistance against a particular disease and they may affect several diseases in opposite ways. I consider this a major drawback of this approach. Natural or experimental challenge inoculation therefore has to confirm that the immune markers are at least correlated with, but preferably causally responsible for resistance to one or more infectious diseases. Another drawback may be negative genetic correlations among various immune response functions as phagocytosis, cell-mediated and humoral immunity (Gavora and Spencer 1983). For example, pigs selected for high combined antibody and cell-mediated response developed more severe arthritis, but less pleuritis and peritonitis upon inoculation with *Mycoplasma hyorhinis*. Thus this selection did not appear to offer a consistent line-related health advantage (Manusson et al 1998, Wilkie and Malard 1999). However, such genetic correlations need not necessarily be absolute, leaving room to simultaneously improve negatively correlated traits.

- Combined approaches. Gavora and Spencer (1983) have showed the advantage of a combined approach of using genetically resistant strains and vaccination against Marek's disease in the chicken, the oncogenic herpesvirus of chickens.

Before considering a specific selection program, relevant heritabilities should be estimated preferably after experimental challenge inoculation. Heritabilities should be sufficiently high, for example higher than 0.3, to allow within breed selection. A subsequent step is to determine that selection for resistance has no or only limited effects on productivity of uninfected animals. These constraints for example hamper the genetic improvement of bovine mastitis. While mastitis has a very large economic impact, its heritability is low (0.01-0.05) and there is a negative genetic correlation between mastitis and milk production traits (Madsen 1989). It is important to try to measure heritability carefully. Large measurement errors and low repeatability of measurements lower heritability estimates and hence the potential effects of breeding programs. One should further keep in mind that heritability estimates may be different during different stages of infection, at different ages, or for different disease phenotypes. The heritability of *Ostertagia circumcincta* infection in lambs increases for example from nearly zero at one and two months of age to 0.33 at six months of age. Somewhat surprisingly, the heritabilities of egg counts following experimental infection are similar to those after natural infection (Stear and Wakelin 1998).

When one considers introducing or selecting for specific alleles that influence complex diseases, it may be advisable to calculate marginal allelic effects. The marginal allelic effect of an allele is a statistical measure of the impact of that particular allele averaged over the phenotypes of all individuals who have a genotype that includes that allele. Most complex diseases, including infectious diseases, are expected to involve only a small number of genes with large marginal allelic effects and a large number of genes with small marginal allelic effects (Hartmann 1989, Sing et al 1996, Weiss 1995).

A next step in genetic improvement programs is to identify the mode of inheritance. This is not only helpful for developing simple procedures to identify genetically resistant animals, but also to estimate the speed of progress. Evidently the rate of progress is much faster when a single resistance gene is involved than when resistance is polygenically regulated. This has for example been examined in some detail for the resistance of chicken lines against Salmonellosis. Resistance to *Salmonella typhimurium* in chickens appears inherited as a complex trait. Two chicken homologs of *Nramp1* and *Tnc* (a locus closey linked to *Lps*) appeared to explain 33 % of the differential resistance to infection with *S. typhimurium* of two chicken lines (Hu et al 1997). Albers and Gray (1986) give the example of a sixty percent increase in resistance of young lambs to the nematode *Haemonchus contortus* achieved after one year of using rams homozygous for a presumed dominant resistance gene. In contrast, thirty per cent increase would be obtained after 12 years of screening if resistance were polygenically controlled with a heritability of 0.29.

Furthermore it should be evaluated whether the effects of a resistance allele are independent of the genetic background into which it is introduced, or not. Ignoring this could lead to false decisions when evaluating the potential benefit of resistance alleles. For example, a large number of genes that each has a small average effect could cumulatively have a very large effect on health through interaction effects. Interaction effects have been observed in poultry with a recessive gene that confers resistance to infection with avian leukosis virus and to the negative effect on egg production that is associated with the infection. In crossbreeding experiments the effects of this gene on egg production and mortality associated with avian leukosis virus depended largely on characteristics of the recipient line (Hartmann 1989). Thus before introducing a gene with a large marginal allelic effect into a purebred breeding line, one should examine whether the gene is also expressed satisfactorily in the crossbred progeny from this line.

Because selection for disease resistance to one disease favors certain alleles and eliminates others (Briles et al 1977, Kuhnlein et al 1989, Simonson et al 1989), it is important to estimate the effects of the selection program on these alleles and their effects in other traits. Moreover, because infectious disease is dynamic, a decrease in polymorphisms induced by the selection program should be considered with regard to potential future diseases as well. It may therefore be imperative to maintain animals with heterozygous combinations that show heterozygous advantage, and to establish gene banks.

Finally, a selection strategy should be devised. As indicated above, selection for disease resistance may either be direct selection towards lower disease incidence, or by indirect selection based on indicator traits or markers. Selection experiments using experimental challenge inoculations may especially be helpful, but observations in naturally exposed breeding stock may suffice. Experimentally induced challenge inoculations are absolutely required when there is no useful marker or when natural infection occurs infrequently.

Comparative genomics
Genetic maps of different mammals are very similar at the fine scale. The mouse genome resembles a reassembled human genome and individual genes within a region are very similar in DNA sequence and function in different mammals. This will allow studies of comparative mammalian genomic organization and will provide many potential candidate genes that may be examined for disease resistance in farm animals. Genetic mapping studies are further underway in many species, including Sheep (SheepBase) and pigs (PigBase). Databases for species-specific or comparative mapping studies are available (see chapter 5).

Successful genetic improvement programs
Initially many efforts have been devoted to finding and exploiting MHC-disease associations. However, as indicated in chapter 9, there have been very few direct, significant and consistent relationships between particular MHC alleles and resistance to infectious disease, limiting its usefulness as marker system in animal breeding. In different populations different alleles may be associated with disease. Moreover, a disease association is in essence a statistical phenomenon that does not indicate a causal relationship between the allelic MHC gene product and the disease. Furthermore, if the MHC polymorphism has been generated or maintained during many generations, for example due to small heterozygous advantages, then the differences between allelic effects may be small (Klein and Figueroa 1986). Thus animal populations may unintentionally have been selected for disease resistance for many generations, making resistance genes with major effects very rare. When the disease is influenced by several loci possibly showing epistatic interaction, or when MHC loci are in linkage disequilibrium with a major disease locus, studies of inheritance are needed to eluci-

date the contribution of the different loci. Using experimental inoculations of groups of animals with limited numbers of different alleles may optimize the statistical power of tests for MHC associations (Ostergård 1989). Yet, despite these drawbacks finding MHC alleles strongly associated with disease may give basic insight in disease mechanisms and about the etiology, and may help in improving genetic resistance in livestock.

A veterinary example of disease resistance conferred by specific MHC alleles is Marek's disease in poultry. Marek's disease is caused by a herpesvirus and is manifested by the development of lymphomas. Heritability estimates for different Marek's disease manifestations range from 0.1 – 0.2. All strains of chickens are susceptible to the virus, but not all succumb to lymphoma development. Certain chicken MHC haplotypes, notably B21 and to a lesser extent B2, B6, and B14, which probably act alone as well as synergistically with each other and with "background" genes, correlate with resistance against morbidity and mortality from Marek's disease virus. Chickens with the B19 haplotype are remarkably susceptible (Gavora et al 1986, Hala et al 1981, Hansen et al 1967, Hartmann 1989). So far, the mechanism of resistance conferred by B21 against Marek's disease is unknown. Blind selection for resistance dramatically increases the frequency of B21 (Briles et al 1977).

The strong association between the chicken's MHC and susceptibility to Marek's disease (and other infections) is likely explained by the simple and compact structure of the MHC in this species, resulting in a low level of recombination, and by the presence of NK receptor gene(s) in the MHC region. Interestingly, another important (non-MHC) locus that determines resistance to Marek's disease in chickens is syntenic with the mouse NK receptor locus (Kaufman et al 1999). As mentioned in chapter 8, resistance to herpesviruses in mouse is determined by the NK cell receptor. Another proposed mechanism by which the chicken's MHC affects Marek's disease resistance is by regulating the cell surface expression of class I molecules. Cells of the B21 haplotype have the lowest and cells of the B19 have the highest expression of these molecules, with cells of the other haplotypes at various levels in between (Kaufman and Salomonson 1997). Interestingly, selection for Marek's disease resistance in White Leghorn chickens also resulted in selection of mitochondrial genome variants, indicating that mitochondrial variants may contribute to variation in Marek's disease resistance (Li et al 1998). (Note that mitochondrial genes are involved in energy generation; variants of mitochondrial genes are often associated with degenerative disorders).

Evidently selection for disease resistance may compete with production parameters. Gavora and Spencer (1983) showed that maximum resistance against Marek's disease was obtained in vaccinated chickens of the most resistant strain, while maximum productivity was obtained in vaccinated chickens of a strain selected for production.

Gregory already reported differential susceptibility of sheep to gastrointestinal nematodes in 1937, and he also suggested selective breeding for resistance to these parasites (Wakelin and Blackwell 1988). Shepherds, however, had already recognized the existence of resistant and susceptible sheep in their flocks for centuries. Whitlock (1955) subsequently showed the heritability of resistance to *Haemonchus contortus* and the feasibility of increasing flock resistance to this worm by breeding from resistant parents. Animals that show greater resistance develop smaller worm burdens, pass fewer eggs in their feces, expel worms earlier and more effectively, suffer less pathology, and respond better to vaccination. Greater resistance is also associated with higher levels of T cell responses and increased inflammatory responses, particularly those involving eosinophils and mast cells (Stear and Wakelin 1998). Subsequent workers from Australia estimated the inheritance of resistance to experimental *H. contortus* infection among sheep (Albers et al 1984, Gray 1987). They estimated a heritability for resistance of 0.29, as reflected in fecal egg output, and of 0.18, as reflected in growth. Both were highly significant and were correlated, suggesting that they partly shared a similar genetic basis. Research in several species of hosts has demonstrated that genes in or around the MHC are associated with resistance to nematode infection. For example, six-month-old Scottish Blackface lambs with the G2 allele of the DRB1 locus have egg counts 50 times lower than lambs without this allele. This locus alone therewith accounted for two-fifths of the genetic variation in fecal egg counts in that population (Stear and Wakelin 1998).

Genetic improvement programs may also be beneficial in providing resistance to coccidiosis, a disease of great economic importance in poultry. Outbred chicken lines clearly differed in their resistance to experimental inoculation with *Eimeria tenella,* as assessed by lesion score, mortality and body weight gain. Moreover, selective breeding of Auburn White Leghorn chickens resulted in a sixfold difference in resistance to coccidia (*Eimeria tenella*) between a resistant and a susceptible line (Pinard-Van der Laan et al 1998, Quist et al 1993).

References

Albers, G.A.A., Gray, G.D. (1986). Breeding for worm resistance: a perspective. In: "Parasitology Quo Vadit?" Ed. by M.J. Howell, Australian Academy of Science, Canberra, 559-566.

Albers, G.A.A., Burgess, S.K., Adams, D.B., Barker, J.S.F., Le Jambre, L.F., Piper, L.R. (1984). Breeding *Haemonchus contortus* resistant sheep – problems and prospects. In: Immunogenetic approaches to the control of endoparasites. Ed. J.K. Dineen and P.M. Outteridge, CSIRO, Australia, pp. 41-52.

Biozzi, G., Mouton, D., Sant'Anna, O.A., Passos, H.D., Gennari, M., Resi, M.H., Ferreira, V.C.A, Huemann, Y., Bouthillier, Y., Ibanez, O.M., Stiffel, C., Sigueira, M. (1979). Genetics of immunoresponsiveness to natural antigens in the mouse. Current Topics in Microbiology and Immunology 85, 31-99.

Briles, W.E., Stone, H.A., Cole, R.K. (1977). Marek's disease: Effects of B histocompatibility alloalleles in resistant and susceptible chicken lines. Science 195, 193-195.

Frisch, J.E., O'Neill, C.J., Kelly, M.J. (2000). Using genetics to control cattle parasites-the Rockhampton experience. Int. J. Parasitol. 30, 253-264.

Gavora, J.S., Simonson, M., Spencer, J.L., Fairfull, R.W., Gowe, R.S. (1986). Changes in the frequency of major histocompatibility haplotypes in chickens under selection for both high egg production and resistance to Marek's disease. Zeitschrift für Tierzüchtung und Züchtungsbiologie 103, 218-226.

Gavora, J.S., Spencer, J.L. (1983). Breeding for immune responsiveness and disease resistance. Anim. Blood Groups and Biochem. Genet. 14, 159-180.

Gray, G.D. (1987). Genetic resistance to haemonchosis in sheep. Parasitology Today 3, 253-255.

Hala, K., Boyd, R., Wick, G. (1981). Chicken major histocompatibility complex and disease. Scand. J. Immunol. 14, 607-616.

Hansen, M.P., Van Zandt, J.N., Law, G.R.J. (1967). Differences in susceptibility to Marek's disease in chickens carrying two different B blood group alleles. Poultry Sci. 46, 1268.

Hartmann, W. (1989). Evaluation of "major genes" affecting disease resistance in poultry in respect to their potential for commercial breeding. Prog. Clin. Biol. Res. 307, 221-231.

Hu, J., Bumstead, N., Barrow, P., Sebastiani, G., Olien, L., Morgan, K., Malo, D. (1997). Resistance to Salmonellosis in the chicken is linked to *NRAMP1* and *TNC*. Genome Res. 7, 693-704.

Kaufman, J., Salomonsen, J. (1997). The "minimal essential MHC" revisited: Both peptide-binding and cell surface expression level of MHC molecules are polymorphisms selected by pathogens in chickens. Hereditas 127, 67-73.

Kaufman, J., Milne, S., Göbel, T.W., Walker, B.A., Jacob, J.P., Auffray, C., Zoorob, R., Beck, S. (1999). The chicken B locus is a minimal essential major histocompatibility complex. Nature 401, 923-925.

Klein, J., Figueroa, F. (1986). Evolution of the major histocompatibility complex. CRC Crit. Rev. Immun. 6, 295-386.

Kuhnlein, U., Sabour, M., Gavora, J.S., Fairfull, R.W., Bernon, D.E. (1989). Influence of selection for egg production and Marek's disease resistance on the incidence of endogenous viral genes in White Leghorns. Poult. Sci. 68, 1161-1167.

Li, S., Aggrey, S.E., Zadworney, D., Fairfull, W., Kuhnlein, U. (1998). Evidence for a genetic variation in the mitochondrial genome affecting traits in White Leghorn chickens. J. Hered. 89, 222-226.

Madsen, P. (1989). Genetic resistance to bovine mastitis. In: Improving genetic disease resistance in farm animals. Ed. by Van der Zijpp and Sybesma, Kluwer Academic Publishers, Dordrecht, The Netherlands, 169-177.

Magnusson, U., Wilkie, B., Mallard, B., Rosendal, S., Kenedy, B. (1998). *Mycoplasma hyorhinis* infection of pigs selectively bred for high and low immune response. Vet. Immunol. Immunopathol. 61, 83-96.

Nestor, K.E., Saif, Y.M., Zhu, J., Noble, D.O. (1996). Influence of growth selection in turkeys on resistance to *Pasteurella multocida*. Poult. Sci. 75, 1161-1163.

Ostergård, H. (1989). Statistical aspects of cattle MHC (BOLA) and disease associations exemplified by an investigation of subclinical mastitis. In: Improving genetic disease resistance in farm animals. Ed. by Van der Zijpp and Sybesma, Kluwer Academic Publishers, Dordrecht, The Netherlands, 115-123.

Pinard-Van der Laan, M.H., Monvoisin, J.L., Pery, P., Hamet, N., Thomas, M. (1998). Comparison of outbred lines of chickens for resistance to experimental infection with coccidiosis (*Eimeria tenella*). Poult. Sci. 77, 185-191.

Quist, K.L., Taylor, R.L. Jr. Johnson, L.W., Strout, R.G. (1993). Comparative development of *Eimeria tenella* in primary chick kidney cell cultures derived from coccidia-resistant and -susceptible chickens. Poult. Sci. 72, 82-87.

Qureshi, M.A., Miller, L. (1991). Comparison of macrophage function in several commercial broiler genetic lines. Poult. Sci. 70, 2094-2101.

Simonson, M., Arnul, M., SØrensen, P. (1989). The chicken MHC and its importance. In: Improving genetic disease resistance in farm animals. Ed. by Van der Zijpp and Sybesma, Kluwer Academic Publishers, Dordrecht, The Netherlands, 42-62.

Sing, C.F., Haviland, M.B., Reilly, S.L. (1996). Genetic architecture of common multifactorial diseases. In: Variation in the human genome. Ciba Foundation Symposium 197. John Wiley, Chisester, p.p. 211-232.

Stear, M.J., Wakelin, D. (1998). Genetic resistance to parasitic infection. Rev. sci. tech. Off. Int. Epiz. 17, 143-153.

Wakelin, D., Blackwell, J.M. (1988). Genetic resistance to bacterial and parasitic infection. Taylor and Francis, London, Philadelphia, New York.

Weiss, K.M. (1995). Genetic variation and human disease. Principles and evolutionary approaches. Cambridge University Press.

Whitlock, J.H. (1955). A study of inheritance of resistance to Trichostrongyloides. Cornell Veterinarian 45, 422-439.

Wilkie, B., Mallard, B. (1999). Selection for high immune response: an alternative approach to animal health maintenance? Vet. Immunol. Immunopathol. 72, 231-235.

Zijpp, A.J. van der, Sybesma, W. (1989). Improving genetic disease resistance in farm animals. Kluwer Academic Publishers, Dordrecht, The Netherlands.

Chapter 30. Infectious disease genetics in laboratory animal science

Knowledge of genetic determinants of susceptibility to infectious diseases may be important for scientists working with animal models in the study of both infectious and non-infectious responses. Some considerations on these aspects follow.

First, particular strains, lines, breeds or species of laboratory animals may be more or less susceptible to contaminating infections of the animal facilities or breeding colonies. Researchers therefore have to consider whether particular contaminating infections may be confounding factors in their study or not. This may prompt them to take measures in order to clear the infection from their facilities, or to use lines or breeds more resistant to the accidental infection.

Second, the pathogenesis and immune response to infection and vaccination may show both qualitative and quantitative differences to experimental infection. Strains may for example be responders or non-responders to a particular stimulus. Depending on their specific research question, researchers may choose a strain particular good responding to their stimulus of interest, or they may for example choose a strain particularly good representing humans in a particular response.

Third, animal models may be modified by transgenesis in order to make them more susceptible to infection, to modify their response towards that seen in humans, or to study the specific influence of a gene on the response.

Finally, genetic differences in response to infection and vaccination may be used to dissect the genetic basis of these differences and to study the physiologic mechanisms involved. While the literature is abundant on all these matters, I shall give but a few examples to illustrate them.

Susceptibility of laboratory animals to natural infection.
Differences in susceptibility/resistance to a large number of naturally occurring infections between laboratory animal strains have been reported. Some of the mechanisms explaining these differences have been elucidated. A summary of some examples is given in Table 10.

Genetic differences in response to infection and vaccination.
Very often animal experiments are designed to mimic the natural infection in humans. However animal species and strains may differ in the way they respond to experimentally induced infections and these differences should therefore taken into account when choosing the strains that are used for experimentation. For example, mice have been used to study *Chlamydia trachomatis*-induced salpingitis and subsequent infertility, but susceptibility to chlamydial salpingitis is clearly under genetic control in mice. No salpingitis and infertility are observed in BALB mice, but lesions are apparent in CBA and C3H strains. This control is not simply associated with the major H-2 gene complex, as mouse strains of the same haplotype (H-2k) differ in susceptibility. The fertility of inoculated BALB/c (H-2d) and BALB/K (H-2k) is no different from that of non-inoculated controls, while congenic C3H mice of differing H-2 haplotypes (H-2k and H-2o) show reduced fertility. Reduced fertility is paralleled by the extent of histological oviductal inflammation in mice of each strain (Tuffrey et al 1992).

Table 10. Genetic differences in susceptibility to naturally occurring infections in laboratory animals.

Pathogen	Disease	Species	Susceptible/ resistant strains	Mechanism	Ref.
Citrobacter rodentium	Transmissible murine colonic hyperplasia	Mouse	NIH Swiss are susceptible. C57BL/6J are resistant.		Barthold et al 1977
Cilia-associated respiratory (CAR) bacillus	Respiratory tract disease	Rat	F344, LEW and SD rats are similarly susceptible.		Schoeb et al 1997
Clostridium piliforme	Tyzzer's disease	Mouse	CBA/N, C3 are susceptible. Other strains are resistant.	IgM defect	
Corynebacterium kutscheri		Mouse	BALB/c-nu/nu, A/J, CBA/N, MPS, and BALB/cCr are susceptible. C3H/He are intermediate. C57BL/6Cr, B10.BR/SgSn, ddY, and ICR are resistant.	polygenic control	Amao et al 1993; Hirst et al 1976
E. coli O115a,c;K(B)	Megaentron	Mouse	CF1 and C3H/He mice are susceptible. NC and C57BL/6 are intermediate. BALB/c are resistant.	Bacterial adherence (and other function)	Itoh et al 1988
Helicobacter hepaticus	Inflammatory large bowel disease	Mouse	Immunodeficient NCr-nu/nu, BALB/c AnNCr-nu/nu, C57BL/6NCr-nu/nu, and C.B17/Icr-scid/NCr are susceptible.	Immuno-deficiency	Ward et al 1996
Mycoplasma pulmonis	Chronic respiratory disease	Mouse	C57BL/6 and B10.BR are resistant. A.By, C3H/Bi, C3H/HeJ, and BALB.B are susceptible.	Macrophage bactericidal function	Lai et al 1993

continue table 10. Genetic differences in susceptibility to naturally occurring infections in laboratory animals.

Pathogen	Disease	Species	Susceptible/ resistant strains	Mechanism	Ref.
Pseudomonas aeruginosa	Endobronchial inflammation	Mouse	DBA/2 mice are susceptible. BALB/c mice are resistant. A/J and C57BL/6 are intermediate.	Recruitment of inflammatory cells.	Morissette et al 1995.
Streptobacillus moniliformis	Cervical lymphadenitis	Mouse	C57BL/6J mice are susceptible. BALB/cJ, C3H/He, DBA/2J, CB6F1, and B6D2F1 mice are resistant.		Wullenweber et al 1991

Another example is the pulmonary infection caused by the yeast *Paracoccidioides brasiliensis*, a disease of man in the tropical and subtropical areas of South and Central America. Pathologic and immunologic responses may be studied in susceptible Swiss mice, while BALB/c mice are non-susceptible. In Swiss mice the pulmonary infection is progressive, while it is regressive in BALB/c mice. Progression of the infection is associated with depressed cellular immune responses (Vilani-Morena et al 1998).

A good alternative for studying experimental infections of mice or rats with human pathogens is to inoculate animals with the natural homolog of the human pathogen. For example, experimentally-induced infection of pigs with the natural herpesvirus of pigs, Aujeszky's disease virus (syn. pseudorabies virus), is a very good model to study many aspects of herpes simplex type 1 infection of humans. These include cell-to-cell transmission, neuropathogenesis, latency, reactivation from latency, immune mechanisms associated with recovery, mucosal immunity, vaccine-induced immunity, and the role of viral proteins in these responses (De Wind et al 1994, Kimman et al 1995, Mulder et al 1996, Mulder et al 1997, Bouma et al 1997).

Genetic modification
The literature is abundant on deletion or insertion mutagenesis of mice to study the role of specific molecules in the pathogenesis of infectious disease (knockout mice), or to modify them in such manner that studying specific infections becomes feasible (transgenic mice). For example, mice are not naturally susceptible to poliovirus infection. Hence pathogenesis and vaccine studies had to be conducted in higher primates. The development of transgenic mice bearing the human receptor for poliovirus (tgPVR) has provided a non-primate animal model to examine neurovirulence of poliovirus (vaccine) strains, and to examine mechanisms of infection and immunity against poliovirus *in vivo*. For example, using these mice we were able to demonstrate that expression of the poliovirus receptor is necessary for a protective virus-specific IgA response at mucosal surfaces (Buisman et al 2000).

Transgenesis of the human *apoE* gene makes mice susceptible to diet-induced atherosclerosis. This and several other mouse models of atherosclerosis may therefore be very suitable to study the additional influence of infection with *Chlamydia pneumoniae* on the development of atherosclerosis and coronary disease, a hotly debated issue at present.

Dissecting genetic differences in response to infection and vaccination
It is evident from many examples that mice that differ in their response to infection may be used to identify genetic factors and physiologic mechanisms that determine this differential response. For example, mutant mice that differ in their response to lipopolysaccharide (LPS) have allowed identifying a gene regulating this response. Gram-negative bacterial LPS evokes a protective proinflammatory response in the normal infected host. This LPS sensitivity is experimentally measured by TNF activity. When the response is strong, TNF and other toxic cytokines may ultimately lead to shock during sepsis. Nonetheless, recognition of LPS by cells of the innate immune system promotes effective clearance of an infection with Gram-negative bacteria before it becomes disseminated. Through genetic analysis of mutant mice, the gene encoding Toll-like receptor 4 (*Tlr4*) was identified as a critical component of this host defense mechanism. The gene is localized on mouse chromosome 4. Several inbred mouse strains, including C3H/HeJ, C57BL/10ScNCr and C57BL/10ScCr bear mutations

at the *Lps* locus (*Lpsd*) that confers hyporesponsiveness to the immunostimulatory properties of LPS and susceptibility to overwhelming gram-negative bacterial infection. These mice also exhibit a natural tolerance to the lethal effects of LPS. The phenotypic expression of *Lpsd* is pleiotropic, affecting several cell types crucial to host defense, including the macrophage. By positional cloning *Tlr4* has been identified as the gene encoded by *Lps*. It has further been established that a point mutation in the *Tlr4* gene in C3H/HeJ mice (*Lpsd*) underlies their hyporesponsiveness to LPS. The mutation results in a nonconservative substitution of a highly conserved proline by histidine at codon 712 (Poltorak et al 1999). F1 progeny from crosses between mice that carry a 9-cM deletion of chromosome 4 (including deletion of *LpsTlr4*) and C3H/HeJ mice (Lps0 x Lpsd F1 mice) exhibit a pattern of LPS sensitivity that is indistinguishable from that exhibited by *Lpsn* x *Lpsd* F1 progeny and whose average response is intermediate to parental responses. These experiments indicate that the C3H/HeJ defect exerts a dominant negative effect on LPS sensitivity and that expression of a normal Toll-like receptor 4 is apparently not required.

Tlr4 is a transmembrane protein with a cytoplasmic domain that bears homology to the interleukin-1 receptor. The gene is a member of an ancient gene family that plays a central role in the signaling pathways that regulate antimicrobial host defense in plants, invertebrates and mammals. In mammals *Tlr4* may be activated by the interaction of LPS with CD14 via LPS-Binding Protein, while *Tlr2* recognizes components of cell walls from gram-positive bacteria such as *Listeria monocytogenes*. The role of Toll-like receptors in innate immunity was first discovered in the fruitfly *Drosophila melanogaster*, and the subsequent identification of its homolog in mammals thus illustrates that studying host responses in *Drosophila* may give us new clues about resistance to infection in mammals. In mammals there are at least nine members of the Toll family. In mammals Toll-like receptors are expressed on the apical surface of gut mucosal epithelial cells and on macrophages. They enable these cells to detect bacterial products (i.e. "pathogen-associated molecular patterns") and to initiate an innate immune response (Beutler and Poltorak 2001, Qureshi et al 1999a, Qureshi et al 1999b, Qureshi et al 1999c, Vogel et al 1999).

To illustrate the usefulness of a mouse-to-human approach it is interesting to note that very recently two common co-segregating missense mutations (Asp299Gly and Thr 399Ile), affecting the extracellular domain of the TLR4, have been demonstrated to cause hyporesponsiveness to inhaled LPS in humans. The frequency of the mutant allele is approximately 7 % (Ardour et al 2000).

References

Amao, H., Komakai, Y., Sugiyama, M., Saito, T.R., Takahashi, K.W., Saito, M. (1993). Differences in susceptibility of mice among various strains to oral infection with *Corynebacterium kutscheri*. Experimental Animals 42, 539-545.

Ardour, N.C., Lorenz, E., Schutte, B.C., Zabner, J., Kline, J.N., Jones, M., Frees, K., Watt, J.L., Schwartz, D.A. (2000). TLR4 mutations are associated with endotoxin hyporesponsiveness in humans. Nat. Genet. 25, 187-191.

Barthold, S.W., Osbaldiston, G.W., Jonas, A.M. (1977). Dietary, bacterial, and host genetic interactions in the pathogenesis of transmissible murine colonic hyperplasia. Lab. Anim. Sci. 27, 938-945.

Beutler, B., Poltorak, A. (2001). The sole gateway to endotoxin response: how *lps* was identified as *tlr4*, and its role in innate immunity. Drug Metab. Dispos. 29, 474-478.

Bouma, A., Zwart, R.J., de Bruin, M.G.M., de Jong, M.C.M., Kimman, T.G., Bianchi, A.T.J. (1997). Immunohistologic characterization of the local cellular response directed against pseudorabies virus in pigs. Veterinary Microbiology 58, 145 – 154.

Buisman, A.M., Sonsma, J.A., Kimman, T.G., Koopmans, M.P. (2000). Mucosal and systemic immunity against poliovirus in mice transgenic for the poliovirus receptor: The poliovirus receptor is necessary for a virus-specific mucosal IgA response. J. Infect. Dis. 181, 815-823.

De Wind, N., Peeters, B.P.H., Zijderveld, A., Gielkens, A.L.J., Berns, A.J.M., Kimman, T.G. (1994). Mutagenesis and characterization of a 41-kilobase-pair region of the pseudorabies virus genome: transcription map, search for virulence genes, and comparison with homologs of herpes simplex virus type 1. Virology 200, 784-790.

Hirst, R.G., Wallace, M.E. (1976). Inherited resistance to *Corynebacterium kutscheri* in mice. Infect. Immun. 14, 475-482.

Itoh, K., Matsui, T., Tsuji, K., Mitsuoka, T., Ueda, K. Genetic control in the susceptibility of germfree inbred mice to infection by *Escherichia coli* O115a,c:K(B). Infect. Immun. 56, 930-935.

Kimman, T.G., de Bruin, T.M.G., Voermans, J.J.M., Peeters, B.P.H., Bianchi, A.T.J. (1995). Development and antigen specificity of the lymphoproliferation response of pigs to pseudorabies virus. Dichotomy between secondary B and T cell responses. Immunology, 86, 372-378.

Lai, W.C., Linton, G., Bennett, M., Pakes, S.P. (1993). Genetic control of resistance to *Mycoplasma pulmonis* infection in mice. Infect. Immun. 61, 4615-4621.

Morisette, C., Skamene, E., Gervais, F. (1995). Endobronchial inflammation following *Pseudomonas aeruginosa* infection in resistant and susceptible strains of mice. Infect. Immun. 63, 1718-1724.

Mulder, W.A.M., Pol, J., Kimman, T.G., Kok, G., Priem, J., Peeters, B. (1996). Glycoprotein gD-negative pseudorabies can spread transneuronally via direct neuron-to-neuron transmission in its natural host, the pig, but not after additional inactivation of gE or gI. Journal of Virology, 70, 2191-2200.

Mulder, W.A.M., Pol, J.M.A., Gruys, E., Jacobs, L., de Jong, M.C.M., Peeters, B., Kimman, T.G. (1997). Pseudorabies virus infections in pigs. Role of viral proteins in virulence, pathogenesis and transmission. Veterinary Research, 28, 1 - 17.

Poltorak, A., He, X., Smirnova, I., Liu, M.Y., Huffel, C.V. et al (1999). Defective LPS signaling in C3H/HeJ and C57BL/10ScCr mice: mutations in *Tlr4* gene. Science 282, 2085-2088.

Qureshi, S.T., Gros, P., Malo, D. (1999a). Host resistance to infection: genetic control of lipopolysacccharide responsiveness by Toll-like receptor genes. Trends Genet. 15, 291-294.

Qureshi, S.T., Gros, P., Malo, D. (1999b). The *Lps* locus: genetic regulation of host response to bacterial lipopolysaccharide. Inflamm. Res. 48, 613-620.

Qureshi, S.T., Lariviere, L., Leveque, G., Clermont, S., Moore, K.J., Gros, P., Malo, D. (1999c). Endotoxin-tolerant mice have mutations in Toll-like receptor 4. J. Exp. Med. 189, 615-625.

Schoeb, T.R., Davidson, M.K., Davis, J.K. (1997). Pathogenicity of cilia-associated respiratoy (CAR) bacillus isolates for F344, LEW, and SD rats. Vet. Pathol. 34, 263-270.

Tuffrey, M., Alexander, F., Woods, C., Taylor-Robinson, D. (1992). Genetic susceptibility to chlamydial salpingitis and subsequent infertility in mice. J. Reprod. Fertil. 95, 31-38.

Vilani-Moreno, F., Fecchio, D., de Mattos, M.C., Moscardi-Bacchi, M., Defaveri, J., Franco, M. (1998). Study of pulmonary experimental paracoccidioidomycosis by analysis of bronchoalveolar lavage cells: resistant vs. susceptible mice. Mycopathologia 141, 79-91.

Vogel, S.N., Johnson, D., Perera, P.Y., Medvedev, A., Lariviere, L., Qureshi, S.T., Malo, D. (1999). Cutting edge: functional characterization of the effect of the 3H/HeJ defect in mice that lack an *Lpsn* gene: *in vivo* evidence for a dominant negative mutation. J. Immunol. 162, 5666-5670.

Ward, J.M., Anver, M.R., Haines, D.C., Melhorn, J.M., Gorelick, P., Yan, L., Fox, J.G. (1996). Inflammatory large bowel disease in immunodeficient mice naturally infected with *Helicobacter hepaticus*. Lab. Anim. Sci. 46, 15-20.

Wullenweber, M., Kaspareit-Rittinghausen, J., Farouq, M. (1990). *Streptobacillus moniliformis* epizootic in barrier-maintained C57BL/6J mice and susceptibility to infection of different strains of mice. Lab. Anim. Sci. 40, 608-612.

Epilogue

As we know there are 46 human chromosomes, which contain approximately 3,000 million base pairs of DNA. The human genome may encode up to 38.000 genes. While the coding regions make up only about 2 % of the genome, the function of the remaining 98 % is unknown. Many genes are variable and some of these gene variants likely affect the outcome of infectious disease. A complete understanding of infectious disease requires knowledge of the complex interplay between host and pathogen genomes, as well as of socioeconomic and other environmental factors that influence this interaction. A thorough understanding of the pathogenesis of infectious disease would further require complete knowledge of the time, location, and level of expression of genes from host and pathogen, their function, and the influence of gene variants on these parameters. Unfortunately, it should be clear that at this moment we have fragmentary knowledge of these aspects only. Given the rapid developments in high throughput array-technology and bioinformatics we may however anticipate many studies examining these aspects in a more systematic genome-wide fashion in the near future. Apparently large sample collections for genetic epidemiological and pathophysiological studies of infectious diseases are needed for that purpose. Other studies may restrict themselves to study certain pathways in the pathogenesis of a disease and give more detailed information on such a pathway. Examples include, among others, genes functioning in the Th1 or Th2 pathway, in macrophage function, or in attachment and entry of pathogens.

The relation between host genetics and susceptibility to infectious disease has many aspects. Some gene variants reflect seldom mutations, against which there has been strong selection in the past. Other polymorphisms, which may be frequent, do not or hardly appear to affect the normal physiologic functioning of the host. Some of the polymorphisms have only been detected by carefully examining the course of infections in genetically diverse populations. In addition, genetic variability in susceptibility to infection may or may not be advantageous for the population at present. Some polymorphisms may enhance susceptibility to infection while others may afford protection. Likewise, polymorphisms may have small, major, or all-or-nothing effects on the course of infectious disease. The latter appear to function mainly at the early stages of entry and penetration. For example, the ΔCCR5 mutation or lack of the Duffy blood group antigen results in almost complete protection against HIV infection and malaria caused by *Plasmodium vivax* respectively. Other polymorphisms, notably in the MHC system, appear to have subtle effects only. Such polymorphisms likely affect the quantity or configuration of immune factors operating during the later stages of infection. Polymorphisms may further affect the outcome of different infections in different ways. Certain genes could have a beneficial effect in some backgrounds and a detrimental effect in others. In any case, as a result of evolution every individual has a set of interacting genes that positively affect the outcome of some diseases and negatively affect the outcome of others.

The general principle behind the search for relationships between genetic polymorphisms and susceptibility is that it will likely identify relevant genes. Despite this truism, non-polymorphic genes may also severely affect the course of infection, for example by coding for pathogen receptors or for important immune functions. This situation may be clear when such genes are mutated and give rise to seldom occurring disease manifestations. In such cases the fraction of disease attributable to a gene variant may be high, while heritability in the population is low. As mentioned, the interaction of susceptibility genes with other genes and environmental factors may further affect the final influence of a host gene on the course of infection. Different pathogen strains, in itself a major reason for variation in the course of infectious disease, may also interact differently with the host genome. These pathogen-host interactions will likely be a major field of research in the near future.

The final outcome of infectious disease may depend on many factors and may be very difficult to predict, especially wen host-pathogen-interactions are dynamic due to ongoing molecular evolutionary changes. When, for example, the impact of host genetic variation on the course of disease is small, and when the impact of many interactions between host and pathogen genes and environmental factors is larger, the final prediction of disease course may be rather uncertain. Thus it may paradoxically be that the more knowledge of genes, both from host and pathogen, their variation and effects, and environmental factors is acquired, the more difficult and uncertain prediction of the final outcome of the interactions will be. Gaining more knowledge in this field requires large population studies and advanced statistical techniques. We may fail to detect subtle effects when even large studies fail to afford the necessary power. Thus another paradox is that evolution has shaped many subtle gene variants adapted to specific challenges, which effects nowadays are perhaps even too subtle to be discriminated or exploited in medicine. Nonetheless, by studying even minor gene effects, important basic functional knowledge on infectious disease may be acquired. Less subtle methods, such as gene knockout experiments in laboratory animals, may also be useful for answering such questions. Notwithstanding these remarks, it seems appropriate that in complex diseases, such as infectious diseases, genes that have a major impact on the course of disease merit most attention. These may be of major clinical relevance, as they may influence diagnosis, therapy and prevention in individual patients, and they may also have a major impact on public health.

Estimates of the contribution of genetic factors to infectious diseases, including attributable risks, heritability in broad sense (all genetic differences) or in a narrow sense (reflecting one or more loci only) are not easy to obtain, but they are important to give further insight in the success of further studies into genetic mechanisms affecting the disease. There are no general rules, but we may speculate that the more complex the pathogenesis is and the more immune factors contribute to pathology, the larger the influence of host genetic variation on the finale disease outcome is. In addition, the stage and intensity of selection, as well as environmental complexity, are likely important determinants of heritability. Heritability estimates are more difficult to achieve in humans compared to veterinary or laboratory animal species. They are nonetheless essential to estimate the genetic contribution to the variation in disease expression and to estimate the chances to find relevant genes and important disease mechanisms under specific circumstances. Although comparisons on the genetic contribution to disease susceptibility between species are very difficult and should be made with very much caution, heritability estimates in veterinary or laboratory animal species may therefore deserve special attention. Another field of comparative medicine, which may gain further significance in the near future, is that of studying infections in organisms with a simple, well-characterized genome, such as *Caenorhabditis elegans* and *Drosophila melanogaster*.

Demonstration of a genetic basis for differences in response to infection requires large sample sizes and carefully chosen populations, in which the influences of nutrition, social and economic factors can be carefully controlled. Data from large vaccination surveys might often supply that information. Such an approach may also provide information on genetically determined variation in the host response to vaccination, which may be crucial information for the design of new vaccines.

From the examples given in the text, it is hopefully obvious that studying genetic variation may give clues to really every step in the pathogenesis of infectious disease. In this book we examined attachment and entry of micro-organisms, innate immunity, antigen processing and presentation, immune and inflammatory responses, lymphocyte function, accessory cell function, and the normal barrier function of tissues. Functional studies of infections in individuals with different susceptibility to infectious disease are thus able to provide important insights into the complex pathogenesis of the disease. The study of host genetic factors influencing infectious diseases therefore has its own scientific purposes. It shows that really in every step of the pathogenesis host molecules may play important roles in the interaction between host and pathogen, and they are thus identified for further in-dept studies. However, perhaps more important are the practical implications of infectious disease genetics for treatment and preferably prevention. Genetic studies may identify potential targets for drug therapy, which may be useful to treat all affected persons. An example is vitamin D, which appears beneficial in treating tuberculosis, and which role has been supported by genetic studies of the vitamin D receptor. Another potential target for therapy identified by genetics is the chemokine receptor CCR5, which also functions as receptor for HIV. Drugs interfering with CCR5 may be useful in treating HIV/AIDS in patients regardless of their genetic make-up. However, therapies or vaccinations may also be tailored to the genetic make-up of the individual. For example, persons with increased susceptibility for bacterial infections may be offered additional vaccinations or early treatments. Future vaccines may also not be efficacious in the general population, but in certain individuals only. This may especially be the case for "difficult" vaccines, such as vaccines against HIV/AIDS or respiratory syncytial virus.

Acknowledgements

I am very grateful to many very skilled colleagues for much fruitful discussions and collaboration. These have resulted in ideas, suggestions for this book, and even concrete experiments.

So, for providing stimulating discussions on the role of cytokine polymorphisms in respiratory tract infection, I thank Barbara Hoebee and Anita Boelen. Peter Demant, Henk van Kranen, Sander Banus, and Leo Schouls were all stimulating in thinking about the role of animal models for studying genetic aspects of infectious disease. Ron Boot was stimulating in showing me the way to much literature on naturally occurring infections in laboratory animals. Rob Vandebriel was very helpful in writing the chapter on the genetics and the response to vaccination.

Much of the work has been done during a sabbatical leave. I therefore thank the institute for giving me the opportunity to consume my off-days for this purpose. I am also much indebted to Jan van Embden, who has been so kind to isolate me from the responsibilities of the everyday world and to take over my administrative duties during this period.

Last, but certainly not least, I thank my greatest supporters, Annette, Maarten and Wouter, for providing stimulating help of a most agreeable kind. Thanks.

Subject Index

MIX
Papier aus verantwortungsvollen Quellen
Paper from responsible sources
FSC® C105338

If you have any concerns about our products,
you can contact us on
ProductSafety@springernature.com

In case Publisher is established outside the EU,
the EU authorized representative is:
Springer Nature Customer Service Center GmbH
Europaplatz 3, 69115 Heidelberg, Germany

Printed by Libri Plureos GmbH
in Hamburg, Germany